Lecture Notes in Computer Science 8305

Commenced Publication in 1973
Founding and Former Series Editors:
Gerhard Goos, Juris Hartmanis, and Jan van Leeuwen

T0236356

Editorial Board

David Hutchison
Lancaster University, UK
Takeo Kanade
Carnegie Mellon University, Pittsburgh, PA, USA
Josef Kittler
University of Surrey, Guildford, UK
Jon M. Kleinberg
Cornell University, Ithaca, NY, USA
Alfred Kobsa
University of California, Irvine, CA, USA
Friedemann Mattern
ETH Zurich, Switzerland
John C. Mitchell
Stanford University, CA, USA
Moni Naor
Weizmann Institute of Science, Rehovot, Israel
Oscar Nierstrasz
University of Bern, Switzerland
C. Pandu Rangan
Indian Institute of Technology, Madras, India
Bernhard Steffen
TU Dortmund University, Germany
Madhu Sudan
Microsoft Research, Cambridge, MA, USA
Demetri Terzopoulos
University of California, Los Angeles, CA, USA
Doug Tygar
University of California, Berkeley, CA, USA
Gerhard Weikum
Max Planck Institute for Informatics, Saarbruecken, Germany

Anthony Bonato Michael Mitzenmacher
Paweł Prałat (Eds.)

Algorithms and Models for the Web Graph

10th International Workshop, WAW 2013
Cambridge, MA, USA, December 14-15, 2013
Proceedings

Springer

Volume Editors

Anthony Bonato
Paweł Prałat
Ryerson University
Department of Mathematics
Toronto, ON, Canada
E-mail: {abonato, pralat}@ryerson.ca

Michael Mitzenmacher
Harvard University
School of Engineering and Applied Sciences
Cambridge, MA, USA
E-mail: michaelm@eecs.harvard.edu

ISSN 0302-9743 e-ISSN 1611-3349
ISBN 978-3-319-03535-2 e-ISBN 978-3-319-03536-9
DOI 10.1007/978-3-319-03536-9
Springer Cham Heidelberg New York Dordrecht London

Library of Congress Control Number: 2013952734

CR Subject Classification (1998): F.2, G.2, H.3, H.2.8, I.2.6

LNCS Sublibrary: SL 1 – Theoretical Computer Science and General Issues

© Springer International Publishing Switzerland 2013
This work is subject to copyright. All rights are reserved by the Publisher, whether the whole or part of the material is concerned, specifically the rights of translation, reprinting, reuse of illustrations, recitation, broadcasting, reproduction on microfilms or in any other physical way, and transmission or information storage and retrieval, electronic adaptation, computer software, or by similar or dissimilar methodology now known or hereafter developed. Exempted from this legal reservation are brief excerpts in connection with reviews or scholarly analysis or material supplied specifically for the purpose of being entered and executed on a computer system, for exclusive use by the purchaser of the work. Duplication of this publication or parts thereof is permitted only under the provisions of the Copyright Law of the Publisher's location, in ist current version, and permission for use must always be obtained from Springer. Permissions for use may be obtained through RightsLink at the Copyright Clearance Center. Violations are liable to prosecution under the respective Copyright Law.
The use of general descriptive names, registered names, trademarks, service marks, etc. in this publication does not imply, even in the absence of a specific statement, that such names are exempt from the relevant protective laws and regulations and therefore free for general use.
While the advice and information in this book are believed to be true and accurate at the date of publication, neither the authors nor the editors nor the publisher can accept any legal responsibility for any errors or omissions that may be made. The publisher makes no warranty, express or implied, with respect to the material contained herein.

Typesetting: Camera-ready by author, data conversion by Scientific Publishing Services, Chennai, India

Printed on acid-free paper

Springer is part of Springer Science+Business Media (www.springer.com)

Preface

The 10th Workshop on Algorithms and Models for the Web Graph (WAW 2013) took place at Harvard University in Cambridge, MA, USA, December 14–15, 2013. This is an annual meeting, which is traditionally co-located with another, related, conference. WAW 2013 was co-located with the 9th Conference on Web and Internet Economics (WINE 2013). Co-location of the workshop and conference provided opportunities for researchers in two different but interrelated areas to interact and to exchange research ideas. It was an effective venue for the dissemination of new results and for fostering research collaboration.

The World Wide Web has become part of our everyday life, and information retrieval and data mining on the Web are now of enormous practical interest. The algorithms supporting these activities combine the view of the Web as a text repository and as a graph, induced in various ways by links among pages, hosts, and users. The aim of the workshop was to further the understanding of graphs that arise from the Web and various user activities on the Web, and stimulate the development of high-performance algorithms and applications that exploit these graphs. The workshop gathered the researchers who are working on graph-theoretic and algorithmic aspects of related complex networks, including citation networks, social networks, biological networks, molecular networks, and other networks arising from the Internet.

This volume contains the papers presented during the workshop. Each submission was reviewed by Program Committee members. Papers were submitted and reviewed using the EasyChair online system. The committee members decided to accept 17 papers.

December 2013

Anthony Bonato
Michael Mitzenmacher
Paweł Prałat

Preface

The Tenth Workshop on Algorithms and Models for the Web Graph (WAW 2013) took place at Harvard University in Cambridge, MA, USA, during December 14–15, 2013. This is an annual meeting, which is traditionally co-located with another, related, conference. WAW 2013 was co-located with the Workshop on Internet and Network Economics (WINE 2013). Co-location of the workshop and conference provided opportunities for researchers in two different but interrelated areas to interact and to exchange research ideas. It was an effective venue for the dissemination of new results and for fostering research collaboration.

The World Wide Web has become part of our everyday life, and information retrieval and data mining on the Web are now of enormous practical interest. The algorithms supporting these activities combine the view of the Web as a text repository and as a graph, induced in various ways by links among pages, hosts and users. The aim of the workshop was to further the understanding of graphs that arise from the Web and various user activities on the Web, and stimulate the development of high-performance algorithms and applications that exploit these graphs. The workshop gathered the researchers who are working on graph-theoretic and algorithmic aspects of related complex networks, including citation networks, social networks, biological networks, molecular networks, and other networks arising from the Internet.

This volume contains the papers presented during the workshop. Each submission was reviewed by Program Committee members. Papers were submitted and reviewed using the EasyChair online system. The committee members decided to accept 17 papers.

December 2013 Anthony Bonato
Fan Chung Graham
Paweł Prałat

Organization

General Chairs

Andrei Z. Broder	Google Research
Fan Chung Graham	University of California San Diego, USA

Organizing Committee

Anthony Bonato	Ryerson University, Canada
Michael Mitzenmacher	Harvard University, USA
Paweł Prałat	Ryerson University, Canada

Sponsoring Institutions

Google
Internet Mathematics
Microsoft Research New England
National Science Foundation
Ryerson University

Program Committee

Dimitris Achlioptas	UC Santa Cruz, USA
Konstantin Avratchenkov	Inria, France
Paolo Boldi	University of Milano, Italy
Anthony Bonato	Ryerson University, Canada
Milan Bradonjic	Bell Laboratories, USA
Fan Chung Graham	UC San Diego, USA
Colin Cooper	King's College, UK
Anirban Dasgupta	Yahoo! Research
Luca de Alfaro	UC Santa Cruz, USA
Raissa D'Souza	UC Davis, USA
Alan Frieze	Carnegie Mellon University, USA
David Gleich	Purdue University, USA
Adam Henry	University of Arizona, USA
Paul Horn	Harvard University, USA
Jeannette Janssen	Dalhousie University, Canada
Ravi Kumar	Google Research
Stefano Leonardi	University of Rome, Italy
Nelly Litvak	University of Twente, The Netherlands

Lincoln Lu	University of South Carolina, USA
Oliver Mason	National University of Ireland
Michael Mitzenmacher	Harvard University, USA
Peter Morters	University of Bath, UK
Mariana Olvera-Cravioto	Columbia University, USA
Allon Percus	Claremont Graduate University, USA
Pawel Pralat	Ryerson University, Canada
D. Sivakumar	Google Research
Stephen Young	University of Louisville, USA

Table of Contents

Asymmetric Distribution of Nodes in the Spatial Preferred Attachment Model

Jeannette Janssen[1], Paweł Prałat[2], and Rory Wilson[1]

[1] Dalhousie University, Halifax, Canada
{jeannette.janssen,rory.ross.wilson}@dal.ca
[2] Ryerson University, Toronto, Canada
pralat@ryerson.ca

Abstract. In this paper, a spatial preferential attachment model for complex networks in which there is non-uniform distribution of the nodes in the metric space is studied. In this model, the metric layout represents hidden information about the similarity and community structure of the nodes. It is found that, for density functions that are locally constant, the graph properties can be well approximated by considering the graph as a union of graphs from uniform density spatial models corresponding to the regions of different densities. Moreover, methods from the uniform case can be used to extract information about the metric layout. Specifically, through link and co-citation analysis the density of a node's region can be estimated and the pairwise distances for certain nodes can be recovered with good accuracy.

Keywords: Spatial Random Graphs, Spatial Preferred Attachment Model, Preferential Attachment, Complex Networks, Web Graph, Co-citation, Common Neighbours.

1 Introduction

There has been a great deal of recent interest in modelling complex networks, a result of the increasing connectedness of our world. The hyperlinked structure of the Web, citation patterns, friendship relationships, infectious disease spread, these are seemingly disparate collections of entities which have fundamentally very similar natures.

Many models of complex networks—such as copy models and preferential attachment models—have a common weakness: the 'uniformity' of the nodes; other than link structure there is no way to distinguish the nodes. One family of models which overcomes this deficiency is spatial (or geometric) models, wherein the nodes are embedded in a metric space. A node's position—especially in relation to the others—has real-world meaning: the character of the node is encoded in its location. Similar nodes are closer in the space than dissimilar nodes. This distance has many potential meanings: in communication networks, perhaps physical distance; in a friendship graph, an interest space; in the World Wide Web, a topic space. As an illustration, a node representing a webpage on

A. Bonato, M. Mitzenmacher, and P. Prałat (Eds.): WAW 2013, LNCS 8305, pp. 1–13, 2013.
© Springer International Publishing Switzerland 2013

pet food would be closer in the metric space to one on general pet care than to one on travel.

The Spatial Preferred Attachment Model [1], designed as a model for the World Wide Web, is one such spatial model. Indeed, as its name suggests, the SPA Model combines geometry and preferential attachment. Setting the SPA Model apart is the incorporation of 'spheres of influence' to accomplish preferential attachment: the greater the degree of the node, the larger its sphere of influence, and hence the higher the likelihood of the node gaining more neighbours. The SPA model produces scale-free networks, which exhibit many of the characteristics of real-life networks (see [1,4]). In [9], it was shown that the SPA model gave the best fit, in terms of graph structure, for a series of social networks derived from Facebook.

As the motivation behind spatial models is the 'second layer of meaning'—the character of the nodes as represented by their positions in the metric space—we hope to uncover this layer through examination of the link structure. In particular, estimating the distance between nodes in the metric space forms the basis for two important link mining tasks: finding entities that are similar—represented by nodes that are close together in the metric space—and finding communities—represented by spatial clusters of nodes in the metric space. We show how a theoretical analysis of a spatial model can lead to reliable tools to extract the 'second layer of meaning'.

The majority of the spatial models to this point have used uniform random distribution of nodes in the space. However, considering the real-world networks these models represent, this concept is impractical: indeed, on a basic level, if the metric space represents actual physical space, and the nodes people, then we note that people cluster in cities and towns, rather than being uniformly spread across the land. More abstractly, there are more webpages on a popular topic, corresponding to a small area of our metric space, than for a more obscure topic. The development of spatial network models naturally then begins to incorporate varying densities of node distribution: both 'clumps' of higher/lower density, as well as gradually changing densities, are both possibilities.

Of the more important goals is that of community recognition: the discovery and quantification of characteristically (semantically) similar nodes.

In this work we generalize the SPA model to non-homogeneous distribution of nodes within the space. We assume very distinct regions of different densities, 'clusters'. We find they behave almost as independent SPA Models of parameters derived from the densities. Many earlier results from the SPA Model then translate easily to this asymmetric version and we begin the process of uncovering the geometry using link analysis.

1.1 Background and Related Work

Efforts to extract node information through link analysis began with a heuristic quantification of entity similarity: numerical values, obtained from the graph structure, indicating the relatedness of two nodes. Early simple measures of entity similarity, such as the Jaccard coefficient [12], gave way to iterative graph theoretic measures, in which two objects are similar if they are related to similar objects, such as SimRank [10]. Many such measures also incorporate co-citation, the number of common neighbours of two nodes, as proposed in the paper by Small [13].

The development of graph models, in particular spatial models—as explored in [3] using thresholds, in combination with protean graphs [2] and with preferential attachment [5,7]—added another dimension to node information extraction. For example, in [6], the authors make inferences on the social space for nodes in a social network, using Bayesian methods and maximum likelihood. But in particular, the authors' previous paper, [8], used common neighbours in a spatial model of the World Wide Web [1] to explore the underlying geometry and quantify node similarity based on distance in the space. In this paper, we draw heavily from [4], which includes further results on the SPA model, and in particular from [8] and extend its results to a generalization that allows us to overcome the reliance on uniform random distribution of nodes in the space. Non-uniform distributions have also been explored in [11,14], as we move to more realistic models.

1.2 The Asymmetric SPA Model

We begin with a brief description of our Asymmetric SPA model. The model presented here is a generalization of the SPA model introduced in [1], the main difference being that we allow for an inhomogeneous distribution of nodes in the space.

Let S be the unit hypercube in $\mathbb{R}^m$, equipped with the torus metric derived from the Euclidean norm, or any equivalent metric. The nodes $\{v_t\}_{t=1}^n$ of the graphs produced by the SPA model are points in S chosen via an m-dimensional point process. Most generally, the process is given by a probability density function ρ; ρ is a measurable function such that $\int_S \rho d\mu = 1$. Precisely, for any measurable set $A \subseteq S$ and any t such that $1 \leq t \leq n$, $\mathbb{P}(v_t \in A) = \int_A \rho d\mu$.

In fact, we will restrict ourselves to probability functions that are *locally constant*. Precisely, we assume that the space $S = [0,1)^m$ is divided into k^m equal sized hypercubes, where k is a constant natural number. Each hypercube is of the form $I_{j_1} \times I_{j_2} \times \cdots \times I_{j_m}$ $(0 \leq j_1, j_2, \ldots, j_m < k)$, where $I_j = [j/k, (j+1)/k)$. Note that any density function ρ can be approximated by such a locally constant function, so that this restriction is justified.

To keep notation as simple as possible, we assume that each hypercube is labelled $\mathcal{R}_\ell$, $1 \leq \ell \leq k^m$. Let ρ_ℓ be the density of $\mathcal{R}_\ell$, so the density function has value ρ_ℓ on $\mathcal{R}_\ell$. For any node v, let $\mathcal{R}(v)$ be the hypercube containing v, and let $\rho(v)$ be the density of $\mathcal{R}(v)$. Clearly, every hypercube has volume k^{-m}. Then the probability that a node v_t, introduced at time t, falls in $\mathcal{R}_\ell$ equals $q_\ell = \rho_\ell k^{-m}$, and the expected number of points in $\mathcal{R}_\ell$ equals $\rho_\ell k^{-m} n$. It is easy to see that $\sum_\ell q_\ell = 1$. Thus we model the point process as follows: at each time step t, one of the regions is chosen as the destination of v_t; region $\mathcal{R}_\ell$ is chosen with probability q_ℓ. Then, a location for v_t is chosen uniformly at random from $\mathcal{R}_\ell$.

The SPA model generates stochastic sequences for graphs $(G_t : t \geq 0)$ with edge set E_t and node set $V_t \subseteq S$. The in-degree of a node v at time t is given by $\deg^-(v,t)$. Likewise the out-degree is given by $\deg^+(v,t)$. The sphere of influence of a node v at time t is defined as the ball, centred at v, with total volume

$$|S(v,t)| = \frac{A_1 \deg^-(v,t) + A_2}{t},$$

where $A_1, A_2 > 0$ are given parameters. If $(A_1 \deg^-(v,t) + A_2)/t \geq 1$, then $S(v,t) = S$ and so $|S(v,t)| = 1$. We impose the additional restriction that $pA_1 \max_j \rho_j < 1$; this avoids regions becoming too dense. This property will be always assumed. The generation of a SPA model graph begins at time $t = 0$ with G_0 being the null graph. At each time step $t \geq 1$ (defined to be the transition from G_{t-1} to G_t), a node v_t is chosen from S according to the given spatial distribution, and added to V_{t-1} to form V_t. Next, independently, for each node $u \in V_{t-1}$ such that $v_t \in S(u, t-1)$, a directed link (v_t, u) is created with probability p, $p \in (0,1)$ being another parameter of the model.

Let $\delta(v)$ be the distance from v to the boundary of $\mathcal{R}(v)$. Let $r(v,t)$ be the radius of the sphere of influence of node v at time t. So if $r(v,t) \leq \delta(v)$, then $S(v,t)$ is completely contained in $\mathcal{R}(v)$ at time t. We see that

$$r(v,t) = (|S(v,t)|/c_m)^{1/m} = \left(\frac{A_1 \deg^-(v,t) + A_2}{c_m t}\right)^{1/m},$$

where c_m is the volume of the unit ball; for example, in 2-dimensions with the Euclidean metric, $c_2 = \pi$.

Our goal is to investigate typical properties of a graph G_n on n nodes, and to use these to infer the spatial layout of the nodes. As typical in random graph theory, we shall consider only asymptotic properties of G_n as $n \to \infty$. We say that an event in a probability space holds *asymptotically almost surely* (a.a.s.) if its probability tends to one as n goes to infinity.

2 Graph Properties of the SPA Model

In the Asymmetric SPA model with a locally constant density function, the probability of an edge forming from a new node v_t to an existing node v at time t equals

$$\mathbb{P}((v_t, v) \in E(G_n)) = p \int_{S(v,t)} \rho d\mu = p \sum_{\ell} \rho_\ell \, |S(v,t) \cap \mathcal{R}_\ell|.$$

Thus, the stochastic process of edge formation in the Asymmetric SPA model is bounded below by the process in which the edge probability is governed by $p\rho_{\min}$, and bounded above by that with $p\rho_{\max}$, where $\rho_{\min}$ and $\rho_{\max}$ are, respectively, the smallest and the largest densities occurring. The bounds on the link probability $\mathbb{P}((v_t, v) \in E(G_n))$ lead to bounds on the expected value of the degree.

Theorem 1. *Let $\omega = \omega(n)$ be any function tending to infinity together with n. The expected in-degree at time t of a node v_i born at time $i \geq \omega$ is given by*

$$(1+o(1))\frac{A_2}{A_1}\left(\frac{t}{i}\right)^{p\rho_{\min}A_1} - \frac{A_2}{A_1} \leq \mathbb{E}(\deg^-(v_i,t)) \leq (1+o(1))\frac{A_2}{A_1}\left(\frac{t}{i}\right)^{p\rho_{\max}A_1} - \frac{A_2}{A_1}.$$

In the analysis of the original SPA model, we find that nodes born quite early have their spheres of influence typically shrinking rapidly, and nodes born late start with small spheres of influence. A node would have to be quite close to the boundary of its region with another for the effect of any other region to be felt. It seems reasonable to expect that the graph formed by nodes in a region $\mathcal{R}_\ell$ with local density ρ_ℓ behaves like an independent SPA model of density ρ_ℓ.

To be specific, assume that nodes in the SPA model do not arrive at fixed time instances t, but instead arrive according to a homogeneous Poisson process with rate 1. (This will not significantly change the analysis.) Then, the process inside a region $\mathcal{R}$ with density ρ will behave like a SPA model with the same parameters A_1, A_2 and p, but with points arriving according to a Poisson process with rate ρ. This means that in each time interval we expect ρ points to arrive, and the expected time interval between arrivals equals $1/\rho$. If we use v_t to denote the t-th node arriving, then the arrival time $a(t)$ of v_t is approximately t/ρ, and thus the volume of the sphere of influence of an existing node v at the time that v_t is born equals

$$|S(v,a(t))| = \frac{A_1 \deg^-(v,a(t)) + A_2}{a(t)} \approx \frac{\rho A_1 \deg^-(v,a(t)) + \rho A_2}{t}.$$

Thus, in the analysis of the degree of an individual node, we expect a node v in the asymmetric SPA model to behave like a node in the original SPA model with parameters $\rho(v)A_1$, $\rho(v)A_2$ instead of A_1, A_2, where the degree of node v at time t in the Asymmetric SPA model corresponds to the degree of a node at time $a(t)$ in the corresponding SPA model. The following theorems show that this is indeed the case.

Theorem 2. *Let $\omega = \omega(n)$ be any function tending to infinity together with n. The expected in-degree at time t of a node v_i born at time $i \geq \omega \log n$, with $\delta(v) \gg (\log n/i)^{1/m}$ is given by*

$$\mathbb{E}(\deg^-(v_i,t)) = (1+o(1))\frac{A_2}{A_1}\left(\frac{t}{i}\right)^{p\rho(v)A_1} - \frac{A_2}{A_1}.$$

Theorem 3. *Let $\omega = \omega(n)$ be any function tending to infinity together with n, and let $\varepsilon > 0$. The following holds a.a.s. For every node v for which* $\deg^-(v,n) =$ $k = k(n) \geq \omega \log n$ *and for which*

$$\delta(v) \geq (1+\varepsilon)\left(\frac{A_1 k + A_2}{c_m n}\right)^{1/m},$$

it holds that for all values of t such that $\max\{t_v, T_v\} \le t \le n$,

$$\deg^-(v,t) = (1+o(1))k\left(\frac{t}{n}\right)^{p\rho(v)A_1}.$$

Times T_v and t_v are defined as follows:

$$T_v = n\left(\frac{\omega \log n}{k}\right)^{p\rho(v)A_1}, \quad t_v = (1+\varepsilon)\left(\frac{A_1 k}{\delta^m c_m n^{p\rho(v)A_1}}\right)^{\frac{1}{1-p\rho A_1}}.$$

The statement of the theorem is rather technical, so we lay it out conceptually:

- the condition on $\delta(v)$ ensures that at time n, $S(v,n)$ is completely contained in $\mathcal{R}(v)$ (the factor of $(1+\varepsilon)$ gives some extra room for argument),
- time T_v is the time node v has $\omega \log n$ neighbours, provided that the process behaves as we expect,
- time t_v is the time when the sphere of influence has shrunk to the point where it became completely contained in $\mathcal{R}(v)$, provided the process behaves well (again, with extra room due to the factor $(1+\varepsilon)$). This occurs at the moment when the expected radius of the sphere of influence is smaller than $\delta(v)$.

The implication of this theorem is that once a node accumulates $\omega \log n$ neighbours and its sphere of influence has shrunk so that it does not intersect neighbouring regions, its behaviour can be predicted with high probability until the end of the process, and is completely governed by its region, and no others.

We note that if $\max\{T_v, t_v\} = t_v$ for a node v, then at time T_v—the time v first reaches in-degree $\omega \log n$—its sphere of influence extends beyond the region of v. However, since a.a.s. no node has degree $\omega \log n$ at time $O(\omega \log n)$, it must be that $T_v \gg \omega \log n$. Thus at time T_v the radius of the sphere of influence of v is $O\left((\omega \log n/T_v)^{1/m}\right) = o(1)$. The implication is that, in order for $\max\{T_v, t_v\}$ to be equal to t_v, a node would have to be very close to the border, that is, $\delta(v) = o(1)$. So for most nodes under consideration, $\max\{T_v, t_v\} = T_v$, and they behave like in a uniform SPA model of density $\rho(v)$ as soon as their degree reaches $\omega \log n$. Further, of these nodes, those with $\deg(v,n) \ge \omega^2 \log n$ reach degree $\omega \log n$ at time $o(n)$, and so have $o(\deg(v,n))$ neighbours outside $\mathcal{R}(v)$.

We can use the results on the degree to show that each graph induced by one of the regions $\mathcal{R}_\ell$ has a power law degree distribution. Let $N_i(j,n)$ denote the number of nodes of degree j at time n in the region $\mathcal{R}_i$ and let $j_f = j_f(n) = (n/\log^8 n)^{\frac{p\rho_{\max} A_1}{4p\rho_{\max} A_1 + 2}}$.

Theorem 4. *A.a.s. the graph induced by the nodes in region $\mathcal{R}_\ell$ has a power law degree distribution with coefficient $1 + 1/p\rho_i A_1$. Precisely,* a.a.s. *for any $1 \le i \le k^m$ there exists a constant c_i such that for any $1 \ll j \le j_f$,*

$$N_i(j,n) = (1+o(1))c_i j^{-(1+\frac{1}{p\rho_i A_1})} q_i n.$$

Moreover, a.a.s. *the entire graph generated by the Asymmetric SPA model has a degree distribution whose tail follows a power law with coefficient $1 + 1/p\rho_{\max} A_1$.*

The number of edges also validates our hypothesis that a region of a certain density behaves almost as a uniform SPA model with adjusted parameters. In the SPA model with parameters $\rho_\ell A_1$, $\rho_\ell A_2$ and p, the average out-degree is approximately $\frac{p\rho_\ell A_2}{1-p\rho_\ell A_1}$, as per [1, Theorem 1.3]. The following theorem shows that the subgraph induced by one of the regions has the equivalent expected number of edges, and most edges have both endpoints in the same region. Moreover, this result shows why we need the condition $p\rho_{\max} A_1 < 1$. In fact, if $p\rho_{\max} A_1 \geq 1$, then the number of edges will grow superlinearly.

Theorem 5. *For a region $\mathcal{R}_\ell$ of density ρ_ℓ,* a.a.s. $|V(G_n) \cap \mathcal{R}_\ell| = (1+o(1))q_\ell n$. *Moreover,*

$$\mathbb{E}(\{(u,v) \in E(G_n) \,|\, u,v \in \mathcal{R}_\ell\}|) = (1+o(1))\frac{p\rho_\ell A_2}{1-p\rho_\ell A_1} q_\ell n.$$

Furthermore, a.a.s.

$$|\{(u,v) \in E(G_n) : \mathcal{R}(u) \neq \mathcal{R}(v)\}| = o(n),$$

i.e. the number of edges that cross the boundary of $\mathcal{R}_\ell$ is of smaller order than the number of edges completely contained in the region.

Our ultimate goal is to derive the pairwise distances between the nodes in the metric space through an analysis of the graph. The following theorem, obtained using the approach of [8], provides an important tool. Namely, it links the number of common in-neighbours of a pair of nodes to their (metric) distance. Using this theorem, we can then infer the distance from the number of common in-neighbours. Let $cn(u,v)$ denote the number of common in-neighbours of two nodes u and v.

The theorem distinguishes three cases. If u and v are relatively far from each other, then a.a.s. they will have no common neighbours. If the nodes are very close, then the number of common neighbours is approximately equal to a fraction p of the degree of the node of smallest degree. The third case provides a 'sweet spot' where the number of common neighbours is a direct function of the metric distance and the degrees of the nodes. For any two nodes u and v, let $cn(u,v,t)$ denote the number of common in-neighbours of u and v at time t.

Theorem 6. *Let $\omega = \omega(n)$ be any function tending to infinity together with n, and let $\varepsilon > 0$. The following holds* a.a.s. *Let u and v be nodes of final degrees* $\deg(u,n) = k$ *and* $\deg(v,n) = j$ *such that $\mathcal{R}(u) = \mathcal{R}(v)$, and $k \geq j \geq \omega^2 \log n$.*

Let $\rho = \rho(v)$ and let $T_v = n\,(\omega \log n/j)^{p\rho A_1}$, and assume that

$$\delta(v)^m \geq cj \text{ and } \delta(u)^m \geq ck, \text{ where } c = (1+\varepsilon)\left(\frac{A_1}{c_m n^{p\rho A_1} T_v^{1-p\rho A_1}}\right).$$

Let $d(u,v)$ be the distance between u and v in the metric space. Then, we have the following result about the number of common in-neighbours of u and v:

Case 1. If for some $\varepsilon > 0$

$$d(u,v) \geq \varepsilon \left(\frac{\omega \log n(k/j)}{T_v} \right)^{1/m}$$

then $cn(u,v,n) = O(\omega \log n)$.

Case 2. If $k \geq (1+\varepsilon)j$ *for some* $\varepsilon > 0$ *and*

$$d(u,v) \leq \left(\frac{A_1 k + A_2}{c_m n} \right)^{1/m} - \left(\frac{A_1 j + A_2}{c_m n} \right)^{1/m} = O\left(\left(\frac{k}{n} \right)^{1/m} \right),$$

then $cn(u,v,n) = (1+o(1))pj$. *If* $k = (1+o(1))j$ *and* $d(u,v)^m \ll (k/n) = (1+o(1))(j/n)$, *then* $cn(u,v,n) = (1+o(1))pj$ *as well.*

Case 3. If $k \geq (1+\varepsilon)j$ *for some* $\varepsilon > 0$ *and*

$$\left(\frac{A_1 k + A_2}{c_m n} \right)^{1/m} - \left(\frac{A_1 j + A_2}{c_m n} \right)^{1/m} < d(u,v) \ll \left(\frac{\omega \log n(k/j)}{T_v} \right)^{1/m},$$

then

$$cn(u,v,n) = C i_k^{-\frac{(p\rho A_1)^2}{1-p\rho A_1}} i_j^{-p\rho A_1} d(u,v)^{-\frac{mp\rho A_1}{1-p\rho A_1}} \left(1 + O\left(\left(\frac{i_k}{i_j} \right)^{p\rho A_1/m} \right) \right), \tag{1}$$

where $i_k = n\left(\frac{A_1}{A_2}k\right)^{-\frac{1}{p\rho A_1}}$ *and* $i_j = n\left(\frac{A_1}{A_2}j\right)^{-\frac{1}{p\rho A_1}}$, *and* $C = pA_1^{-1} A_2^{\frac{1}{1-p\rho A_1}} c_m^{-\frac{p\rho A_1}{1-p\rho A_1}}$. *If* $k = (1+o(1))j$ *and* $\varepsilon(k/n)^{1/m} < d(u,v) \ll (\omega \log n / T_v)^{1/m}$ *for some* $\varepsilon > 0$, *then*

$$cn(u,v,n) = \Theta\left(i_k^{-\frac{(p\rho A_1)^2}{1-p\rho A_1}} i_j^{-p\rho A_1} d(u,v)^{-\frac{mp\rho A_1}{1-p\rho A_1}} \right).$$

3 Reconstruction of Geometry

We set out to discover the character of nodes in a network purely through link structure, and to quantify the similarities. Spatial models allow us a convenient definition of similarity: distances between nodes. In examining the SPA model, the number of common neighbours allows us to uncover pairwise distances, a first step in the reconstruction of the geometry.

Description of Model Used. For simulations, we use an Asymmetric SPA model we call the *diagonal layout*, which has 4 'clusters' of identical high density, with $m = 2$. In the diagonal layout, $k = 4$ and the 4 regions (x, x), $1 \leq x \leq 4$, are dense, with the others sparse. We will use 'dense region' and 'sparse region' to denote the union of all regions with densities ρ_d and ρ_s, respectively. For ease of notation, we note that $\rho_s = 4/3 - \rho_d/3$ so it is enough to provide the value

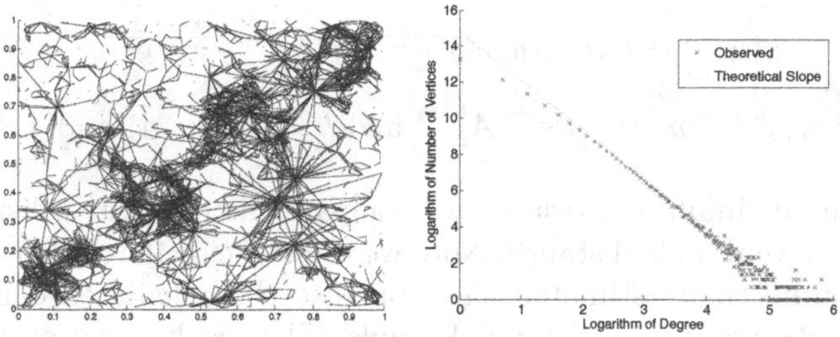

Fig. 1. Left: diagonal layout, $n = 1,000$, $p = 0.6$, $\rho_d = 1.6$, $A_1 = 0.7$, $A_2 = 2.0$; Right: degree distribution $n = 1,000,000$, $p = 0.7$, $\rho_d = 1.2$, $A_1 = 0.7$, $A_2 = 1.0$

of ρ_d only. In Figure 1 we see an example of the diagonal layout with nodes and edges, and we also see evidence that the densest region does dominate the power law degree distribution.

First we assume uniform density and apply the original estimator (Equation 7 from [8]) to our diagonal layout; the results are shown in the left in Figure 2. We eliminate those pairs we assume are in Case 1 (too close) and those in Case 2 (too far), by limiting our pairs to those with more than 10 common neighbours and fewer than $p/2\deg(v,n)$. This leaves 2270 pairs. The figure shows that the approach fails, and that it leads to a consistent overestimate of the distance for the nodes. This is somewhat counterintuitive, but the trouble lies with the estimator for a node's age based on in-degree: a node in $\mathcal{R}_d$ is thought to be much older than it actually is and confounds the distance estimator.

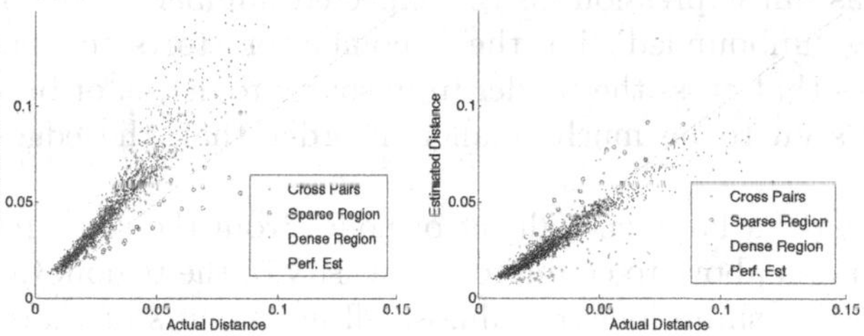

Fig. 2. SPA model, $n = 100,000$, diagonal layout, $p = 0.7$, $\rho_d = 1.6$, $A_1 = 0.7$, $A_2 = 2.0$, actual vs. estimated distances for pairs of nodes; Left: using original estimator; Right: using new estimator, density known

More precision is needed to take into account the varying densities. Examining Theorem 6, we note that for Case 3, equation (1) can be used to obtain an estimate $\hat{d}$ of the distance between a pair of nodes. For a pair of nodes u, v which are both in a region of density ρ, and their distance is such that Case 3 applies, this estimate is given by:

$$\hat{d}(u,v) = C_1(cn(u,v))^{-\frac{(1-p\rho A_1)}{mp\rho A_1}} k^{1/m} j^{\frac{(1-p\rho A_1)}{p\rho A_1 m}} \qquad (2)$$

where $C_1 = (nc_m)^{-1/m} p^{\frac{(1-p\rho(v_1))A_1)}{mp\rho(v_1)A_1}} A_1^{1/m}$ and $k = \deg(u,n)$ and $j = \deg(v,n)$, with $k \geq j$.

Using the same simulation results, we compare the estimated distance using Equation 2 vs. actual node distance. Note we use our calculated density for each node to determine their estimated ages, but use the calculated density of the node of higher degree in the distance formula. The results seen on the right in Figure 2 indicate that our new estimator is quite accurate in predicting distances for some pairs of nodes, given all the parameters of the model, except for the cross-border pairs.

3.1 Estimating the Density

In real-world situations, we cannot assume to know the density of the region containing a given node. In fact, the density of the region containing a node is an important part of the 'second layer of meaning' which we aim to extract from the graph. Therefore, in order to use our estimator for the distances between the nodes, we need to be able to use the graph structure to estimate the densities.

Using the theoretical results obtained from the previous section, we see that we can use the out-degrees of the in-neighbours of v to estimate the density of $\mathcal{R}(v)$. As per Theorem 5, the average out-degree in $\mathcal{R}_\ell$ is approximately $\frac{p\rho_\ell A_2}{1-p\rho_\ell A_1}$. Simulations confirm this expected value. Running sets of parameters 10 times each, we observe that if $p\rho_{\max}A_1 \leq 0.75$, the number of edges per region are within 90% of the expected value, on average. For $0.75 < p\rho_{\max}A_1 \leq 0.8$, the number of edges is within 75% of expected. For $p\rho_{\max}A_1 > 0.8$ we start to see deviation, as our expression for the expected number of edge in the densest region becomes 'unbounded', i.e. the denominator starts to approach 0. The number of edges that cross the border from sparse to dense, or between clusters, is consistently seen to be much smaller in order than the edges within each region.

Thus, if we have a large enough set of nodes from the same region, then we can use the formula above to estimate the density of the region. Consider a node v, and make two assumptions: (i) almost all neighbours of v are contained in $\mathcal{R}(v)$, and (ii) the neighbours of v form a representative sample of all nodes of $\mathcal{R}(v)$. Simulations show that these assumptions are justified and allow us to make an estimate for $\rho(v)$.

Set $\overline{\deg^+}(N^-(v))$ to be the average out-degree of the in-neighbours of v. Assuming the in-neighbours of v are also in $\mathcal{R}(v)$ (a fair assumption, given our earlier theorems), an estimator for the density can be derived from the average out-degree:

$$\hat{\rho(v)} = \frac{\overline{\deg^+}(N^-(v))}{pA_2 + pA_1\overline{\deg^+}(N^-(v))}.$$

We see in the histograms in Figure 3 (Left and Center) that the average out-degree of a node's in-neighbours in the dense region of the diagonal layout is quite accurate, but for the sparse region, the average out-degree is higher than expected. Displayed are the results for nodes with $\deg^-(v) \geq 10$. The calculated theoretical value for the average out-degree of in-neighbours for a node in the dense region is 5.85, and in the sparse, 1.45. This translates in $\rho_d = 1.6$ and $\rho_s = 0.8$. We see peaks that are quite accurate for the dense region, but translated to the right for the sparse region. Likely, those are sparse region nodes located close to the border; our condition on the minimum degree favours the 'rich' sparse region nodes.

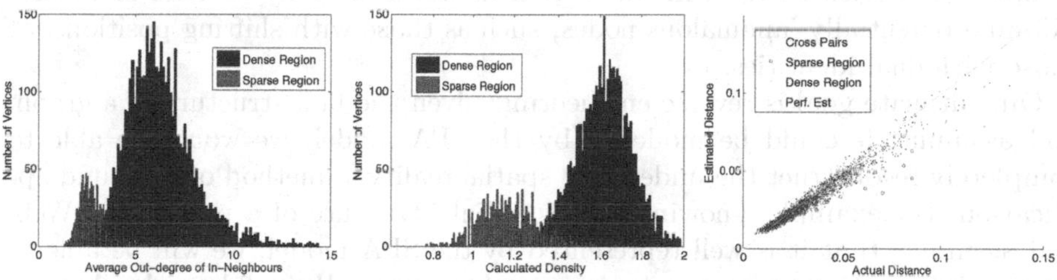

Fig. 3. Diagonal layout, $n = 100,000$, $\rho_d = 1.6$, $A_1 = 0.7$, $A_2 = 2.0$; Left: $p = 0.6$, average out-degree of the in-neighbours; Center: $p = 0.6$, calculated density from average out-degree; Right: $p = 0.7$, using estimated density from the node of greater final degree, all other parameters known

Finally, we use $\hat{\rho}$, and knowing all other parameters, to calculate the distance between the nodes based on number of common neighbours, Equation 2, using the same simulation results as earlier. Again note we use our calculated density for each node to determine their estimated ages, but use the calculated density of the node of higher degree in the distance formula. Using the lower degree node gives similar results. The results are seen in Figure 3 (Right). We obtain very good agreement between calculated and estimated densities.

4 Conclusion

Our analysis of a SPA model with non-uniform random distribution of nodes reveals almost independent clusters of nodes. Expected degree, degree distribution and number of edges behave as they would in localized SPA models with 'adjusted' parameters. It is not examined here, but it is suspected that these adjusted parameters extend to other existing results on the SPA Model such as the small world property and spectral properties: this is a goal of future work. The main result of the paper is that, by using the average out-degree of the in-neighbours of a node, an estimate of its region's density can be obtained. With this density in hand, the examination of common neighbours for pairs of nodes allows us to find their distances in the metric space. Currently, the number of pairs of nodes for which we have distances is quite limited due to the nature of the spheres of influence: that they either start small or shrink rapidly results in only those pairs that are very close having a significant number of common neighbours. Attempts to increase our information could include the use of path lengths, second neighbourhoods, etc.

Although the theoretical results are interesting in and of themselves, further work can be done in examining their validity in the context of real networks, i.e. recovering *meaningful* distances for pairs of nodes. Early results using machine learning and graphlets show that the SPA Model can be an accurate representation of social networks [9]; it would be ideal to extend our knowledge of the accuracy of the SPA Model, in particular the Asymmetric SPA Model, to other complex networks. In the context of real networks we may be able to further examine potentially 'anomalous nodes, such as those with shifting positions, or those with dual identities.

Our ultimate goal is reverse engineering: given the link structure of a graph, and assuming it could be modelled by the SPA model, we would be able to completely reconstruct the underlying spatial reality, a method of profound application. For example, knowing the hyperlink structure of a part of the Web, and assuming that it is well represented by the SPA model, we will be able to use this information to create a topic map of the pages. We will have developed a very powerful tool for prediction in the Web, with both economic and sociological benefits, such as improved web search and the discovery of cyber-communities.

References

1. Aiello, W., Bonato, A., Cooper, C., Janssen, J., Prałat, P.: A spatial web graph model with local influence regions. Internet Mathematics 5(1-2), 175–196 (2008)
2. Bonato, A., Janssen, J., Prałat, P.: Geometric protean graphs. Internet Mathematics 8(1-2), 2–28 (2012)
3. Bradonjić, M., Hagberg, A., Percus, A.G.: Giant component and connectivity in geographical threshold graphs. In: Bonato, A., Chung, F.R.K. (eds.) WAW 2007. LNCS, vol. 4863, pp. 209–216. Springer, Heidelberg (2007)
4. Cooper, C., Frieze, A.M., Prałat, P.: Some typical properties of the spatial preferred attachment model. In: Bonato, A., Janssen, J. (eds.) WAW 2012. LNCS, vol. 7323, pp. 29–40. Springer, Heidelberg (2012)
5. Flaxman, A.D., Frieze, A.M., Vera, J.: A geometric preferential attachment model of networks. In: Leonardi, S. (ed.) WAW 2004. LNCS, vol. 3243, pp. 44–55. Springer, Heidelberg (2004)
6. Hoff, P., Raftery, A., Handcock, M.: Latent space approaches to social network analysis. Journal of the American Statistical Association 97, 1090–1098 (2001)
7. Jacob, E., Mörters, P.: Spatial preferential attachment networks: Power laws and clustering coefficients. ArXiv e-prints (October 2012)
8. Janssen, J., Prałat, P., Wilson, R.: Geometric graph properties of the spatial preferred attachment model. Advances in Applied Mathematics 50(2), 243–267 (2013)
9. Janssen, J., Hurshman, M., Kalyaniwalla, N.: Model selection for social networks using graphlets. Internet Math. 8(4), 338–363 (2012)
10. Jeh, G., Widom, J.: SimRank: a measure of structural-context similarity. In: Knowledge Discovery and Data Mining, pp. 538–543 (2002)

11. Jordan, J.: Geometric preferential attachment in non-uniform metric spaces. ArXiv e-prints (August 2012)
12. Kobayakawa, M., Kinjo, S., Hoshi, M., Ohmori, T., Yamamoto, A.: Fast computation of similarity based on jaccard coefficient for composition-based image retrieval. In: Muneesawang, P., Wu, F., Kumazawa, I., Roeksabutr, A., Liao, M., Tang, X. (eds.) PCM 2009. LNCS, vol. 5879, pp. 949–955. Springer, Heidelberg (2009)
13. Small, H.: Co-citation in the scientific literature: A new measure of the relationship between two documents. Journal of the American Society for Information Science 24(4), 265–269 (1973)
14. Zhang, J.: Growing Random Geometric Graph Models of Super-linear Scaling Law. ArXiv e-prints (December 2012)

A Spatial Preferential Attachment Model with Local Clustering

Emmanuel Jacob[1] and Peter Mörters[2]

[1] École Normale Supérieure de Lyon
[2] University of Bath

Abstract. A class of growing networks is introduced in which new nodes are given a spatial position and are connected to existing nodes with a probability mechanism favouring short distances and high degrees. The competition of preferential attachment and spatial clustering gives this model a range of interesting properties. Most notably, empirical degree distributions converge to a limit law, which can be a power law with any exponent $\tau > 2$, and the average clustering coefficient converges to a positive limit. Our main tool to show these and other results is a weak law of large numbers in the spirit of Penrose and Yukich, which can be applied thanks to a novel rescaling idea. We also conjecture that the networks have a robust giant component if τ is sufficiently small.

Keywords: Scale-free network, Barabasi-Albert model, preferential attachment, dynamical random graph, geometric random graph, power law, degree distribution, edge length distribution, clustering coefficient.

1 Introduction

Many of the phenomena in the complex world in which we live have a rough description as a large network of interacting components. It is therefore a fundamental problem to derive the global structure of such networks from basic local principles. A well established principle is the *preferential attachment paradigm* which suggests that networks are built by adding nodes and links successively, in such a way that new nodes prefer to be connected to existing nodes if they have a high degree [3]. The preferential attachment paradigm offers, for example, a credible explanation of the observation that many real networks have degree distributions following a *power law behaviour*. On the global scale preferential attachment networks are *robust* under random attack if the power law exponent is sufficiently small, and have logarithmic or doubly logarithmic diameters depending on the power law exponent. These features, together with a reasonable degree of mathematical tractability, have all contributed to the enormous popularity of these models. Unfortunately, the local structure of preferential attachment networks significantly deviates from that observed in real networks. In preferential attachment models the neighbourhoods of typical nodes have a tree-like topology [11], [4], which is a crucial feature for their mathematical analysis, but is not in line with the behaviour of many real world networks.

A. Bonato, M. Mitzenmacher, and P. Prałat (Eds.): WAW 2013, LNCS 8305, pp. 14–25, 2013.
© Springer International Publishing Switzerland 2013

The most popular quantity to measure the local clustering of networks are the *clustering coefficients*, which are measured to be positive in most real networks, but which invariably vanish in preferential attachment models that do not incorporate further effects [2], [6]. A possible reason for the clustering of real networks is the presence of a hidden variable assigned to the nodes, such that similarity of values is a further incentive to form links. For the class of protean graphs this idea has allowed Bonato et al. in [7] to generate power law networks with spatial clustering. Several authors have also proposed models combining preferential attachment with spatial features. Among the mathematically sound attempts in this direction are the papers of Flaxman et al. [12], [13], Jordan [16], Aiello et al. [1] and Cooper et al. [8]. These papers show that combining preferential attachment and spatial dependence can retain the global power law behaviour while changing the local topology of the network, for example by showing that the resulting graphs have small separators [12], [13]. None of these papers discusses clustering by analysing the clustering coefficients.

In this paper, we study a generalisation and variant of the spatial preferred attachment (SPA) model introduced in Aiello et al. [1] and further studied in Janssen et al. [15] and Cooper et al. [8]. The original model is based on the idea that a vertex at position x has a ball of influence centred in x. A new vertex can only be connected to it, if it falls within this ball, in which case it does so with a given probability p_0. The preferential attachment effect is modelled through the fact that the size of the ball depends on the degree of the vertex. In our model, this ball of influence is replaced by a profile, rotationally symmetric around x, with the probability of a connection given by the height of the profile. This allows us to relax the spatial rigidity of the model, so that for example two vertices always have a positive probability of being connected, whatever their positions. This generalisation induces a richer phenomenology, in particular when it comes to more complex statistics such as the edge length distribution or the existence of a giant component.

Our analysis of this model is using methods developed originally for the study of random geometric graphs, see Penrose and Yukich [19] for a seminal paper in this area and [18] for an exhibition. Our approach is based on a rescaling which transforms the growth in time into a growth in space. This transformation stabilises the neighbourhoods of a typical vertex and allows us to observe convergence of the local neighbourhoods of typical vertices in the graph to an infinite graph. This infinite graph, which is not a tree, is locally finite and can be described by means of a Poisson point process. We establish a weak law of large numbers, similar to the one given in [19], which allows us to deduce convergence results for a large class of functionals of the graph. For example, we show that the average clustering coefficient always converges to a positive constant for the scale-free networks given by SPA models. We also observe interesting phase transitions in the behaviour of the global clustering coefficient and the empirical edge length distribution. Finally, we informally discuss the existence of a robust giant component, one of the key features of preferential attachment networks which we would like to see retained in our model.

2 The Model

The generalized SPA model may be defined in a variety of metric spaces S. In this paper, we work in dimension $d \geq 1$, and we choose a distance d on $\mathbb{R}^d$ derived from any of the l_p norms. Similarly as in [8], [15] we let S be the unit hypercube in $\mathbb{R}^d$, centred at 0, equipped with its own torus metric d_1, i.e. for any two points $(x,y) \in S$, we set $\mathsf{d}_1(x,y) = \min\{\mathsf{d}(x, y+u) : u \in \{-1,0,1\}^d\}$. Note that S equipped with the torus metric has no boundary and is spatially homogeneous, which avoids some technical difficulties. Let $\mathcal{X}$ denote a Poisson point process of unit intensity on $S \times (0,\infty)$. A point $\mathbf{x} = (x,s)$ in $\mathcal{X}$ is a vertex $\mathbf{x}$, born at time s and placed at position x. Observe that, almost surely, two points of $\mathcal{X}$ neither have the same birth time nor the same position. We say that (x,s) is *older* than (y,t) if $s < t$. An edge is always oriented from the younger to the older vertex. For $t > 0$, write $\mathcal{X}_t$ for $\mathcal{X} \cap (S \times (0,t])$, the set of vertices already born at time t. We construct a growing sequence of graphs $(G_t)_{t>0}$, starting from the empty graph, and adding successively the vertices in $\mathcal{X}$ when they are born (so that the vertex set of G_t is $\mathcal{X}_t$), and connecting them to some of the older vertices. The rule is as follows:

> **Construction Rule**: Given the graph G_{t-} and $\mathbf{y} = (y,t) \in \mathcal{X}$, we add the vertex $\mathbf{y}$ and, independently for each vertex $\mathbf{x} = (x,s)$ in G_{t-}, we insert the edge $(\mathbf{y},\mathbf{x})$, independently of $\mathcal{X}$, with probability
>
> $$\varphi\left(\frac{t^{1/d}\mathsf{d}_1(x,y)}{f(\deg^-(\mathbf{x},t-))^{1/d}}\right). \tag{1}$$
>
> The resulting graph is denoted by G_t.

Here the following definitions and conventions apply:

(i) $\deg^-(\mathbf{x},t-)$ (resp. $\deg^-(\mathbf{x},t)$) denotes the indegree of vertex $\mathbf{x}$ at time $t-$ (respectively t), that is, the total number of incoming edges for the vertex $\mathbf{x}$ in G_{t-} (resp. G_t). Similarly, we denote by $\deg^+(\mathbf{y})$ the outdegree of vertex $\mathbf{y}$, which remains the same at all times $u \geq t$.

(ii) $f\colon \mathbb{N} \cup \{0\} \to (0,\infty)$ is the *attachment rule*. Informally, $f(k)$ quantifies the preferential 'strength' of a vertex of current indegree k, or likelihood of attracting new links. For simplicity, we suppose

$$f(k) = \gamma k + \beta, \quad \gamma \in (0,1), \beta > 0,$$

just as in [1], [8], but most of the results hold unchanged if f is only supposed to be increasing with asymptotic slope $\lim_{k\to\infty} f(k)/k = \gamma$.

(iii) φ: $[0,\infty) \to [0,1]$ is the *profile function*. It is non-increasing and satisfies

$$\int_{\mathbb{R}^d} \varphi(\mathsf{d}(0,y))\, dy = 1. \tag{2}$$

Informally, the profile function describes the spatial dependence of the probability that the newborn vertex $\mathbf{y}$ is linked to the existing vertex $\mathbf{x}$.

Loosely speaking, this form of the construction rule is the only one that ensures that we have a genuine interaction of the spatial and the preferential attachment effects, as a vertex is likely to be connected to a finite number of vertices within distance of order $t^{-1/d}$ and indegree of order 1, as well as to a finite number of vertices at distance $\gg t^{-1/d}$ and indegree $\gg 1$. If φ is integrable, the condition (2) is no loss of generality, as otherwise one can modify φ and f without changing the construction rule. Under (2) one can see that the interesting range of f (leading to degree distributions following an approximate power law) is characterised by asymptotic linearity. If $\gamma = pA_1$, $\beta = pA_2$ and $\varphi = p\mathbb{1}_{[0,r]}$, where r is chosen so that (2) is satisfied, we essentially get the original SPA model of [1], with the slight modification that we work in continuous time with random birth times. With this choice of profile function, the model can be interpreted as follows: Each vertex $\mathbf{x}$ is surrounded by a ball of influence, a ball centered at x and of volume $f(\deg^-(\mathbf{x},t-))/(pt)$. If the new vertex $\mathbf{y}$ falls within this ball of influence, then $\mathbf{y}$ and $\mathbf{x}$ are connected with probability p, otherwise they cannot be connected.

A general profile function can be seen as a mixture of indicator functions, where any values $p \in (0,1]$ are allowed. We are particularly interested in the case of profile functions φ with support the whole $\mathbb{R}_+$, in which case two vertices $\mathbf{x}$ and $\mathbf{y}$ always have a positive probability of being connected. In particular, we will discuss the choice of a polynomially decaying profile function, that is

$$\varphi(x) \asymp (1+x)^{-\delta}, \quad \delta > d,$$

where $g \asymp h$ is the commonly used notation for g/h bounded away from zero and infinity. The condition $\delta > d$ is needed for the integrability condition to be satisfied.

We now illustrate the connection between non-spatial preferential and spatial attachment models. Suppose the graph G_{t-} is given, and a vertex is born at time t, but we do not know its position, which is therefore uniform on S_1. Then, for each vertex $\mathbf{x} = (x,s) \in G_{t-}$, the probability that it is linked to the newborn vertex is equal to

$$\int_{S_1} \varphi\left(K\mathsf{d}_1(x,y)\right)\,\mathrm{d}y = K^{-d} \int_{(-\frac{K}{2},\frac{K}{2}]^d} \varphi(\mathsf{d}(0,y))\,\mathrm{d}y,$$

where we have written $K = t^{1/d}/(f(\deg^-(\mathbf{x},t-))^{1/d})$. As a consequence, the *indegree evolution* process $(\deg^-(\mathbf{x},t))_{t\geq s}$ is a time-inhomogeneous pure birth process, starting from 0 and jumping at time t from state k to state $k+1$ with intensity

$$\frac{f(k)}{t} \int_{\left(-\frac{t^{1/d}}{2f(k)^{1/d}},\frac{t^{1/d}}{2f(k)^{1/d}}\right]^d} \varphi(\mathsf{d}(0,y))\,\mathrm{d}y.$$

We can show that $(\deg^-(\mathbf{x},t))_{t\geq s}$ grows roughly like t^γ, so that the integral is asymptotically 1. Hence the jumping intensity of our process is the same as in the Barabási-Albert model of preferential attachment [3], [20], or its variant studied by Dereich and Mörters [9], [10], [11].

As soon as one deepens the study of the graph further than the first moment calculations, the essential difference with the non-spatial models appears. The presence of edges is now strongly correlated through the spatial positions of the vertices. These correlations both make the model much harder to study, and allow the network to enjoy interesting clustering properties. These are the main concern of this paper.

3 Rescaling the Graph

This section has been simplified from the full-version of this article, see [14]. The interested reader will find in [14] all the details, and the proof of Proposition 1 and Theorem 1 in the one-dimensional case. Everything holds *mutatis mutandis* in the higher dimensional cases

3.1 The Rescaled Picture

In the graph sequence $(G_t)_{t>0}$, the degree of any given vertex goes almost surely to $+\infty$. In this section we introduce a different graph sequence $(G^t)_{t>0}$ such that for every fixed t the graphs G_t and G^t have the same law. The new sequence has a different dynamics in which growth in time is replaced by growth in space, and the degrees of fixed vertices remain finite. Loosely speaking the sequence $(G^t)_{t>0}$ represents the graphs as seen from a typical vertex in the original graph sequence $(G_t)_{t>0}$, and hence a fixed point in $(G^t)_{t>0}$ does not age whereas a fixed point in $(G_t)_{t>0}$ does. The graph sequence $(G^t)_{t>0}$ will be easier to analyse, in particular it will converge and this goes along with convergence results for a large class of statistics derived from $(G_t)_{t>0}$.

To be more precise, let $\mathcal{Y}$ be a Poisson point process with intensity 1 on $\mathbb{R}^d \times (0,1]$. We interpret the first coordinate as space and the second as time, which is now restricted to the unit interval. For $t > 0$, we define S_t to be the hypercube

$$S_t = \left(-\frac{t^{1/d}}{2}, \frac{t^{1/d}}{2}\right]^d$$

of volume t. It is seen as a subspace of $\mathbb{R}^d$ but it is endowed with its own torus distance d_t. Observe that for any x and y of $\mathbb{R}^d$, for t large enough, x and y will be in S_t and satisfy $\mathsf{d}_t(x, y) = \mathsf{d}(x, y)$. For $t > 0$, define $\mathcal{Y}_t = \mathcal{Y} \cap (S_t \times (0, 1])$, and construct a graph G^t on $\mathcal{Y}_t$ with the same construction rule as before, with the new understanding that time now belongs to $(0, 1]$, and the distance is now replaced by d_t in (1). It is easily seen that the graphs G^t and the original graph G_t have the same law. Just multiply the time coordinate by t^{-1}, the space coordinates by $t^{1/d}$, and observe that the point process is still a Poisson point process of intensity 1, while the construction rule (1) is unchanged. It will turn out that there is a limiting graph $G^\infty = \lim G^t$, which can be obtained directly by applying our construction to the point set $\mathcal{Y}$ endowed with the distances in $\mathbb{R}^d$ in the construction rule (1).

3.2 Convergence

Proposition 1.

(i) The graph G^∞ is almost surely a well-defined locally finite graph, in the sense that each vertex has finite degree.
(ii) Almost surely, the graph G^t converges locally to G^∞, in the sense that for each $\mathbf{x} \in \mathcal{Y}$, for sufficiently large t, the neighbours of $\mathbf{x}$ in G^t and in G^∞ coincide.

The proposition states local convergence of G^t to G^∞ around any given vertex $\mathbf{x} \in \mathbb{R}^d \times (0,1]$. Its proof is based on a study of the indegree evolution process and bounds on the probability that a vertex has an exceptionally high indegree, or outdegree, or connects to an exceptionally distant vertex.

The following theorem completes Proposition 1 by describing the local structure of G^t around a *randomly chosen* vertex $\mathbf{x} \in G^t$. It is also the key to proving global results for the graphs G_t, see the following sections. It can be seen as a geometric law of large numbers in the spirit of Yukich and Penrose, the proof using that distant regions of space are asymptotically independent. For $t \in (0,\infty]$ let U be uniform on $(0,1]$ and $G^t(U)$ be the graph obtained by adding the point $(0,U)$ to $\mathcal{Y}$ before the construction of the graph G^t. Let $\xi(\mathbf{x}, G)$ be a 'local' function on a graph G around a distinguished point $\mathbf{x}$. For the purpose of this article, we can simply define such a local function to be a function on the neighbourhood of $\mathbf{x}$ up to graph distance a given finite value.

Theorem 1 (Weak law of large numbers). *Suppose, for some $a > 1$, the following uniform moment condition holds,*

$$\sup_{t>0} \mathbb{E}[\xi((0,U), G^t(U))^a] < \infty.$$

Then the following convergence in probability is satisfied,

$$\frac{1}{|\mathcal{Y}_t|} \sum_{\mathbf{x}\in\mathcal{Y}_t} \xi(\mathbf{x}, G^t) \longrightarrow \mathbb{E}[\xi((0,U), G^\infty(U))]. \tag{3}$$

In other words, the law of the local structure of the graph G^t around a randomly chosen vertex is the same as the law of the local structure of the infinite graph G^∞, conditioned[1] to have a vertex with position 0 and birth time U uniform in $(0,1]$, around this vertex. The next sections provide various applications to Proposition 1 and Theorem 1.

4 Results

Indegree. Denote by μ the law of the indegree of $(0,U)$ in $G^\infty(U)$ defined by

$$\mu(k) = \mathbb{P}(\deg^-((0,U), G^\infty(U)) = k).$$

[1] We recall here that for a Poisson point process, adding a point at $(0,U)$ is equivalent to conditioning the Poisson point process on having a point at $(0,U)$.

The local finiteness of $G^\infty(U)$ ensures that it is a probability law on $\mathbb{N} \cup \{0\}$. Applying the law of large numbers to the functionals $\xi_k(\mathbf{x}, G) = \mathbb{1}\{\deg^-(\mathbf{x}, G) = k\}$, we get that the empirical indegree distribution of G_t, defined by

$$\mu_t(k) = \frac{1}{|\mathcal{X}_t|} \sum_{\mathbf{x} \in \mathcal{X}_t} \mathbb{1}\{\deg^-(\mathbf{x}, G_t) = k\},$$

converges to μ in probability. Given the construction of $G^\infty(U)$, it is remarkable that the law of the indegree μ can be calculated explicitly. It relies on the study of the *indegree evolution* process, which we omit here. We find

$$\mu(k) = \frac{1}{\gamma} \frac{\Gamma(k + \frac{\beta}{\gamma}) \Gamma(\frac{\beta+1}{\gamma})}{\Gamma(k + \frac{\beta+\gamma+1}{\gamma}) \Gamma(\frac{\beta}{\gamma})} \sim \frac{\Gamma(\frac{\beta+1}{\gamma})}{\gamma \Gamma(\frac{\beta}{\gamma})} k^{-(1+1/\gamma)} \qquad \text{as } k \uparrow \infty,$$

which is in line with Theorem 1.1 of [1], and verifies the scale-free property of the network with power law exponent $\tau = 1 + 1/\gamma \in (2, \infty)$.

Actually, we can prove a stronger convergence result:

Theorem 2. *For any nondecreasing function $g: \mathbb{N} \cup \{0\} \to [0, \infty)$, we have the following convergence in probability, as $t \to \infty$,*

$$\sum \mu_t(k) g(k) \longrightarrow \sum \mu(k) g(k).$$

Applying this result with $g(k) = k$, we get that the total number of edges is always asymptotically of the same order as the number of vertices. More interestingly, applying it with $g(k) = k^2$, we get that

$$\frac{1}{|\mathcal{X}_t|} \sum_{\mathbf{x} \in \mathcal{X}_t} \deg^-(\mathbf{x}, G_t)^2$$

converges to a finite constant if $\gamma < 1/2$ and to infinity if $\gamma \geq 1/2$.

Outdegree and Total Degree. Similarly to the empirical indegree distribution we define the empirical outdegree distribution ν_t of G_t, and let ν the law of the outdegree of $(0, U)$ in $G^\infty(U)$. As before, Theorem 1 yields convergence in probability of ν_t to ν. We do not have an explicit expression for ν, in particular it is *not* a Poisson distribution as in the model of Dereich-Mörters [9]. By a study of the infinite picture, we can however prove that ν is light-tailed and get the following theorem.

Theorem 3. *For any $0 < \alpha < 1 - \gamma$, we have*

$$\nu([k, +\infty)) = o(e^{-k^\alpha}).$$

Moreover, for any function $g: \mathbb{N} \cup \{0\} \to \mathbb{R}$ satisfying $g(k) = o(e^{k^\alpha})$ for some $0 < \alpha < 1 - \gamma$, we have the following convergence in probability, as $t \to \infty$,

$$\sum \nu_t(k) g(k) \longrightarrow \sum \nu(k) g(k).$$

These results complete Theorem 1.5 in [1], which controls the maximum degree in G_t. Further, it is not hard to see that the law of the outdegree of $(0,u)$ in $G^\infty(u)$ is independent of u, and that the law of the total degree of $(0,U)$ in $G^\infty(U)$ is the convolution $\mu * \nu$. Hence, the empirical total degree distribution in G_t converges to $\mu * \nu$, which is also decaying polynomially with parameter τ. We can check that the mean degree is

$$\sum k\ \mu * \nu(k) = 2\sum k\nu(k) = 2\sum k\mu(k) = \frac{2\beta}{1-\gamma}.$$

Clustering. In this section, we forget the orientation of the edges of G_t to define its clustering coefficients. These coefficients are based on the number of triangles and open triangles in the graph. An open triangle of G_t with tip x is simply a subgraph of the form $(\{\mathbf{x},\mathbf{y},\mathbf{z}\},\{\{\mathbf{x},\mathbf{y}\},\{\mathbf{x},\mathbf{z}\}\})$, where $\mathbf{y}$ and $\mathbf{z}$ could either be connected in G_t and hence form a triangle, or not. Note that every triangle in G contributes three open triangles.

The *global clustering coefficient* of G is defined as

$$c^{\text{glob}}(G_t) := 3\,\frac{\text{Number of triangles included in } G_t}{\text{Number of open triangles included in } G_t}.$$

Note that always $c^{\text{glob}}(G_t) \in [0,1]$. The local clustering coefficient of G_t at a vertex $\mathbf{x}$ with degree at least two is defined by

$$c^{\text{loc}}_{\mathbf{x}}(G_t) := \frac{\text{Number of triangles included in } G_t \text{ containing vertex } \mathbf{x}}{\text{Number of open triangles with tip } \mathbf{x} \text{ included in } G_t},$$

which is also an element of $[0,1]$. Finally, the *average clustering coefficient* is defined as

$$c^{\text{av}}(G) := \frac{1}{|V_2|}\sum_{\mathbf{x}\in V_2} c^{\text{loc}}_{\mathbf{x}}(G),$$

where $V_2 \subset V$ is the set of vertices with degree at least two in G. The global and average clustering coefficients have the following probabilistic interpretation:

- *Pick a vertex uniformly at random and condition on the event that this vertex has degree at least two. Pick two of its neighbours, uniformly at random. Then the probability that these two vertices are linked is equal to* $c^{\text{av}}(G_t)$.
- *Pick two edges sharing a vertex, uniformly from all such pairs of edges in the graph. Then the probability that the two other vertices bounding the edges are connected is equal to* $c^{\text{glob}}(G_t)$.

Theorem 4.

(i) The average clustering coefficient of G_t *converges in probability to a strictly positive number* c^{av}_∞.

(ii) The global clustering coefficient of G_t *converges in probability to nonnegative number* c^{glob}_∞*, which is nonzero if and only if* $\gamma < 1/2$.

The first part of this theorem is easy to prove by considering the functional which, to a vertex, associates its local clustering coefficient. It is clear that the expected local clustering coefficient of $(0, U)$ in $G^\infty(U)$ belongs to $(0, 1)$, and there is nothing more to argue. For the second part, we estimate both the number of triangles and the number of open triangles in G_t thanks to two applications of Theorem 1. We choose to count the triangles *from their youngest vertex* and consider the functional which associates to a vertex $\mathbf{x}$ the number of triangles containing $\mathbf{x}$ and having $\mathbf{x}$ as youngest vertex. The light-tail of the outdegree distribution ensures that this functional satisfies the uniform moment condition. We can apply Theorem 1 and deduce that the number of triangles is always asymptotically proportional to the number of vertices, that is of order t. To estimate the number of open triangles, we should in particular estimate the number of open triangles with tip in $\mathbf{x}$ the *oldest vertex*. But this number is simply

$$\sum_{\mathbf{x} \in \mathcal{X}_t} \deg^-(\mathbf{x}, G_t)(\deg^-(\mathbf{x}, G_t) - 1),$$

and we know, thanks to the work on the indegree, that it is linear if $\gamma < 1/2$ and superlinear if $\gamma \geq 1/2$. This discussion is almost enough to prove Theorem 4.

An interesting extension, suggested by an anonymous referee, is to look at the average local clustering coefficient of vertices with a fixed degree k. Our methods are expected to show that this quantity converges to a deterministic limit, which decays of order $1/k$, as $k \to \infty$. Details will be discussed elsewhere.

The phase transition in the global clustering coefficient has been observed in a similar form for random intersection graphs [5]. The behaviour of the clustering coefficients in the case $\gamma \geq 1/2$ matches the behaviour expected in the world wide web: if you pick a webpage at random, and click on two hyperlinks, it is likely that the two pages you get have actually a direct hyperlink. However, if you pick two webpages which both have a hyperlink to the Google homepage, it is not likely that these two pages have a direct link.

Edge Length Distribution. In the rescaled graphs G^∞ or in G^t, we expect a typical edge to have geometric length (in $\mathbb{R}^d$ or in S_t) of order 1. Therefore, in the original graph G_t, we expect edges to have length of order $t^{-1/d}$. Write $E(G_t)$ for the set of the edges of the graph G_t. Define λ, the (rescaled) empirical edge length distribution, by

$$\lambda_t = \frac{1}{|E(G_t)|} \sum_{(\mathbf{x},\mathbf{y}) \in E(G_t)} \delta_{t^{1/d} \mathsf{d}_1(\mathbf{x},\mathbf{y})}.$$

Similarly, write $E(G^\infty(U))$ for the set of the edges of the graph $G^\infty(U)$, and define a probability distribution λ on $\mathbb{R}$ by

$$\lambda(A) = \frac{1-\gamma}{2\beta} \, \mathbb{E} \left[\sum_{((0,U),(x,s)) \in E(G^\infty(U))} \delta_{\mathsf{d}(0,x)}(A) \right].$$

Another application of Theorem 1 enables us to prove convergence of λ_t to λ, in probability. It is of course not possible to have an explicit expression for λ. However, we can estimate its tail behaviour in the case of a polynomially decaying profile function.

Theorem 5. *Suppose that there exists $\delta > d$ such that the profile function satisfies $\varphi(x) \asymp (1+x)^{-\delta}$. Then*

$$\lambda([K, +\infty)) \asymp (1+K)^{-\eta},$$

where $\eta \in (0, d]$ is the smallest of the three constants d, $d(\frac{1}{\gamma} - 1)$ and $\delta - d$.

The proof is the most technical of our work and is omitted here. Note that if $d = 1$ or if $\gamma \geq \frac{d}{1+d}$ or if $\delta \leq d + 1$, then λ does not have a first moment, and the *mean* edge length is not of order $t^{-1/d}$. The heavy tails of the empirical edge length distribution highlight the nature of our networks as *small worlds.*

The empirical edge length distribution for the original SPA model, corresponding roughly to the case $\delta = \infty$, is also studied in Janssen et al. [15]. They show that if $\gamma > \frac{1}{2}$ and $\frac{3\gamma+2}{4\gamma+2} < \alpha < 1$, then

$$\left|\{\text{edges of length longer than } t^{-\alpha/d}\}\right| \sim C\, t^{(2-\alpha)+\frac{1}{\gamma}(\alpha-1)}$$

for an explicit constant $C > 0$. Our result uses a different order of limits, but leads to the same order of growth for the comparable quantity $t\lambda[t^{(1-\alpha)/d}, \infty)$. Note that the general form of the profile functions allows for a genuinely richer phenomenology in our case.

5 Further Work

In [8], the authors find small separators for the SPA model. They deduce that the spectral gap of the normalised Laplacian of the graph G_t converges to 1, yielding bad expansion properties for G_t. The separators they found are simply obtained by cutting the hypercube in half. We expect that the same strategy would yield similar results for our generalised model, with the slight difference that the separators will not be as small, depending on the tail of the profile function φ.

Existence of a Giant Component. Let us forget about the orientation of the edges of G_t, and simply consider it as an unoriented graph. Note that, as $\mu * \nu(0) > 0$, the graph has a number of isolated vertices growing linearly in time, and is therefore not connected. Before using G_t as a model for the world wide web, we would like to ensure the existence of a giant component of G_t, i.e. a connected component of linear size. Moreover, we would be interested in the robustness of a giant component under random attack. Robustness is defined by the existence, for every positive value of p, of a giant component in the graph obtained by removing every edge independently with probability $1 - p$.

Proposition 1 suggests that the existence of a giant component for G_t should be related to percolation in G^∞, that is, the existence of an *infinite* connected component in G^∞. As the construction of the graph G^∞ is based (at least in the case of an indicator profile functions) on balls with random positions and random sizes, it resembles the construction of random geometric graphs, and so it seems plausible to use methods from this field as surveyed, for example, in Meester and Roy [17]. Just like in continuum percolation, we cannot really hope for a precise criterion deciding whether there is or is no percolation in G^∞ for any attachment rule f and any profile function φ. However, we hope to identify the domain of existence of a robust giant component. At this point we conjecture the following results, based on preliminary calculations, which highlight interesting phase transitions not occurring for non-spatial models, and show the crucial role of the tail of the profile function (at least in dimension 1). The phase corresponding to our first conjecture seems the best candidate for a model of the world wide web. In (1) and (3) we assume the profile function satisfies

$$\varphi(x) \asymp (1+x)^{-\delta}.$$

(1) *There is always a robust giant component if* $\gamma > 1 - \frac{1}{1+\delta/d}$.

In this case, the shortest paths between two typical vertices in the giant component is doubly logarithmic in the number of nodes. Similarly to the non-spatial models there is a 'core' of old vertices connected to each other in no more than a finite number of steps. Short paths between remote vertices typically connect via this core. The condition on γ, and in particular a finite value of δ, is necessary for the formation of a core.

(2) *There is never a robust giant component if* $\gamma < 1/2$.

This conjecture is based on the corresponding result for non-spatial preferential attachment networks, and the idea that the spatial correlations of the model cannot help the construction of a giant component.

(3) *There is never a robust giant component if* $d = 1$ *and* $\gamma < 1 - 1/\delta$.

Here we have a strong concentration of the length of vertices around the typical value, which give the network some characteristics of geometric random graphs.

All the other cases seem to be even tougher open questions. It would be interesting if we could get a robust giant component in a case not covered by our first conjecture, as this giant component would then have a very different topology from the one in the non-spatial models.

Acknowledgments. We gratefully acknowledge support of this project by the *European Science Foundation* through the research network *Random Geometry of Large Interacting Systems and Statistical Physics (RGLIS).*

References

1. Aiello, W., Bonato, A., Cooper, C., Janssen, J., Prałat, P.: A spatial web graph model with local influence regions. Internet Mathematics 5, 175–196 (2009)
2. Albert, R., Barabási, A.-L.: Statistical mechanics of complex networks. Rev. Modern Phys. 74, 47–97 (2002)
3. Barabási, A.-L., Albert, R.: Emergence of scaling in random networks. Science 286, 509–512 (1999)
4. Berger, N., Borgs, C., Chayes, J., Saberi, A.: Asymptotic behavior and distributional limits of preferential attachment graphs. Ann. Prob. (to appear, 2013)
5. Bloznelis, M.: Degree and clustering coefficient in sparse random intersection graphs. Ann. Appl. Prob. 23, 1254–1289 (2013)
6. Bollobás, B., Riordan, O.M.: Mathematical results on scale-free random graphs. In: Handbook of Graphs and Networks, pp. 1–34. Wiley-VCH, Weinheim (2003)
7. Bonato, A., Janssen, J., Pralat, P.: Geometric protean graphs. Internet Mathematics 8, 2–28 (2012)
8. Cooper, C., Frieze, A., Pralat, P.: Some typical properties of the spatial preferred attachment model. In: Bonato, A., Janssen, J. (eds.) WAW 2012. LNCS, vol. 7323, pp. 29–40. Springer, Heidelberg (2012)
9. Dereich, S., Mörters, P.: Random networks with sublinear preferential attachment: degree evolutions. Electron. J. Probab. 14(43), 1222–1267 (2009)
10. Dereich, S., Mörters, P.: Random networks with concave preferential attachment rule. Jahresber. Dtsch. Math.-Ver. 113, 21–40 (2011)
11. Dereich, S., Mörters, P.: Random networks with sublinear preferential attachment: the giant component. Ann. Prob. 41, 329–384 (2013)
12. Flaxman, A.D., Frieze, A.M., Vera, J.: A geometric preferential attachment model of networks. Internet Math. 3, 187–205 (2006)
13. Flaxman, A.D., Frieze, A.M., Vera, J.: A geometric preferential attachment model of networks II. In: Bonato, A., Chung, F.R.K. (eds.) WAW 2007. LNCS, vol. 4863, pp. 41–55. Springer, Heidelberg (2007)
14. Jacob, E., Mörters, P.: Spatial preferential attachment networks: Power laws and clustering coefficients. arXiv:1210.3830 (2012)
15. Janssen, J., Pralat, P., Wilson, R.: Geometric graph properties of the spatial preferred attachment model. Adv. Appl. Math. 50, 243–267 (2013)
16. Jordan, J.: Degree sequences of geometric preferential attachment graphs. Adv. in Appl. Probab. 42, 319–330 (2010)
17. Meester, R., Roy, R.: Continuum percolation. Cambridge Tracts in Mathematics, vol. 119. Cambridge University Press, Cambridge (1996)
18. Penrose, M.: Random geometric graphs. Oxford Studies in Probability, vol. 5. Oxford University Press, Oxford (2003)
19. Penrose, M.D., Yukich, J.E.: Weak laws of large numbers in geometric probability. Ann. Appl. Probab. 13, 277–303 (2003)
20. Rudas, A., Tóth, B., Valkó, B.: Random trees and general branching processes. Random Structures Algorithms 31, 186–202 (2007)

A Local Clustering Algorithm for Connection Graphs

Fan Chung and Mark Kempton

Department of Mathematics,
University of California, San Diego
{fan,mkempton}@ucsd.edu

Abstract. We give a clustering algorithm for connection graphs, that is, weighted graphs in which each edge is associated with a d-dimensional rotation. The problem of interest is to identify subsets of small Cheeger ratio and which have a high level of consistency, i.e. that have small edge boundary and the rotations along any distinct paths joining two vertices are the same or within some small error factor. We use PageRank vectors as well as tools related to the Cheeger constant to give a clustering algorithm that runs in nearly linear time.

1 Introduction

In this paper, we study connection graphs, which are generalizations of weighted graphs in which each edge is associated with both a positive scalar weight and a d-dimensional rotation matrix for some fixed positive integer d. The Laplacian of a connection graphs are higher dimensional versions of the normalized Laplacian matrix, which are linear operators acting on the space of vector-valued functions (instead of the usual real-valued functions).

Connection graphs arise in applications involving high dimensional data sets where some data points are related by rotation matrices. Some early usage of connection graphs can be traced back to work in graph gauge theory for computing the vibrational spectra of molecules and examining spins associated with vibrations [9]. There have been more recent developments of related research in principal component analysis [13], cryo-electron microscopy [11,15], angular synchronization of eigenvectors [10,14], and vector diffusion maps [16]. In computer vision, there has been a great deal of work dealing with the many photos that are available on the web, in which information networks of photos can be built. The edges of the associated connection graphs correspond to the rotations determined by the angles and positions of the cameras used [1]. Recently, related work has been done on a synchronization problem, for which the connection Laplacian acts on the space of functions which assign an orthogonal matrix to each vertex [4].

For high dimensional data sets, a central problem is to uncover lower dimensional structures in spite of possible errors or noises. An approach for reducing the effect of errors is to consider the notion of inconsistency, which quantifies the

A. Bonato, M. Mitzenmacher, and P. Prałat (Eds.): WAW 2013, LNCS 8305, pp. 26–43, 2013.
© Springer International Publishing Switzerland 2013

difference of accumulated rotations while traveling along distinct paths between two vertices. In many applications, it is desirable to identify edges causing the inconsistencies, or to identify portions of the graph that have relatively small inconsistency. In [8], an algorithm is given, utilizing a version of effective resistance from electrical network theory, that deletes edges of a connection graph in such a way that reduces inconsistencies. In this paper, rather than deleting edges, our focus is on identifying subsets of a connection graph with small inconsistency. The notion of ϵ-consistency of a subset of the vertex set of a connection graph will be introduced, which quantifies the amount of inconsistency for the subset to within an error ϵ. This can be viewed as a generalization of the notion of consistency.

One of the major problems in computing is to design efficient clustering algorithms for finding a good cut in a graph. That is, it is desirable to identify a subset of the graph with small edge boundary in comparison to the overall volume of the subset. Many clustering algorithms have been derived including some with quantitative analysis (e.g., [2,3]). As we are looking for ϵ-consistent subsets, it is natural that that clustering and the Cheeger ratio should arise in examining local subsets of a graph. In this paper, we will combine the clustering problem and the problem of identifying ϵ-consistent subsets. In particular, we will give an algorithm that uses PageRank vectors to identify a subset of a connection graph which both has a small cut, and is ϵ-consistent.

The notion of PageRank was first introduced by Brin and Page [5] in 1998 for Google's web search algorithms. It has since proven useful in graph theory for quantifying relationships between vertices in a graph. Algorithms from [2] and [3] utilize PageRank vectors to locally identify good cuts and cluster in a graph. In [8], a vectorized version of PageRank is given for connection graphs. Here we use these connection PageRank vectors in a manner similar to [3] to find good cuts which are ϵ-consistent.

A Summary of the Results. The results in this paper can be summarized as follows:

- We define the notion of ϵ-consistency and establish several inequalities relating ϵ-consistency with the smallest eigenvalue of the connection Laplacian and the Cheeger ratio of subsets of a connection graph.
- We define connection PageRank vectors and establish several inequalities relating the sharp drops in the connection PageRank vectors to the Cheeger ratio and the ϵ-consistency of the subsets.
- We give an algorithm that outputs a subset of the vertices (if one exists) which is a good cut and is ϵ-consistent. The run time of the algorithm is $O\left(d^2 x \frac{\log^2 m}{\phi^2}\right)$, where m is the number of edges, d is the dimension of the rotations, ϕ is the target Cheeger ratio, and x is the target volume.

The remainder of the paper is organized as follows: In Section 2, we give some of the basic definitions of a connection graph, the connection Laplacian, and the notion of consistency, as well as some useful facts on consistency from [8].

In Section 3 we introduce the notion of ϵ-consistency which generalizes the notion of consistency, and gives some results relating ϵ-consistency of a connection graph to the spectrum of the normalized connection Laplacian. In Section 4 we examine subsets of a connection graph that are ϵ-consistent, and relate the spectrum of the normalized Laplacian to the Cheeger ratio of such subsets. In Section 5, we utilize connection PageRank vectors in the study of ϵ-consistent subsets, and present a local partition algorithm for a connection graph, completed with complexity analysis.

2 Preliminaries

2.1 The Normalized Connection Laplacian

Suppose $G = (V, E, w)$ is an undirected graph with vertex set V, edge set E, and edge weights $w_{uv} = w_{vu} > 0$ for edges (u, v) in E. Let $\mathcal{F}(V, \mathbb{R})$ denote the space of all functions $f : V \to \mathbb{R}$. The usual adjacency matrix A, combinatorial Laplacian matrix L, and normalized Laplacian $\mathcal{L}$, are all operators on the space $\mathcal{F}(V, \mathbb{R})$. (See, for example, [6] for definitions of A, L, and $\mathcal{L}$.) For undefined terminology, the reader is referred to [8].

Now suppose each oriented edge (u, v) is also associated with a rotation matrix $O_{uv} \in \mathsf{SO}\,(d)$ satisfying $O_{uv}O_{vu} = I_{d\times d}$. Here $\mathsf{SO}\,(d)$ denotes the special orthogonal group of dimension d, namely, the group of all $d \times d$ matrices S satisfying $S^{-1} = S^T$ and $\det(S) = 1$. Let O denote the set of rotations associated with all oriented edges in G. The *connection graph*, denoted by $\mathbb{G} = (V, E, O, w)$, has G as the *underlying graph*. The *connection adjacency matrix* $\mathbb{A}$ of $\mathbb{G}$ is defined by:

$$\mathbb{A}(u, v) = \begin{cases} w_{uv}O_{uv} & \text{if } (u, v) \in E, \\ 0_{d\times d} & \text{if } (u, v) \notin E \end{cases}$$

where $0_{d\times d}$ is the zero matrix of size $d \times d$. We view $\mathbb{A}$ as a block matrix where each block is either a $d \times d$ rotation matrix O_{uv} multiplied by a scalar weight w_{uv}, or a $d\times d$ zero matrix. The matrix $\mathbb{A}$ is an operator on the space $\mathcal{F}(V, \mathbb{R}^d) = \{f : V \to \mathbb{R}^d\}$. The matrix $\mathbb{A}$ is symmetric as $O_{uv}^T = O_{vu}$ and $w_{uv} = w_{vu}$.

The *connection Laplacian* $\mathbb{L}$ of a graph $\mathbb{G}$ is defined by

$$\mathbb{L} = \mathbb{D} - \mathbb{A}$$

where $\mathbb{D}$ is the diagonal matrix defined by the diagonal blocks $\mathbb{D}(u, u) = d_u I_{d\times d}$ for $u \in V$. Here d_u is the weighted degree of u in G, i.e., $d_u = \sum_{(u,v)\in E} w_{uv}$. The connection Laplacian is an operator on $\mathcal{F}(V, \mathbb{R}^d)$, where its action on a function $f : V \to \mathbb{R}^d$ is given by

$$\mathbb{L}f(v) = \sum_{u\sim v} w_{uv} \left(f(v) - f(u)O_{uv}\right).$$

(The elements of $\mathcal{F}(V, \mathbb{R}^d)$ is sometimes viewed as a row vector so that $f(u)O_{uv}$ is the product of matrix multiplication of $f(u)$ and O_{uv}.)

Recall that for any orientation of edges of the underlying graph G on n vertices and m edges, the combinatorial Laplacian L can be written as $L = B^T WB$ where W is a $m \times m$ diagonal matrix with $W_{e,e} = w_e$, and B is the edge-vertex incident matrix of size $m \times n$ such that $B(e, v) = 1$ if v is e's head; $B(e, v) = -1$ if v is e's tail; and $B(e, v) = 0$ otherwise. A useful observation for the connection Laplacian is the fact that it can be written in a similar form. Let $\mathbb{B}$ be the $md \times nd$ block matrix given by

$$\mathbb{B}(e,v) = \begin{cases} O_{uv} & v \text{ is } e\text{'s head,} \\ -I_{d\times d} & v \text{ is } e\text{'s tail,} \\ 0_{d\times d} & \text{otherwise.} \end{cases}$$

Let the block matrix $\mathbb{W}$ denote the diagonal block matrix given by $\mathbb{W}(e, e) = w_e I_{d\times d}$ where $\mathbb{W}$ is actually of size $md \times md$. Then it can be verified by direct computation that, given an orientation of the edges, the connection Laplacian also can alternatively be defined as

$$\mathbb{L} = \mathbb{B}^T \mathbb{W}\mathbb{B}.$$

We define the *normalized connection Laplacian* $\hat{\mathcal{L}}$ to be the operator on $\mathcal{F}(V, \mathbb{R}^d)$ given by

$$\hat{\mathcal{L}} = \mathbb{D}^{-1/2}\mathbb{L}\mathbb{D}^{-1/2} = I_{nd\times nd} - \mathbb{D}^{-1/2}\mathbb{A}\mathbb{D}^{-1/2}.$$

We remark that $\mathbb{L}$ and $\hat{\mathcal{L}}$ are symmetric, positive semi-definite matrices. Using the Courant-Fischer Theorem (see, for example, [12]), we can investigate the eigenvalues of $\hat{\mathcal{L}}$ by examining the *Rayleigh quotient*

$$\mathcal{R}(g) = \frac{g\hat{\mathcal{L}}g^T}{gg^T}$$

where $g : V \to \mathbb{R}^d$ is thought of as a $1 \times nd$ row vector. Defining $f = g\mathbb{D}^{-1/2}$, we see that

$$\mathcal{R}(g) = \frac{f\mathbb{L}f^T}{f\mathbb{D}f^T} = \frac{\sum_{(u,v)\in E} w_{uv} \|f(u)O_{uv} - f(v)\|_2^2}{\sum_{v\in V} d_v \|f(v)\|_2^2}.$$

In particular, letting $0 \le \lambda_1 \le \lambda_2 \le \cdots \le \lambda_{nd}$ denote the eigenvalues of $\hat{\mathcal{L}}$. It is not hard to see that $\mathcal{R}(f) \le 2$ which implies $\lambda_k \le 2$ for all k.

2.2 Consistency

For a connection graph $\mathbb{G} = (V, E, O, w)$, we say that $\mathbb{G}$ is *consistent* if

$$\inf_{\substack{f:V\to\mathbb{R}^d \\ f\neq 0}} \sum_{(u,v)\in E} w_{uv}\|f(u)O_{uv} - f(v)\|_2^2 = 0.$$

An equivalent definition for consistency is an assignment of a vector $f(u) \in \mathbb{R}^d$ to each vertex $u \in V$ such that for all vertices v adjacent to u, $f(v) = f(u)O_{uv}$. Therefore for any two vertices u, v in a consistent graph, any two distinct paths starting and ending at u and v, $P_1 = (u = u_1, u_2, ..., u_k = v)$ and $P_2 = (u = v_1, v_2, ..., v_l = v)$, then the product of rotations along either path is the same. That is,

$$\prod_{i=1}^{k-1} O_{u_i u_{i+1}} = \prod_{j=1}^{l-1} O_{v_j v_{j+1}}.$$

For any cycle $C = (v_1, v_2, ..., v_k, v_{k+1} = v_1)$ of the underlying graph, the product of rotations along the cycle C is the identity, i.e. $\prod_{i=1}^{k} O_{v_i v_{i+1}} = I_{d \times d}$.

For ease of notation, given a cycle $C = (v_1, v_2, ..., v_k, v_{k+1} = v_1)$, define $O_C = \prod_{i=1}^{k} O_{v_i v_{i+1}}$, and for a path joining distinct vertices u and v, $P_{uv} = (u = v_1, v_2, ..., v_k = v)$, define $O_{P_{uv}} = \prod_{i=1}^{k-1} O_{v_i v_{i+1}}$. Therefore consistency can be characterized by saying $O_C = I_{d\times d}$ for any cycle C, or given any two vertices u and v of $\mathbb{G}$, then $O_{P_{uv}} = O_{P'_{uv}}$ for any two paths P_{uv}, P'_{uv} connecting u and v.

In [8], a spectral characterization of consistency for a connection graph is given in terms of the eigenvalues of the connection Laplacian $\mathbb{L}$. We note that an easy modification of the argument in [8] yields the similar statements for the normalized connection Laplacian. Namely, let $\hat{\mathcal{L}}$ be the normalized connection Laplacian of the connection graph $\mathbb{G}$, let $\mathcal{L}$ be the normalized Laplacian of the underlying graph G. For a connected connection graph $\mathbb{G}$, the following statements are equivalent:

(i) $\mathbb{G}$ is consistent.
(ii) The normalized connection Laplacian $\hat{\mathcal{L}}$ of $\mathbb{G}$ has eigenvalue 0.
(iii) The eigenvalues of $\hat{\mathcal{L}}$ are the n eigenvalues of $\mathcal{L}$, each of multiplicity d.
(iv) For each vertex u in G, we can find $O_u \in \mathsf{SO}(d)$ such that for any edge (u, v) with rotation O_{uv}, we have $O_{uv} = O_u^{-1}O_v$.

2.3 The Cheeger Ratio

Given a subset of the vertex set, $S \subset V$, we define $E(S, \bar{S})$ to be the set of all edges having one endpoint in S and the other endpoint outside of S. We define the volume of S, denoted vol(S), by vol$(S) = \sum_{v\in S} d_v$. We define the *Cheeger ratio* of S, denoted $h_G(S)$, by

$$h_G(S) = \frac{|E(S, \bar{S})|}{\mathrm{vol}(S)}.$$

The *Cheeger constant* (sometimes called the *conductance*) of a graph G is

$$h_G = \min\left\{h(S) : S \subset V,\ \mathrm{vol}(S) \leq \frac{1}{2}\,\mathrm{vol}(G)\right\}.$$

Determining the Cheeger constant of a graph can be thought of as a discrete version of the classical isoperimetric problem from geometry. One of the classic results in spectral graph theory (see, for example, [6]) is the *Cheeger Inequality*, which relates the Cheeger constant of a graph to the eigenvalues of its normalized Laplacian. Given a graph G with normalized Laplacian $\mathcal{L}$ with eigenvalues $0 = \lambda_1 \leq \lambda_2 \leq \cdots \leq \lambda_n$, the Cheeger Inequality states that

$$\frac{h_G^2}{2} \leq \lambda_2 \leq 2h_G.$$

We will be giving results analogous to the Cheeger inequality for ϵ-consistent connection graphs, and the Cheeger ratio will play a critical role in our algorithm and its analysis in Section 5.

3 ϵ-Consistency

We say a connection graph $\mathbb{G}$ is *ϵ-consistent* if, for every simple cycle $C = (v_1, v_2, ..., v_k, v_{k+1} = v_1)$ of the underlying graph G, we have $\|O_C - I_{d\times d}\|_2 \leq \epsilon$ where $O_C = \prod_{i=1}^{k} O_{v_i v_{i+1}}$. That is, the product of rotations along any cycle is within ϵ of the identity in the 2-norm. An equivalent formulation is as follows. Given vertices u and v, and two distinct paths from u to v, $P_1 = (v_1 = u, v_2, ..., v_k = v)$ and $P_2 = (u_1 = u, u_2, ..., u_l = v)$, define $O_{P_1} = \prod_{i=1}^{k-1} O_{v_i v_{i+1}}$ and $O_{P_2} = \prod_{i=1}^{l-1} O_{u_i u_{i+1}}$. Then $\mathbb{G}$ is ϵ-consistent if and only if $\|O_{P_1} - O_{P_2}\|_2 \leq \epsilon$. This follows from the observation that $O_C = O_{P_1} O_{P_2}^{-1} = O_{P_1} O_{P_2}^T$ and the fact that the 2-norm of a rotation matrix is 1.

We observe that the triangle inequality implies that any connection graph is 2-consistent, and that a consistent connection graph is 0-consistent. We generalize the first part of the above mentioned result from [8] with the following theorem, which bounds the d smallest eigenvalues of the normalized connection Laplacian for an ϵ-consistent connection graph.

Theorem 1. *Let $\mathbb{G}$ be an ϵ-consistent connection graph whose underlying graph is connected. Let $\hat{\mathcal{L}}$ be the normalized connection Laplacian and let $0 \leq \lambda_1 \leq \cdots \leq \lambda_{nd}$ be the eigenvalues of $\hat{\mathcal{L}}$. Then for $i = 1, ..., d$,*

$$\lambda_i < \frac{\epsilon^2}{2}.$$

The proof can be found in the appendix.

The following result concerns the second block of d eigenvalues of $\hat{\mathcal{L}}$ for an ϵ-consistent connection graph, and gives an analog to the upper bound in the Cheeger inequality.

Theorem 2. *Let $\hat{\mathcal{L}}$ be the normalized connection Laplacian of the connection graph $\mathbb{G}$, with eigenvalues $\lambda_1 \leq \cdots \leq \lambda_{nd}$, and let h_G denote the Cheeger constant of the underlying graph. Then for $i = d + 1, ..., 2d$,*

$$\lambda_i \leq 2h_G + \frac{\epsilon^2}{2}.$$

The proof of this result is also in the appendix.

4 Consistent and ϵ-Consistent Subsets

In this section, we will consider the case where a connection graph has been created in which some subset of the data is error-free (or close to it), leading to a consistent or ϵ-consistent induced subgraph. We will define functions on the vertex set in such a way that the Rayleigh quotient will keep track of the edges leaving the consistent subset. In this way, we will obtain bounds on the spectrum of the normalized connection Laplacian involving the Cheeger ratio of such subsets.

Theorem 3. *Let $\mathbb{G}$ be a connection graph of dimension d with normalized connection Laplacian $\hat{\mathcal{L}}$, and $S \subset V$ a subset of the vertex set that is ϵ-consistent for given $\epsilon \geq 0$. Then for $i = 1, ..., d$,*

$$\lambda_i(\hat{\mathcal{L}}) \leq \frac{\epsilon^2}{2} + h_G(S).$$

Proof. Fix a spanning tree T of the subgraph induced by S. Define f as follows. For a fixed vertex u of S, define $f(u) = x$ where $||x|| = 1$, and for $v \in S$, define f to be consistent with the subtree T. For $v \notin S$, define $f(v) = 0$. Fix an edge $uv \in E$ and note that for $u, v \notin S$, $||f(u)O_{uv} - f(v)|| = 0$, for $u, v \in S$, $||f(u)O_{uv} - f(v)|| = ||f(v)\left(O_{P_{vu}}O_{uv} - I\right)|| < \epsilon$, and for $u \in S$, $v \notin S$, $||f(u)O_{uv} - f(v)|| = 1$. Therefore

$$\begin{aligned}
\mathcal{R}(f) &= \frac{\sum\limits_{(u,v)\in E} w_{uv}||f(u)O_{uv} - f(v)||^2}{\sum\limits_{v} d_v||f(v)||^2} \\
&< \frac{\sum\limits_{\substack{uv\in E\\ u,v\in S}} w_{uv}\epsilon^2}{\mathrm{vol}(S)} + \frac{\sum\limits_{\substack{uv\in E\\ u\in S, v\notin S}} w_{uv}}{\mathrm{vol}(S)} \\
&\leq \frac{\epsilon^2}{2} + h_G(S).
\end{aligned}$$

There are d orthogonal choices for the initial choice of x leading to d orthogonal vectors satisfying this bound, so by the Courant-Fisher Theorem, the result follows.

In the next result, we consider the situation where most of the edges are is close to being consistent except for some edges in the edge boundary of a subset.

Theorem 4. *Suppose G is an ϵ_1-consistent graph for some $\epsilon_1 > 0$, and suppose that $S \subset V$ is a set such that the subgraphs induced by S and $\bar{S}$ are ϵ_2-consistent, with $0 \leq \epsilon_2 < \epsilon_1$, and $\mathrm{vol}(S) \leq \frac{1}{2}\mathrm{vol}(G)$. Let $\hat{\mathcal{L}}$ be the normalized connection Laplacian. Then for $i = 1, ..., d$,*

$$\lambda_i(\hat{\mathcal{L}}) < \frac{\epsilon_2^2}{2} + \frac{\epsilon_1^2}{2} h_G(S)$$

Proof. We will construct a function $f : V \to \mathbb{R}^d$ whose Rayleigh quotient will bound λ_1. Fix a spanning tree T of S and T' of $\bar{S}$, and fix a vector $w \in S$. Choose a unit vector $x \in \mathbb{R}^d$, and assign $f(w) = x$. For $v \in S$, assign $f(v)$ for each vertex $v \in S$ such that $f(v) = f(u)O_{uv}$ moving along edges uv of T. Now choose an arbitrary edge $e = yz \in E(S, \bar{S})$ such that $y \in S$ and $z \in \bar{S}$. Assign $f(z) = f(y)O_{yz}$. Assign the remaining vertices of $\bar{S}$ so that $f(v) = f(u)O_{uv}$ moving along edges uv of T'. Note that f is consistent with both T and T'.

Let us examine the Dirichlet sum $\sum_{uv \in E} w_{uv} ||f(u)O_{uv} - f(v)||^2$. Consider an edge $f = uv \in E(S, \bar{S})$, $f \neq e$. We may, without loss of generality, assume that both S and $\bar{S}$ are connected. (If one or both is not, then we may alter our definition of f to be consistent along even more edges). Therefore, there is a cycle, $C = v_1 v_2 ... v_k v_1$ where $v_1 = u$, $v_k = v$, C contains the edges e and f, and all other edges have endpoints lying in either S or $\bar{S}$. By construction, $f(v) = f(u)O_{P_{uv}}$, so by the ϵ-consistency condition, we have

$$\begin{aligned}
||f(u)O_{uv} - f(v)|| &= ||f(v)O_{P_{vu}}O_{uv} - f(v)|| \\
&= \left\| f(v) \left(\prod_{i=1}^{k-1} O_{v_i v_{i+1}} O_{v_k v_1} - I \right) \right\| \\
&\leq \epsilon_1 ||f(v)|| = \epsilon_1.
\end{aligned}$$

In a similar manner, we have that $||f(u)O_{uv} - f(v)|| \leq \epsilon_2$ for each edge uv with both u and v in S or both in $\bar{S}$.

Therefore

$$\begin{aligned}
\lambda_1 \leq \mathcal{R}(f) &= \frac{\sum\limits_{(u,v)\in E} w_{uv} ||f(u)O_{uv} - f(v)||^2}{\sum\limits_v d_v ||f(v)||^2} \\
&\leq \frac{\sum\limits_{(u,v)\in E} w_{uv}\epsilon_2^2}{\sum\limits_v d_v} + \frac{\sum\limits_{\substack{u\sim v \\ u\in S, v\in \bar{S}}} w_{uv}\epsilon_1^2}{\sum\limits_v d_v} \\
&\leq \frac{\epsilon_2^2 |E(G)|}{\mathrm{vol}(G)} + \frac{\epsilon_1^2 |E(S,\bar{S})|}{2\,\mathrm{vol}(S)} \\
&= \frac{\epsilon_2^2}{2} + \frac{\epsilon_1^2}{2} h_G(S).
\end{aligned}$$

We have d orthogonal choices for the initial assignment of x, which leads to d orthogonal vectors satisfying this inequality. Therefore $\lambda_1, ..., \lambda_d$ all satisfy this bound.

5 Identifying Subsets

In this section, we follow ideas from [2] and [3] to relate connection PageRank vectors to the Cheeger ratio of ϵ-consistent subsets of a connection graph.

We will give an algorithm, which runs in time nearly linear in the size of the vertex set, which outputs a subset of the vertex set (if one exists) which has small Cheeger ratio and is ϵ-consistent.

5.1 PageRank Vectors and ϵ-Consistent Subsets

We define, for $S \subset V$, $f(S) = \sum_{v \in S} ||f(v)||_2$. Given a vertex v of $\mathbb{G}$, define a connection characteristic function χ_v to be any vector satisfying $||\chi_v(v)||_2 = 1$ and $\chi_v(u) = 0$ for $u \neq v$. Likewise, for a subset S of V, define a characteristic function χ_S to be a function such that $||\chi_S(v)||_2 = 1$ for $v \in S$, and $\chi_S(v) = 0$ for $v \notin S$.

Recall the definition of connection pagerank given a seed vector $\hat{s} : V \to \mathbb{R}^d$ is the vector $\widehat{\mathrm{pr}}(\alpha, \hat{s}) : V \to \mathbb{R}^d$ that satisfies

$$\widehat{\mathrm{pr}}(\alpha, \hat{s}) = \alpha\hat{s} + (1-\alpha)\,\widehat{\mathrm{pr}}(\alpha, \hat{s})\mathbb{Z}$$

where $\mathbb{Z}$ is the matrix for the random walk. Define $\mathbb{R}_\alpha = \alpha(I - (1-\alpha)\mathbb{Z})^{-1} = \alpha\sum_{t=0}^{\infty}(1-\alpha)^t\mathbb{Z}^t$ and note that $\widehat{\mathrm{pr}}(\alpha, \hat{s}) = \hat{s}\mathbb{R}_\alpha$.

Lemma 1. *Let $S \subset V$ be a subset of the vertex set of a connection graph, and let χ_S be a characteristic function for S. Then*

$$||\chi_S \mathbb{D}\mathbb{R}_\alpha(v)||_2 \leq d_v$$

for all $v \in V$

Proof. First, we will show that

$$\left\|\chi_S \mathbb{D}\mathbb{Z}^k(v)\right\|_2 \leq d_v$$

for all k by induction. For $k = 1$,

$$||\chi_S \mathbb{D}\mathbb{P}(v)||_2 = ||\chi_S \mathbb{A}(v)||_2 \leq \sum_{\substack{u \in S \\ u \sim v}} w_{uv}||\chi_S(u)O_{uv}||_2 \leq d_v.$$

By the induction hypothesis

$$\begin{aligned}\left\|\chi_S \mathbb{D}\mathbb{P}^{k+1}(v)\right\|_2 = \left\|\chi_S \mathbb{D}\mathbb{P}^k\mathbb{P}(v)\right\|_2 &= \left\|\sum_{u \in V}\chi_S \mathbb{D}\mathbb{P}^k(u)\mathbb{P}(u,v)\right\|_2 \\ &\leq \sum_{u \in V}\left\|\chi_S \mathbb{D}\mathbb{P}^k(u)\right\|_2 ||\mathbb{P}(u,v)||_2 \\ &\leq \sum_{u \in V} d_u \left\|\frac{1}{d_u}w_{uv}O_{uv}\right\|_2 = \sum_{u \in V} w_{uv} = d_v\end{aligned}$$

so this claim follows by induction.

Then from this claim,

$$||\chi_S \mathbb{D}\mathbb{R}_\alpha(v)||_2 = \left\|\chi_S \mathbb{D}\alpha\sum_{k=0}^{\infty}(1-\alpha)^k\mathbb{P}^k\right\|_2 \leq \alpha\sum_{k=0}^{\infty}(1-\alpha)^k\left\|\chi_S \mathbb{D}\mathbb{P}^k(v)\right\|_2 \leq d_v.$$

Lemma 2. *Let $S \subset V$ be a subset of the vertices such that the subgraph of $\mathbb{G}$ induced by S is ϵ-consistent. Let χ_S be some connection characteristic function for S that is consistent with some spanning subtree T of S. Define $\hat{f}_S$ by $\hat{f}_S(v) = \frac{d_v}{\mathrm{vol}(S)}\chi_S(v)$. The function $\hat{f}_S$ is the expected value for a characteristic function χ_u when a vertex u is chosen from S at random with probability $d_u/\mathrm{vol}(S)$. Then*

$$\widehat{\mathrm{pr}}(\alpha, \hat{f}_S)(S) \geq 1 - \frac{1-\alpha}{\alpha}(h(S) + \epsilon).$$

Proof. We have

$$\begin{aligned}
\widehat{\mathrm{pr}}(\alpha, \hat{f}_S)(S) &= \sum_{v \in S} ||\widehat{\mathrm{pr}}(\alpha, \hat{f}_S)(v)||_2 = \sum_{v \in S} ||\widehat{\mathrm{pr}}(\alpha, \hat{f}_S)(v)||_2 ||\chi_S(v)||_2 \\
&\geq \sum_{v \in S} \widehat{\mathrm{pr}}(\alpha, \hat{f}_S)(v)\chi_S(v)^T = \widehat{\mathrm{pr}}(\alpha, \hat{f}_S)\chi_S^T = f_S \mathbb{R}_\alpha \chi_S^T \\
&= \hat{f}_S \left(I - \frac{(1-\alpha)(I - \mathbb{Z})}{I - (1-\alpha)\mathbb{Z}}\right)\chi_S^T = 1 - \left(\hat{f}_S \frac{(1-\alpha)(I-\mathbb{Z})}{I - (1-\alpha)\mathbb{Z}}\right)\chi_S^T \\
&= 1 - \left(\frac{(1-\alpha)\chi_S \mathbb{D}}{\alpha\,\mathrm{vol}(S)} \frac{\alpha I}{I - (1-\alpha)\mathbb{Z}}(I - \mathbb{Z})\right)\chi_S^T \\
&= 1 - \frac{1-\alpha}{\alpha\,\mathrm{vol}(S)}\left(\chi_S \mathbb{D}\mathbb{R}_\alpha \mathbb{D}^{-1}\frac{(\mathbb{D} - \mathbb{A})}{2}\right)\chi_S^T \\
&= 1 - \frac{1-\alpha}{2\alpha\,\mathrm{vol}(S)} \sum_{u \sim v} w_{uv}\left(\chi_S \mathbb{D}\mathbb{R}_\alpha \mathbb{D}^{-1}(u)O_{uv} - \chi_S \mathbb{D}\mathbb{R}_\alpha \mathbb{D}^{-1}(v)\right)((\chi_S(u)O_{uv})^T - \chi_S(v)^T).
\end{aligned}$$

Note that χ_S is a characteristic function, so all the terms in the sum corresponding to $u, v \notin S$ are 0, for $v \in S$ and $u \notin S$ we are left with just $\chi_s(v)$, and for $u, v \in S$, since S is ϵ-consistent and χ_S was chosen to be consistent with a spanning subtree of S, then we have $\chi_S(u)O_{uv} - \chi_S(v)$ has norm less than ϵ. Applying this, the Cauchy-Schwarz Inequality, and the triangle inequality to the above, we have

$$\begin{aligned}
\widehat{\mathrm{pr}}(\alpha, \hat{f}_S)(S) \geq& 1 - \frac{1-\alpha}{2\alpha\,\mathrm{vol}(S)}\Big(\sum_{\substack{u \sim v \\ v \in S, u \in \bar{S}}} w_{uv} \left\|\chi_S \mathbb{D}\mathbb{R}_\alpha \mathbb{D}^{-1}(u)O_{uv} - \chi_S \mathbb{D}\mathbb{R}_\alpha \mathbb{D}^{-1}(v)\right\|_2 \\
&+ \sum_{\substack{u \sim v \\ u, v \in S}} w_{uv} \left\|\chi_S \mathbb{D}\mathbb{R}_\alpha \mathbb{D}^{-1}(u)O_{uv} - \chi_S \mathbb{D}\mathbb{R}_\alpha \mathbb{D}^{-1}(v)\right\|_2 \left\|\chi_S(u)O_{uv} - \chi_S(v)\right\|_2\Big) \\
\geq& 1 - \frac{1-\alpha}{2\alpha\,\mathrm{vol}(S)}\Big(\sum_{\substack{u \sim v \\ v \in S, u \in \bar{S}}} w_{uv} \left(||\chi_S \mathbb{D}\mathbb{R}_\alpha \mathbb{D}^{-1}(u)O_{uv}|| + ||\chi_S \mathbb{D}\mathbb{R}_\alpha \mathbb{D}^{-1}(v)||\right) \\
&+ \sum_{\substack{u \sim v \\ u, v \in S}} w_{uv} \left(||\chi_S \mathbb{D}\mathbb{R}_\alpha \mathbb{D}^{-1}(u)O_{uv}|| + ||\chi_S \mathbb{D}\mathbb{R}_\alpha \mathbb{D}^{-1}(v)||\right)\epsilon\Big).
\end{aligned}$$

Using Lemma 1 we can conclude that

$$\widehat{\mathrm{pr}}(\alpha, \hat{f}_S)(S) \geq 1 - \frac{1-\alpha}{\alpha\,\mathrm{vol}(S)}(|\partial S| + \epsilon|E(S, S)|) \geq 1 - \frac{1-\alpha}{\alpha}(h(S) + \epsilon).$$

Theorem 5. *Let $S \subset V$ be a subset of the vertex set such that the subgraph induced by S is ϵ-consistent. Let χ_S be some connection characteristic function* *for S that is consistent with some spanning subtree T of S. For each vertex $v \in S$, define $\chi_v : V \to \mathbb{R}^d$ by $\chi_v(v) = \chi_S(v)$ and $\chi_v(u) = 0$ for $u \neq v$. Then for any $\alpha \in (0,1]$, there is a subset $S_\alpha \subset S$ with volume $\mathrm{vol}(S_\alpha) \geq \mathrm{vol}(S)/2$ such that for any vertex $v \in S_\alpha$, the PageRank vector $\widehat{\mathrm{pr}}(\alpha, \chi_v)$ satisfies*

$$\widehat{\mathrm{pr}}(\alpha, \chi_v)(S) \geq 1 - \frac{2(h(S)+\epsilon)}{\alpha}.$$

Proof. Let v be a vertex of S chosen randomly from the distribution given by $\hat{f}_S$ of the previous result. Define the random variable $X = \widehat{\mathrm{pr}}(\alpha, \chi_v)(\bar{S})$ and note that the definition of PageRank and linearity of expectation implies that $E[X] = \widehat{\mathrm{pr}}(\alpha, \hat{f}_S)$. Therefore, by the preceding result,

$$E[X] = \widehat{\mathrm{pr}}(\alpha, \hat{f}_S)(\bar{S}) \leq \frac{1-\alpha}{\alpha\,\mathrm{vol}(S)}(h(S)+\epsilon) \leq \frac{h(S)+\epsilon}{\alpha}.$$

Then Markov's inequality implies

$$\Pr[v \notin S_\alpha] \leq \Pr[X > 2E[X]] \leq \frac{1}{2}.$$

Therefore $\Pr[v \in S_\alpha] \geq \frac{1}{2}$, so $\mathrm{vol}(S_\alpha) \geq \frac{1}{2}\mathrm{vol}(S)$.

5.2 A Local Partitioning Algorithm

We will follow ideas from [3] to produce an analogue of the Sharp Drop Lemma. Given any function $p : V \to \mathbb{R}^d$, define $q^{(p)} : V \to \mathbb{R}^d$ by $q^{(p)}(u) = p(u)/d_u$ for all $u \in V$. Order the vertices such that $q^{(p)}(v_1) \geq q^{(p)}(v_2) \geq \cdots \geq q^{(p)}(v_n)$. Define $S_j = \{v_1, ..., v_j\}$. The following lemma will be the basis of our algorithm, whose proof we will present in the appendix.

Lemma 3 (Sharp Drop Lemma). *Let $v \in V(\mathbb{G})$ and let $p = \widehat{\mathrm{pr}}(\alpha, \chi_v)$ for some $\alpha \in (0,1)$, let $q = q^{(p)}$ and let $\phi \in (0,1)$ be a real number. Then for any index $j \in [1,n]$, either the edges of S_j satisfy*

$$h(S_j) < 2\phi,$$

or there exists some index $k > j$ such that

$$\mathrm{vol}(S_k) \geq \mathrm{vol}(S_j)(1+\phi) \text{ and } \|q(v_k)\| \geq \|q(v_j)\| - \frac{2\alpha}{\phi\,\mathrm{vol}(S_j)}.$$

For our algorithm, we will employ an efficient algorithm for computing an approximate connection PageRank vector called ApproximatePR. The specifics of the algorithm as well as its run-time analysis can be found in [7] and a version for connection graphs is found in [17]. We will state the relevant result from [17] as the following:

Theorem 6. *For $\alpha, \epsilon \in (0,1)$ and $v \in V$, the algorithm ApproximatePR(v, α, ϵ) outputs a vector $\hat{p} = \widehat{\mathrm{pr}}(\alpha, \chi_v - \hat{r})$ such that*

$$\frac{\|\hat{r}(v)\|_2}{d_v} \leq \epsilon$$

for all vertices v. The running time of the algorithm is $O\left(\frac{d^2}{\epsilon\alpha}\right)$.

We note that

$$\frac{\|\hat{p}(u)\|}{d_u} \geq \frac{\|\widehat{\mathrm{pr}}(\alpha, \chi_v)(u)\|}{d_u} - \epsilon$$

for all u.

We are now ready to present the algorithm ConnectionPartition that utilizes PageRank vectors to come up with an ϵ-consistent subset of small Cheeger ratio.

ConnectionPartition(v, ϕ, x):

The input into the algorithm is a vertex $v \in V$, a target Cheeger ratio $\phi \in (0, 1)$, and a target volume $x \in [0, 2m]$.

1. Set $\gamma = \frac{1}{8} + \sum_{k=1}^{2m} \frac{1}{k}$ where m is the number of edges, $\alpha = \frac{\phi^2}{8\gamma}$, and $\delta = \frac{1}{16\gamma x}$.
2. Compute $p =$ ApproximatePR(v, α, δ) (which approximates $\widehat{\mathrm{pr}}(\alpha, \chi_v)$).
   Set $q(u) = p(u)/d_u$ for each u and order the vertices $v_1, \ldots, v_n$ so that $\|q(v_1)\| \geq \|q(v_2)\| \geq \cdots \geq \|q(v_n)\|$ and for each $j \in [1, n]$ define $S_j = \{v_1, \ldots, v_j\}$.
3. Choose a starting index k_0 such that $\|q(v_{k_0})\| \geq \frac{1}{\gamma \operatorname{vol}(S_{k_0})}$.
   If no such starting vertex exists, output Fail: No starting vertex.
4. While the algorithm is running:
   (a) If $(1 + \phi) \operatorname{vol}(S_{k_i}) > \operatorname{vol}(G)$, output Fail: No cut found.
   (b) Otherwise, let k_{i+1} be the smallest index such that $\operatorname{vol}(S_{k_{i+1}}) \geq (1 + \phi) \operatorname{vol}(S_{k_i})$.
   (c) If $\|q(v_{k_{i+1}})\| \leq \|q(v_{k_i})\| - 2\alpha/\phi \operatorname{vol}(S_{k_i})$, then output $S = S_{k_i}$ and quit.
   Otherwise repeat the loop.

Theorem 7. *Suppose $\mathbb{G}$ is a connection graph with a subset C such that $\operatorname{vol}(C) \leq \frac{1}{2} \operatorname{vol}(\mathbb{G})$, and $h(C) \leq \alpha/64\gamma$ with α as chosen in the algorithm. Assume further that C is ϵ-consistent for some $\epsilon < h(C)$. Let $C_\alpha = \left\{v \in C : \widehat{\mathrm{pr}}(\alpha, \chi_v)(\bar{C}) \leq \frac{2(h(C)+\epsilon)}{\alpha}\right\}$. Then for $v \in C_\alpha$, $\phi < 1/2$, and $x \geq \operatorname{vol}(C)$, the algorithm ConnectionPartition outputs a set S satisfying the following properties:*

1. $h(S) \leq 2\phi$.
2. $\operatorname{vol}(S) \leq (2/3) \operatorname{vol}(G)$.
3. $\operatorname{vol}(S \cap C) \geq (3/4) \operatorname{vol}(S)$.

Proof. Claim. There exist an index j such that $\|q(v_j)\| \geq \frac{1}{\gamma \operatorname{vol}(S_j)}$.

Proof. Suppose that $\|q(v_j)\| < \frac{1}{\gamma \operatorname{vol}(S_j)}$ for every index j. Since $v \in C_\alpha$, $\epsilon < h(C)$, and $h(C) \leq \alpha/64\gamma$ then we know that

$$p(C) \geq \widehat{\mathrm{pr}}(\alpha, \chi_v)(C) - \delta \operatorname{vol}(C) \geq 1 - \frac{2(h(C) + \epsilon)}{\alpha} - \frac{1}{16\gamma x} \operatorname{vol}(C) \geq 1 - \frac{1}{16\gamma} - \frac{1}{16\gamma} = 1 - \frac{1}{8\gamma}$$

since $x \geq \operatorname{vol}(C)$.

On the other hand, under our assumption,

$$\begin{aligned} p(C) \le p(V) = \sum_{i=1}^{n} \|p(v_i)\| &= \sum_{i=1}^{n} \|q(v_i)\| d_{v_i} \\ &< \sum_{i=1}^{n} \frac{d_{v_i}}{\gamma \operatorname{vol}(S_j)} \\ &\le \frac{1}{\gamma} \sum_{k=1}^{2m} \frac{1}{k}. \end{aligned}$$

Putting these together, we have

$$1 - \frac{1}{8\gamma} < \frac{1}{\gamma} \sum_{k=1}^{2m} \frac{1}{k}.$$

With the choice of $\gamma = \frac{1}{8} + \sum_{k=1}^{2m} \frac{1}{k}$ as in the algorithm, this yields a contradiction. Therefore there exists some index j with $\|q(v_j)\| \ge \frac{1}{\gamma \operatorname{vol}(S_j)}$ and the claim is proved.

It follows from Claim 5.2, that the algorithm will not fail to find a starting vertex.

Let k_f be the final vertex considered by the algorithm.

Claim. If $k_0, ..., k_f$ is a sequence of vertices satisfying both

- $\|q(v_{k_{i+1}})\| \ge \|q(v_{k_i})\| - \frac{2\alpha}{\phi \operatorname{vol}(S_{k_i})}$
- $\operatorname{vol}(S_{k_{i+1}}) \ge (1+\phi) \operatorname{vol}(S_{k_i})$

then

$$\|q(k_f)\| \ge \|q(k_0)\| - \frac{4\alpha}{\phi^2 \operatorname{vol}(S_{k_0})}.$$

Proof. We note that $\operatorname{vol}(S_{k_{i+1}}) \ge (1+\phi)^i \operatorname{vol}(S_{k_0})$, and so we have

$$\begin{aligned} \|q(k_f)\| &\ge \|q(k_0)\| - \frac{2\alpha}{\phi \operatorname{vol}(S_{k_0})} - \frac{2\alpha}{\phi \operatorname{vol}(S_{k_1})} - \cdots - \frac{2\alpha}{\phi \operatorname{vol}(S_{k_{f-1}})} \\ &\ge \|q(k_0)\| - \frac{2\alpha}{\phi \operatorname{vol}(S_{k_0})} \left(1 + \frac{1}{1+\phi} + \cdots + \frac{1}{(1+\phi)^{f-1}}\right) \\ &\ge \|q(k_0)\| - \frac{2\alpha}{\phi \operatorname{vol}(S_{k_0})} \frac{1+\phi}{\phi} \\ &\ge \|q(k_0)\| - \frac{4\alpha}{\phi^2 \operatorname{vol}(S_{k_0})} \end{aligned}$$

and the claim follows.

Now we will use this claim, the choice of $\alpha = \phi^2/8\gamma$, and the condition on the starting vertex $\|q(k_0)\| \geq 1/\gamma \operatorname{vol}(S_{k_0})$ to obtain a lower bound on $\|q(k_f)\|$,

$$\begin{aligned}\|q(k_f)\| &\geq \|q(k_0)\| - \frac{4\alpha}{\phi^2 \operatorname{vol}(S_{k_0})} \\ &\geq \frac{1}{\gamma \operatorname{vol}(S_{k_0})} - \frac{1}{2\gamma \operatorname{vol}(S_{k_0})} \\ &\geq \frac{1}{2\gamma \operatorname{vol}(S_{k_0})}.\end{aligned}$$

As in the proof of Claim 5.2, we have that $p(C) \geq 1 - \frac{1}{8\gamma}$, and therefore $p(\bar{C}) \leq \frac{1}{8\gamma}$.

Now observe that $\operatorname{vol}(S_{k_f} \cap \bar{C}) \leq \dfrac{p(\bar{C})}{\|q(k_f)\|}$. This follows since $\|q(k_f)\| \operatorname{vol}(S_{k_f} \cap \bar{C}) = \sum_{v \in S_{k_f} \cap \bar{C}} \|q(k_f)\| d_v \leq \sum_{v \in S_{k_f} \cap \bar{C}} \|q(v)\| d_v \leq \sum_{v \in \bar{C}} \|p(v)\| = p(\bar{C})$. Thus

$$\begin{aligned}\operatorname{vol}(S_{k_f} \cap \bar{C}) &\leq \frac{p(\bar{C})}{\|q(k_f)\|} \\ &\leq \frac{2\gamma \operatorname{vol}(S_{k_f})}{8\gamma} \\ &= \frac{1}{4} \operatorname{vol}(S_{k_f}).\end{aligned}$$

Therefore $\operatorname{vol}(S_{k_f}) \leq \operatorname{vol}(C) + \operatorname{vol}(S_{k_f} \cap \bar{C}) \leq \operatorname{vol}(C) + \frac{1}{4} \operatorname{vol}(S_{k_f})$, implying that $\operatorname{vol}(S_{k_f}) \leq \frac{4}{3} \operatorname{vol}(C)$. Using that fact that $\operatorname{vol}(C) \leq \frac{1}{2} \operatorname{vol}(G)$,

$$\operatorname{vol}(S_{k_f}) \leq \frac{4}{3} \operatorname{vol}(C) \leq \frac{2}{3} \operatorname{vol}(G) \leq \frac{\operatorname{vol}(G)}{1+\phi}.$$

This last step follows under the assumption that $\phi \leq 1/2$. We can do this without loss of generality since the guarantee on $h(S)$ in the theorem is trivial for $\phi > 1/2$.

The above shows that the algorithm will not experience a failure due to the volume becoming too large, and we have seen that conditions (2) and (3) will be satisfied by the output.

Finally, to show condition (1), we apply the Sharp Drop Lemma. We know that k_f is the smallest index such that $\operatorname{vol}(S_{k_f+1}) \geq (1+\phi) \operatorname{vol}(S_{k_f})$, and $\|q(v_{k_f+1})\| \leq \|q(v_{k_f})\| - 2\alpha/\phi \operatorname{vol}(S_{k_i})$. Therefore the Sharp Drop Lemma guarantees that $h(S_{k_f}) < 2\phi$, and the proof is complete.

Theorem 8. *The running time for the algorithm* ConnectionPartition *is* $O\left(d^2 x \dfrac{\log^2 m}{\phi^2}\right)$.

Proof. The running time is dominated by the computation of the PageRank vector. According to Theorem 6, the running time for this is $O\left(\frac{d^2}{\delta\alpha}\right)$. In the algorithm, we have $\alpha = \frac{\phi^2}{8\gamma}$, $\delta = \frac{1}{16\gamma x}$, and $\gamma = \frac{1}{8} + \sum_{k=1}^{2m} \frac{1}{k} = \Theta(\log m)$. Therefore $\alpha = O(\frac{\phi^2}{\log m})$ and $\delta = O(\frac{1}{x \log m})$. Therefore the running time is as claimed.

References

1. Agarwal, S., Snavely, N., Simon, I., Seitz, S.M., Szeliski, R.: Building Rome in a Day. In: Proceedings of the 12th IEEE International Conference on Computer Vision, pp. 72–79 (2009)
2. Andersen, R., Chung, F., Lang, K.: Using PageRank to locally partition a graph. Internet Math. 4(1), 35–64 (2007)
3. Andersen, R., Chung, F.: Detecting sharp drops in PageRank and a simplified local partitioning algorithm. In: Cai, J.-Y., Cooper, S.B., Zhu, H. (eds.) TAMC 2007. LNCS, vol. 4484, pp. 1–12. Springer, Heidelberg (2007)
4. Bandeira, A.S., Singer, A., Spielman, D.A.: A Cheeger Inequality for the Graph Connection Laplacian (2012), `http://arxiv.org/pdf/1204.3873v1.pdf`
5. Brin, S., Page, L.: The anatomy of a large-scale hypertextual Web search engine. Computer Networks and ISDN Systems 30(1-7), 107–117 (1998)
6. Chung, F.: Spectral Graph Theory. AMS Publications (1997)
7. Chung, F., Zhao, W.: A sharp PageRank algorithm with applications to edge ranking and graph sparsification. In: Kumar, R., Sivakumar, D. (eds.) WAW 2010. LNCS, vol. 6516, pp. 2–14. Springer, Heidelberg (2010)
8. Chung, F., Kempton, M., Zhao, W.: Ranking and sparsifying a connection graph. Internet Mathematics (2013)
9. Chung, F., Sternberg, S.: Laplacian and vibrational spectra for homogeneous graphs. J. Graph Theory 16, 605–627 (1992)
10. Cucuringu, M., Lipman, Y., Singer, A.: Sensor network localization by eigenvector synchronization over the Euclidean group. ACM Transactions on Sensor Networks 8(3), No. 19 (2012)
11. Hadani, R., Singer, A.: Representation theoretic patterns in three dimensional cryo-electron microscopy I - the intrinsic reconstitution algorithm. Annals of Mathematics 174(2), 1219–1241 (2011)
12. Horn, R., Johnson, C.: Matrix Analysis. Cambridge University Press (1985)
13. Jolliffe, I.T.: Principal Component Analysis, 2nd edn. Springer Series in Statistics (2002)
14. Singer, A.: Angular synchronization by eigenvectors and semidefinite programming. Applied and Computational Harmonic Analysis 30(1), 20–36 (2011)
15. Singer, A., Zhao, Z., Shkolnisky, Y., Hadani, R.: Viewing angle classification of cryo-electron microscopy images using eigenvectors. SIAM Journal on Imaging Sciences 4(2), 723–759 (2011)
16. Singer, A., Wu, H.-T.: Vector Diffusion Maps and the Connection Laplacian. Communications on Pure and Applied Mathematics 65(8), 1067–1144 (2012)
17. Zhao, W.: Ranking and sparsifying edges of a graph. Ph.D. Thesis, University of California, San Diego (2012)

Appendix

Proof of Theorem 1

Proof. We will define a function $f : V \to \mathbb{R}^d$ whose Rayleigh quotient will bound the smallest eigenvalue. For a fixed vertex $z \in V$, we assign $f(z) = x$, where x is a unit vector in $\mathbb{R}^d$. Fix a spanning tree T of G, and define f to be consistent with T. That is, for any vertex v of G assign $f(v)$ as follows. Let $P_{zv} = (z = v_1 v_2 ... v_k = v)$ be the path from z to v in T. Then let $f(v) = f(z) O_{P_{zv}}$. Notice that $\|f(v)\| = 1$ for all $v \in V$. We will examine the Rayleigh quotient of this function. Notice that for uv an edge of T, we have

$$\|f(u)O_{uv} - f(v)\| = \|f(v) - f(v)\| = 0$$

by construction. For any other edge uv of G, consider the cycle obtained by taking the path $P_{vu} = (v = v_1 v_2 ... v_k = u)$ in T, and adding in the edge uv. Then by construction of f and the ϵ-consistency condition, we have

$$\begin{aligned}\|f(u)O_{uv} - f(v)\| &= \|f(v)O_{P_{vu}}O_{uv} - f(v)\| \\ &= \left\| f(v) \left(\prod_{i=1}^{k-1} O_{v_i v_{i+1}} O_{v_k v_1} - I \right) \right\| \\ &\leq \epsilon \|f(v)\| = \epsilon.\end{aligned}$$

Therefore

$$\begin{aligned}\lambda_1 \leq \mathcal{R}(f) &= \frac{\sum_{(u,v)\in E} w_{uv} ||f(u)O_{uv} - f(v)||^2}{\sum_v d_v ||f(v)||^2} \\ &< \frac{\sum_{(u,v)\in E} w_{uv}\epsilon^2}{\sum_v d_v} = \frac{\epsilon^2}{2}.\end{aligned}$$

The initial choice of the unit vector $x \in \mathbb{R}^d$ in the construction of f was arbitrary. We thus have d orthogonal choices for the initial assignment of x, which leads to d orthogonal functions satisfying this inequality. Therefore, by the Courant-Fischer Theorem, $\lambda_1, ..., \lambda_d$ all satisfy this bound.

Proof of Theorem 2

Proof. Let $f_1, ..., f_d$ be the orthogonal set of vectors defined in the proof of Theorem 1, each with $\mathcal{R}(f) \leq \epsilon^2/2$. Then $\|f(v)\|^2 = 1$ for all v. Given $A \subset V$ and $B = \bar{A}$, define $g_i : V \to \mathbb{R}^d$ by

$$g_i(v) = \begin{cases} \frac{1}{\text{vol}\, A} f_i(v) & \text{for } v \in A \\ -\frac{1}{\text{vol}\, B} f_i(v) & \text{for } v \in B \end{cases}$$

For ease of notation we will simply write g and f for g_i and f_i. Note that if both $u, v \in A$, then $\|g(u)O_{uv} - g(v)\|^2 = \left\|\frac{1}{\operatorname{vol} A} f(u)O_{uv} - \frac{1}{\operatorname{vol} A} f(v)\right\|^2 \leq \frac{1}{(\operatorname{vol} A)^2}\epsilon^2$. Similarly, if both $u, v \in B$, $\|g(u)O_{uv} - g(v)\|^2 \leq \frac{1}{(\operatorname{vol} B)^2}\epsilon^2$. For $u \in A$ and $v \in B$, we have $\|g(u)O_{uv} - g(v)\|^2 = \left\|\frac{1}{\operatorname{vol} A} f(u)O_{uv} + \frac{1}{\operatorname{vol} B} f(v)\right\|^2 \leq \left(\frac{1}{\operatorname{vol} A} + \frac{1}{\operatorname{vol} B}\right)^2$ by the triangle inequality.

Therefore

$$
\begin{aligned}
\mathcal{R}(g) &= \frac{\sum\limits_{(u,v)\in E} w_{uv}\|g(u)O_{uv} - g(v)\|_2^2}{\sum\limits_{v\in V} \|g(v)\|^2 d_v} \\
&\leq \frac{\frac{1}{2}\operatorname{vol} A \frac{1}{(\operatorname{vol} A)^2}\epsilon^2 + \frac{1}{2}\operatorname{vol} B \frac{1}{(\operatorname{vol} B)^2}\epsilon^2 + \left(\frac{1}{\operatorname{vol} A} + \frac{1}{\operatorname{vol} B}\right)^2 |E(A,B)|}{\sum\limits_{v\in A} \frac{1}{(\operatorname{vol} A)^2} d_v + \sum\limits_{v\in B} \frac{1}{(\operatorname{vol} B)^2} d_v} \\
&= \frac{\frac{1}{2}\epsilon^2\left(\frac{1}{\operatorname{vol} A} + \frac{1}{\operatorname{vol} B}\right) + \left(\frac{1}{\operatorname{vol} A} + \frac{1}{\operatorname{vol} B}\right)^2 |E(A,B)|}{\frac{1}{\operatorname{vol} A} + \frac{1}{\operatorname{vol} B}} \\
&\leq \frac{1}{2}\epsilon^2 + 2h_G(A).
\end{aligned}
$$

Therefore we have d orthogonal vectors $g_1, ..., g_d$ satisfying this bound, each orthogonal to $f_1, ..., f_d$ which clearly satisfy the bound, so the result follows.

Proof of Lemma 3

Proof. Let $S \subset V$ be a subset of the vertex set that contains v. We have

$$
\begin{aligned}
p\mathbb{Z}(S) &= \sum_{u\in S} \|p\mathbb{Z}(u)\| = \sum_{u\in S} \left\|\frac{1}{2}p(u) + \frac{1}{2}q\mathbb{A}(u)\right\| \leq \frac{1}{2}\left(\sum_{u\in S}\|p(u)\| + \sum_{u\in S}\left\|\sum_{v\sim u} q(v)O_{uv}\right\|\right) \\
&\leq \frac{1}{2}\left(\sum_{u\in S}\|p(u)\| + \sum_{u\in S}\sum_{v\sim u}\|q(v)\|\right) = \frac{1}{2}\left(2\sum_{u\in S}\|p(u)\| - \sum_{(u,v)\in E(S,\bar{S})}(\|q(u)\| - \|q(v)\|)\right) \\
&= p(S) - \frac{1}{2}\sum_{(u,v)\in E(S,\bar{S})}(\|q(u)\| - \|q(v)\|).
\end{aligned}
$$

Since $p = \widehat{\mathrm{pr}}(\alpha, \chi_v)$, we have that p satisfies $p\mathbb{Z} = \alpha\chi_v + (1-\alpha)p\mathbb{Z}$, therefore

$$
\|p\mathbb{Z}(u)\| = \frac{1}{1-\alpha}\|p(u) - \alpha\chi_v(u)\| \geq \|p(u)\| - \alpha\|\chi_v(u)\|
$$

for any u. Therefore

$$
p\mathbb{Z}(S) \geq p(S) - \alpha.
$$

Combining these, we see that

$$
\sum_{(u,v)\in E(S,\bar{S})} (\|q(u)\| - \|q(v)\|) \leq 2\alpha. \tag{1}
$$

Now we will consider S_j. If $\operatorname{vol}(S_j)(1+\phi) > \operatorname{vol}(G)$, then

$$
|E(S_j, \bar{S}_j)| \leq \operatorname{vol}(\bar{S}_j) \leq \operatorname{vol}(G)\left(1 + \frac{1}{1+\phi}\right) \leq \phi \operatorname{vol}(S_j)
$$

and the result holds. Assume $\mathrm{vol}(S_j)(1+\phi) \leq \mathrm{vol}(G)$. Then there exists a unique index $k > j$ such that

$$\mathrm{vol}(S_{k-1}) \leq \mathrm{vol}(S_j)(1+\phi) \leq \mathrm{vol}(S_k).$$

If $e(S_j, \bar{S}_j) < 2\phi\,\mathrm{vol}(S_j)$, then we are done. If $e(S_j, \bar{S}_j) \geq 2\phi\,\mathrm{vol}(S_j)$, then we note that we can also get a lower bound on $e(S_j, \bar{S}_{k-1})$, namely

$$e(S_j, \bar{S}_{k-1}) \geq e(S_j, \bar{S}_j) - \mathrm{vol}(S_{k-1} \setminus S_j) \geq 2\phi\,\mathrm{vol}(S_j) - \phi\,\mathrm{vol}(S_j) = \phi\,\mathrm{vol}(S_j).$$

Therefore, by Equation 1

$$\begin{aligned} 2\alpha &\geq \sum_{(u,v)\in E(S_j,\bar{S}_j)} (\|q(u)\| - \|q(v)\|) \\ &\geq \sum_{(u,v)\in E(S_j,\bar{S}_{k-1})} (\|q(u)\| - \|q(v)\|) \\ &\geq e(S_j, \bar{S}_{k-1})(\|q(v_j)\| - \|q(v_k)\|) \\ &\geq \phi\,\mathrm{vol}(S_j)(\|q(v_j)\| - \|q(v_k)\|). \end{aligned}$$

This implies that $\|q(v_j)\| - \|q(v_k)\| \leq 2\alpha/\phi\,\mathrm{vol}(S_j)$, and the result follows.

On the Power of Adversarial Infections in Networks

Michael Brautbar[1,*], Moez Draief[2], and Sanjeev Khanna[1]

[1] Computer and Information Science, University of Pennsylvania
{brautbar,sanjeev}@cis.upenn.edu
[2] Electrical and Electronic Engineering, Imperial College London
m.draief@imperial.ac.uk

Abstract. Over the last decade we have witnessed the rapid proliferation of online networks and Internet activity. Such activity is considered as a blessing but it brings with it a large increase in risk of computer malware — malignant software that actively spreads from one computer to another. To date, the majority of existing models of malware spread use stochastic behavior, when the set of neighbors infected from the current set of infected nodes is chosen obliviously. In this work, we initiate the study of *adversarial* infection strategies which can decide intelligently which neighbors of infected nodes to infect next in order to maximize their spread, while maintaining a similar "signature" as the oblivious stochastic infection strategy as not to be discovered. We first establish that computing an optimal and near-optimal adversarial strategies is computationally hard. We then identify necessary and sufficient conditions in terms of network structure and edge infection probabilities such that the adversarial process can infect polynomially more nodes than the stochastic process while maintaining a similar "signature" as the oblivious stochastic infection strategy. Among our results is a surprising connection between an additional structural quantity of interest in a network, the network *toughness*, and adversarial infections. Based on the network toughness, we characterize networks where existence of adversarial strategies that are pandemic (infect all nodes) is guaranteed, as well as efficiently computable.

1 Introduction

Over the last decade we have witnessed the rapid proliferation of online networks and Internet activity. While such a proliferation is considered by many as a blessing, it brings with it an increase in risk of computer malware — malignant software that actively spreads from one computer to another. Indeed, a long thread of research has been devoted to understanding the spread of malicious malware such as computer viruses, computer worms and other malignant forms of computer infection cf. [3, 16, 17, 19, 21]. However, such work has, so far,

* Now in The Laboratory for Information and Decision Systems, Massachusetts Institute of Technology, brautbar@mit.edu.

A. Bonato, M. Mitzenmacher, and P. Prałat (Eds.): WAW 2013, LNCS 8305, pp. 44–55, 2013.
© Springer International Publishing Switzerland 2013

only considered oblivious malware propagation, where the spreading malware does not behave strategically in its choice of which nodes to spread to. What now is a standard way of containing the spread of malware is the control of the amount of information that spreads from one computer to others [3, 19]. This is also known as throttling. Under rate control, a malware that does not want to get detected spreads obliviously to a small set of neighbors while abiding by the rate constraint. Inspired by this fact, in this work we initiate the study of *intelligent* malware propagation, where the spreading malware can strategically decide which neighboring nodes to infect under rate constraints, in order to maximize the total number of infections over time. Each edge (u, v) is equipped with an edge weight $p(u, v)$ representing the amount of a typical and normal communication between the two nodes u and v. Typical examples of such networks include email networks, instant messaging networks, and online social networks, among others. Under the rate constraint, a malware spreading from an infected node u can infect at most a number of nodes that is not more than the typical "signature" of communication, namely,

$$\left\lceil \sum_{\{v \text{ neighbor of } u\}} p(u, v) \right\rceil .$$

More generally, we will demand that for any subset X, the malware must not infect more nodes than the amount of traffic involving X permits it to, namely,

$$\left\lceil \sum_{u \in X} \sum_{\{v \text{ neighbor of } u\}} p(u, v) \right\rceil .$$

In particular, such malware can use the structure of the network in order to choose which of the neighboring computers to infect from a newly infected node. We initiate a detailed comparison of the behavior of an adversarial infection, that can use the network structure, to that of an oblivious stochastic one, that spreads according to the standard Independent Cascades diffusion model. In order to defend against future malware it is of primary importance to first analyse the adversarial infections strategy and to contrast its behavior with that of the oblivious stochastic strategy.

We would like to further emphasize the need to understand the behavior of adversarial infections with the following example, comparing the behavior of a well-planned adversarial infection to that of a simple heuristic. Consider a path connected at one of its ends to the root of a two level binary tree. Set all edge weights to 1/2. A greedy strategy may consist, for each newly infected node u, to infect its r_u neighbors of highest degree, where r_u is the rate constraint of u. However, such a strategy would miss the path altogether. In contrast, an adversary with a global knowledge of the graph can plan ahead and infect the whole path, by starting from the root of the tree but spending its budget to infect the next node on the path (and the extra budget to infect part of tree). While greedy would infect only $O(1)$ nodes, a well-planned adversary would infect $n - O(1)$ nodes.

1.1 Our Results

We first show that the problem of computing an optimal and near-optimal adversarial strategies under typical constraints is computationally hard. We then identify necessary and sufficient conditions in terms of network structure and edge infection probabilities such that the adversarial process can infect polynomially more nodes than the stochastic process while maintaining a similar "signature" as the oblivious stochastic infection strategy. Our first set of results show that when the minimum weighted graph cut value is $\Omega(\log n)$ (on a network with n nodes), the standard oblivious stochastic infection strategy can essentially infect all nodes. Thus the interesting regime to analyze is when the minimum weighted cut value is $o(\log n)$. In this regime we demonstrate that the optimal adversarial infection can be pandemic (namely infect all nodes) while the oblivious stochastic strategy infects, in expectation, a constant number of nodes. We then identify a surprising connection between an additional structural quantity of interest in a network, the network *toughness*, and adversarial infections. Based on the network toughness, we characterize conditions guaranteeing the existence of a pandemic adversarial strategy as well as its efficient computation.

1.2 Related Work

Most work on computer malware has been focused on virus and worm propagation [17, 19]. The vast majority of the literature has focused on modelling and simulations of the behavior of a stochastic malware which spreads according to the Independent Cascades model or its extension to repeated infection attempts, the Susceptible-Infected model [7, 19–22]. However, none of these papers consider intelligently-designed malware that can choose which nodes to infect based on some prior computation or knowledge of the network.

Another line of research that we would like to mention is the one devoted to error and attack tolerance in online networks; see the seminal work of [1] and its long line of follow-up research such as [9]. The main thread of research is devoted to understanding how attacking and removing, in an unweighted graph, nodes of high-degree results in more parts of the network becoming disconnected than attacking and removing the same number of nodes, obliviously at random. In contrast to this work, we are interested in analyzing cascading effects that spread through the network, rather than a single attack and removal of nodes (and edges). Furthermore, our main interest is in coordinated attacks that are strategically designed, and so the type of node first targeted is chosen as to maximize a global effect in a provable way (resulting in infecting as many nodes as possible) rather than based on local heuristics (choosing high degree nodes).

1.3 Outline

In section 2 we provide a detailed definition of what comprises an adversarial infection and the description of the behavior of the oblivious stochastic infection strategy. In section 3 we discuss the computational complexity of computing an optimal, as well as near optimal, adversarial infection strategies. In section 4 we provide necessary conditions in terms of the network cuts such that the adversarial process can infect polynomially more nodes than the stochastic process. In section 5 we provide necessary and sufficient conditions for the existence of a pandemic adversarial infection strategy and provide efficient algorithms for computing such a strategy. Finally, in section 6 we summarize our contributions and list several intriguing directions for future research.

2 Model and Preliminaries

The Input Network. We are given an undirected, edge-weighted network $G = (V, E, p)$ on $|V| = n$ nodes, $|E| = m$ edges and an edge weight function $p : E \to (0, 1]$. We think of the weight $p(u, v)$ as the average amount of communication between neighbors u and v over a typical period of time. We will be particularly interested in the behavior of adversarial infections and the independent cascade model on the input network.

Adversarial Diffusion. An adversary strategy A is a rooted tree T_A, rooted at its seed node of choice (namely, the source of infection), that spans some arbitrary subset S of nodes. Each node u in T_A is responsible for infecting its children in T_A. For any subset X of nodes in T_A, let

$$\text{infect}(A, X) = \{v \in T_A\colon v\text{'s parent in } T_A \text{ belongs to } X\}.$$

We say that an adversary strategy A (with its rooted tree T_A) infects a set S of nodes, while obeying the first order constraints, if its rooted tree T_A spans S such that for every subset $X \subseteq S$ of nodes we have,

$$|\text{infect}(A, X)| \leq \left\lceil \sum_{u \in X} \sum_{v:(u,v) \in E} p(u, v) \right\rceil .$$

Namely, the number of node infections attributed to X is constrained by the ceiling of the total weight adjacent to the set X.

In this paper we only consider adversarial infections that obey the first-order constraints; unless stated explicitly otherwise, an adversary would be assumed to obey the first-order constraints. We will call the problem of finding a first-order constrained adversarial infection that maximizes number of infections *the adversarial infection problem*[1]. An adversarial strategy will be called *pandemic* if it infects all nodes.

Independent Cascades Diffusion Model. The Independent Cascades (IC) model of diffusion was formalized by Kempe et al. [13] and is by now a standard model to describe infection propagation in social and other contact networks [10].

[1] It is not hard to see that the optimal adversary is deterministic: any stochastic adversary is a convex combination of trees so one can take the best one w.l.o.g.

The IC model can be thought of as a discrete version of the known *Susceptible-Infected-Removed (SIR) model.* The IC diffusion spreads via a random process, beginning at its seed node of choice from V. The process proceeds in rounds. In each round, a node u that got infected in the previous round gets a chance to subsequently infect each healthy neighbor v, with probability equal to the weight of the edge (u, v), namely $p(u, v)$. The node u becomes then recovered and does not spread the virus any further. If multiple nodes try to infect a new node in the current round, then each succeeds, independently, according to the corresponding edge weight. In this paper, we will focus on the the IC process only on undirected graphs. Throughout the text, we shall also refer to the stochastic infection strategy, as defined by the IC process, as the *oblivious stochastic strategy* or sometime as the *stochastic diffusion.*

Quantities of Interest. All graphs considered in the paper are undirected graphs. Throughout our analysis we will frequently refer to the minimum weighted cut value and maximum weighted cut value in the weighted input graph: the minimum (resp. maximum) weighted cut value, denoted $\Phi_G^{\min}$ (resp. $\Phi_G^{\max}$), is the value of the cut $(S, \overline{S})$ that minimizes (resp. maximizes) the sum

$$\Phi_G(S) := \sum_{\{u\in S, v\in\overline{S}, (u,v)\in E\}} p(u, v) \,.$$

When the identity of G is clear, we will often omit the subscript G.

We denote by $|S|$ the size of a set S. We denote the degree of a node u in a graph G by $d_G(u)$.

All logarithms in the paper are in base 2.

3 Hardness of Maximizing Adversarial Infection

In this section, we show that in general, the adversarial infection problem is hard to approximate under first order constraints.

Theorem 1. *The adversarial maximization problem is* $2^{(\log^{1-\epsilon} n-1)}$*-hard to approximate, for any* $\epsilon > 0$, *unless* $\boldsymbol{NP} \subseteq \boldsymbol{DTIME}(2^{O(\log^{1/\epsilon} n)})$.

Proof. The proof is by reduction from the longest path problem on undirected graphs. Let $G(V, E)$ be the input graph for the longest path problem. We create an instance of the adversarial infection problem from G as follows. Starting with the graph G, we attach $(n - d_G(u))$ auxiliary vertices to each vertex $u \in V$. Assign a weight of $1/n$ to each edge in the resulting graph. Let H be the resulting graph.

Note that any infection strategy in H obeying the first-order constraints must be a path, since the total incident weight on any node in H is at most 1 (and is exactly 1 for nodes from G). If the path length is ℓ in H it translates to a path of length at least $\ell - 2$ in G (which is at least $\ell/2$ for $\ell \geq 4$), and a path of length ℓ in G translates to an infection strategy following a path of length ℓ in H. The longest path problem is $2^{(\log^{1-\epsilon} n)}$-hard to approximate in undirected graphs, unless $\mathbf{NP} \subseteq \mathbf{DTIME}(2^{O(\log^{1/\epsilon} n)})$ [12], and the result easily follows.

We next show that the problem remains hard to approximate to within any constant factor even when the input instances are restricted to regular graphs with uniform infection probabilities.

Theorem 2. *The adversarial maximization problem does not admit a constant factor approximation in undirected regular graphs with uniform edge weights, unless* $\mathbf{P} = \mathbf{NP}$*.*

Proof. Let G be an undirected k-regular graph, $k \geq 2$ with uniform infection probability of $1/k$ on edges. Now the problem of finding a good strategy obeying first-order constraints becomes exactly the problem of finding a long path in the graph. Thus the problem of maximizing adversarial infection is as hard to approximate as the longest path on regular graphs. Even in 3-regular Hamiltonian graphs, the longest-path problem is known not to have any constant factor approximation, unless $\mathbf{P} = \mathbf{NP}$ [5].

4 A Cut-Based Analysis

We next proceed to exploring networks where an adversarial infection can infect polynomially more nodes than the oblivious stochastic strategy. We will show that two important parameters in understanding this goal is the value of the minimum weighted cut, $\Phi_G^{\min}$, and the value of maximum weighted cut, $\Phi_G^{\max}$, in the input graph G.

We first show that if $\Phi_G^{\min}$ is at least logarithmically large the oblivious stochastic strategy is essentially pandemic.

Theorem 3 (Theorem 1 of [2]). *Let* $G = (V, E, p)$*. For every positive constant* b*, there exists a constant* $c = c(b) > 0$ *so that if* $\Phi_G^{\min} \geq c\log(n)$*, then the probability that a realization of* G *is disconnected, where each edge* $(u, v) \in E$ *is kept with probably* $p(u, v)$*, is at most* $\frac{1}{n^b}$*.*

By the theorem we conclude,

Corollary 1. *For* c *large enough, the oblivious stochastic strategy would infect at least* $n - o(1)$ *nodes in expectation.*

Thus in this parameter regime, an improvement using adversarial infection strategies would be non-significant over the oblivious stochastic strategy. We note, however, that having a logarithmically large minimum cut is a quite stringent condition; in particular, any graph that has even one node with degree of size $o(\log(n))$ would violate the minimum cut condition. It is thus of high interest to analyze the regimes when this condition is violated. For this purpose we next show that no adversarial strategy can infect more than $\Theta(\Phi_G^{\max})$ nodes, and that when $\Phi_G^{\max}$ is large enough, a polynomial gap between the adversary to the oblivious stochastic strategy is feasible.

Theorem 4. *For any graph* G *no adversarial infection strategy obeying first-order constraint, as well as the oblivious stochastic strategy, can infect more than* $\Theta(\Phi_G^{\max})$ *nodes.*

On the other hand, for any value n there exists a graph G on n nodes with $\Phi_G^{\max} = \Theta(n)$ such that the optimal adversarial strategy obeying first order constraint infects $\Theta(\Phi_G^{\max})$ nodes while the oblivious stochastic strategy infects only $O(1)$ nodes in expectation. Moreover, the gap result holds on graphs where the edge infection probability is uniform and a constant (independent of n).

Proof. To prove the first part we make use of the following known fact that can be easily proven using the probabilistic method.

Fact 1. *For any weighted undirected graph $G = (V, E, p)$,*

$$\Phi_G^{\max} \geq 1/2 \sum_{(u,v)\in E} p(u,v) .$$

Now consider any adversarial infection strategy A, and let T_A be its tree of infections; let S be the set of nodes infected. Ignoring the root of T_A, either the total number of nodes at the odd levels of T_A must be at least $(|S|-1)/2$ or the total number of nodes at the even levels of T_A must be at least $(|S|-1)/2$. Since all infections at odd levels have to be attributed to nodes at the even levels (and vice versa), if the tree T_A conforms to first-order constraints, then we must have

$$(|S|-1)/2 \leq \lceil 2\Phi_G^{\max} \rceil .$$

Since this holds for any choice of tree T_A (and hence any adversarial strategy), the claim follows for any adversary. A simple argument shows the claim for the oblivious stochastic process: any node u can infect at most $\sum_{v:(u,v)\in E} p(u,v)$ new nodes in expectation, and thus the total number of infections is, in expectation, at most

$$\sum_{u\in V} \sum_{v:(u,v)\in E} p(u,v) = \Theta(\Phi_G^{\max}).$$

We now prove the other part of the theorem. To show this we need to provide a graph G on n nodes such that the optimal adversary can infect $\Phi_G^{\max} = \Theta(n)$ nodes while the oblivious stochastic strategy infects $O(1)$ in expectation.

For simplicity of exposition assume that n is even. Take a cycle on $n/2$ nodes and set each edge probability on the cycle to be $1/2$. Now attach to each node u_i on the cycle, an auxiliary nodes v_i using also an edge of probability $1/2$. Note that by fact 1, the value of the maximum weighted cut $\Phi_G^{\max}$ is $\Omega(n)$. Clearly, $\Phi_G^{\max} \leq |E| = n+1$ and so $\Phi_G^{\max} = \Theta(n)$, as required.

The adversary chooses all n nodes on the cycle (infection tree is a path), and no auxiliary nodes. This satisfies first-order constraints because each node u_i has an infection budget of at least 1 and it needs to infect exactly one node.

However, for any choice of the seed vertex, a stochastic strategy obtains only $O(1)$ nodes in expectation. To see this note that the infection survives for k steps on the cycle with probability at most $2/2^k$ and so it can infect, in expectation, at most $\sum_k 8k/2^k = 16 = O(1)$ nodes.

5 Pandemic Infections

In this section, we further explore the setting where the value of the minimum weighted cut is $o(\log n)$. As we have seen earlier, in this setting the gap between an adversarial infection (obeying first-order constraints) and oblivious stochastic diffusion can be as large as $\Omega(n)$. We obtain here sufficient and necessary conditions for an adversarial infection to become pandemic (i.e., infect all nodes) by relating existence of such strategies to the notion of *toughness* of the graph.

The notion of graph toughness was first introduced in order to study conditions for the existence of Hamiltonian cycles in graphs, see [4]. Given an undirected graph $G = (V, E)$ and a subset of nodes S, let $|S|$ be the size of S and $c_G(S)$ be the number of connected components in the graph induced on $V \setminus S$ obtained from G by deleting all nodes in the set S. The toughness of the graph G, where G is not the complete graph, denoted by $\tau(G)$ is defined as follows

$$\tau(G) = \min_{S \subset V,\, c_G(S) > 1} \frac{|S|}{c_G(S)} . \tag{1}$$

The toughness of the complete graph is defined to be infinity. Toughness of a cycle, on the other hand, is 1 since by deleting any subset of k nodes, we can create at most k connected components, and removing two nodes with no edge between creates exactly two components. As another example, it is easy to verify that the toughness of a tree with maximum degree Δ (and at least three nodes) is $1/\Delta$. It is also easy to verify that the toughness of a graph is positive if and only if the graph is connected.

There is a vast literature on the connections between graph toughness and spanning trees; for a recent survey see [15]. In what follows, we will show a close connection between existence and algorithms for adversarial infections that are pandemic to the toughness of the underlying connections graph.

We start by developing sufficient conditions under which there exists a spanning tree that can be exploited by an adversary obeying the first-order constraints.

Theorem 5. *For any connected, weighted undirected graph $G = (V, E, p)$ such that*

$$\forall u \in V, \quad \sum_{v:(u,v) \in E} p(u,v) \geq \left\lceil \frac{1}{\tau(G)} + 1 \right\rceil \tag{2}$$

there is an adversary strategy that infects all nodes in G and obeys the first-order constraints. Moreover, assume that

$$\forall u \in V, \quad \sum_{v:(u,v) \in E} p(u,v) \geq \left\lceil \frac{1}{\tau(G)} + 2 \right\rceil . \tag{3}$$

Then, there is a polynomial-time computable *adversary strategy that infects all nodes in G and obeys the first-order constraints.*

Proof. By Win's Theorem [18], every undirected graph G with toughness $\tau(G)$ has a spanning tree with maximum degree bounded by $d = \lceil (1/\tau(G) + 2) \rceil$. Let T be such a tree. Root this tree at a leaf node s and let this node be the seed. The adversary strategy is to infect all nodes of G, starting with node s, by using the edges of T. Now no node is responsible for infecting more than $d-1$ children and so the first-order constraints are obeyed: for each node u, its degree minus one is at most $\lceil (1/\tau(G)+2) \rceil - 1 = \lceil (1/\tau(G)+1) \rceil \leq \sum_{v:(u,v)\in E} p(u,v)$. To show the other part of the theorem we make use of an algorithmic result by Fürer and Raghavachari [11] that states that if there exists a spanning tree of degree at most Δ than one can construct in polynomial time a spanning tree of degree at most $\Delta + 1$. The assertion of the theorem then follows similarly to the previous case, where now for each node u, $\lceil 1/\tau(G) + 2 \rceil \leq \sum_{v:(u,v)\in E} p(u,v)$.

Example 1. To illustrate the theorem, consider the following example. Take a random graph $G \in \mathrm{G}(n,p)$ with $p \geq 2\log n/n$. With probability of $1 - o(1)$, G contains a Hamiltonian cycle (see for instance [14]). So $\tau(G) \geq 1$, and thus to apply Theorem 5, all we need is that for every node u, $\sum_{v:(u,v)\in E} p(u,v)$ is at least 2. By the Chernoff bound, all vertex degrees in G are at least $pn/2$ with high probability. Hence by setting $p(u,v) = 4/pn$ on each edge (u,v), we satisfy the conditions of Theorem 5, and can conclude existence of an adversarial strategy that leads to pandemic infection. Note that as network degrees increase in this example, smaller and smaller edge infection probabilities suffice to get adversarial pandemic infection.

Discussion:
We note that if we slightly weaken the condition stated in Theorem 5, the result no longer holds. Specifically, if we allow just $O(1)$ nodes to violate condition (2) by only a slack of 1, that is, allow $O(1)$ nodes u with $\sum_{u:(u,v)\in E} p(u,v) = \lceil (1/\tau(G) \rceil$, then for infinitely many toughness values there will exist infinitely many networks such that the optimal adversarial strategy is not pandemic (namely, infects all nodes). To see this, let $k \geq 2$ be an integer and consider the following family of graphs $H_{n,k}$, where $n \geq 3k(k+3)+3$ is integral. Take three nodes v_1, v_2, v_3 as connect them as a clique. In addition create $3k$ vertex-isolated cliques, indexed $C(i,j)$ for $1 \leq i \leq 3, 1 \leq j \leq k$, each on $(n-3)/3k$ nodes. To complete the construction connect each node v_i to some node in each clique $C(i,j)$ ("a representative"), where $1 \leq j \leq k$. Thus the degree of each node v_i in the construction is $k+2$, in addition to being connected to k representatives it is also connected to the other v_i nodes. Set all edge weights to 1, except for each i the edges connecting v_i to the cliques $C(i,1)$ and $C(i,2)$; the weight on each of these edges is set to $1/2$.

We now observe a few simple facts about the graph $H_{n,k}$. First, its toughness is $1/(k+1)$. Second, it has a spanning tree (tree spanning all graph nodes). Third, in each of its spanning trees each v_i must be connected to a representative from each $C(i,j)$ for $1 \leq j \leq k$. In addition, one of the v_is must be connected to the two other v_is (otherwise the spanning tree is disconnected), and so its degree in the spanning tree must be $k+2$. Last, except for the three v_is, the sum of edge-weights touching any node is at least $k+2$ (as $\frac{n-3}{3k} \geq k+3$ condition (2) holds for all clique nodes). However, the sum of edge-weights touching a v_i, namely $\sum_{u:(v_i,u)\in E} p(v_i,u)$, is $2+k-2+1/2\cdot 2 = k+1$. Thus the degree of each of the v_is in a tree representing an adversarial infection can be at most $k+1$, which is strictly smaller than $k+2$, and so any such adversary cannot infect all nodes in the graph. In fact, a constant fraction of the graph nodes will not be infected — all the nodes belonging to one of the cliques $C(i,j)$. □

We now show that for a given value of the toughness τ, there exist infinitely many graphs such that the stochastic diffusion can only infect $O(\tau)$ nodes, for $\tau \geq 4$. In light of theorem 5, on such graphs the gap between the stochastic diffusion and the adversarial diffusion is large.

Theorem 6. *For any value of the toughness $\tau \geq 4$ and positive integer $\ell \geq 3$, there exists a weighted undirected graph $G = (V, E, p)$ on $n = \tau\ell$ nodes and toughness τ, such that condition (2) is satisfied (hence the adversarial process can infect all n nodes), yet the stochastic diffusion infects only $O(\tau)$ nodes in expectation.*

Proof. Take a cycle with ℓ nodes, say, $v_0, v_1, ..., v_{\ell-1}$. Now replace each v_i by a clique C_i on $\tau \geq 4$ nodes. Now for each i, connect vertices in clique C_i to vertices in C_{i+1} by a complete bipartite graph. Note that the total number of nodes n equals $\tau\ell$. Also, one can verify that the toughness of this modified cycle is τ, since the toughness of a simple cycle is 1.

Assign a probability of 1 to edges inside each C_i, and probability $\frac{1}{2\tau^2}$ to edges between the cliques. Note that the probability that any edge between two adjacent cliques gets realized is less than 1/2. Indeed, the probability that no edge between two adjacent cliques gets realized is

$$(1 - 1/(2\tau^2))^\tau \geq 1 - \frac{1}{2\tau} > 1/2\,,$$

where we used the inequality $(1+x)^r \geq 1 + rx$, for $x \geq -1$ and $r \in \mathbb{R} \setminus (0,1)$. In particular, the behavior of the stochastic diffusion on this network is essentially as it is on a cycle with edge probability less than 1/2 that was analyzed in the proof of Theorem 4; the only difference is that now each node on the cycle infects as well all the nodes in its clique. Thus the stochastic diffusion infects at most $O(\tau)$ nodes in expectation. Finally, condition (2) trivially holds since $\tau \geq 4$ and so each vertex has a probability mass of at least 3 incident on it, while $1/\tau \leq 1$.

6 Conclusions and Future Work

In this work we initiated the study of *adversarial* infection strategies which can decide intelligently which nodes to infect next in order to maximize their spread, while obeying first-order constraints as to not get discovered. We have demonstrated that a well-planned adversarial infection can substantially increase the number of nodes infected with respect to the standard Independent Cascades infection strategy. We designed necessary and sufficient conditions to understand when this is possible. Based on novel connection to the network toughness, we characterize networks where existence of adversarial strategies that are pandemic (infect all nodes) is guaranteed, as well as efficiently computable.

Our results have focused on first order constraints: keeping the traffic involving any set X lower than the ceiling of its expected value, namely,

$$\left\lceil \sum_{u \in X} \sum_{v:(u,v) \in E} p(u,v) \right\rceil .$$

An interesting future direction is to consider flow constraints where the number of infections caused by any set X is at most the flow leaving X, namely,

$$\left\lceil \sum_{u \in X} \sum_{v \notin X:(u,v) \in E} p(u,v) \right\rceil .$$

Moreover, it would be interesting to consider directed graphs as well.

Another interesting avenue for further exploration is to analyze the effectiveness of vaccination strategies designed for controlling stochastic epidemics in limiting the spreading of the adversarial epidemic. In particular, one could consider immunization strategies such as immunizing high degree nodes or acquaintance immunization based on the immunization of a small fraction of random neighbors of randomly selected nodes. Such strategies are known to be effective at controlling stochastic epidemics [6, 8] but might be ineffective containing the first-order constrained adversarial infection. To demonstrate this, let us go back to the example of the introduction, namely a path connected at one of its ends to the root of a two-level binary tree. Consider vaccinating the node with the highest weighted degree, which is in this example the root of tree. Vaccinating the root of the tree will not help in containing adversarial infections, which would still infect $n - O(1)$ nodes (the path nodes). Yet vaccinating the middle node of the path is much better for containment — yielding at most $n/2 + O(1)$ infected nodes by an adversarial epidemic. It is therefore interesting to understand and develop vaccination schemes that aim to minimize the number of infections under the adversarial setting.

Acknowledgements. We would like to thank the anonymous reviewers for their helpful comments.

Moez Draief was supported by QNRF grant NPRP-09-1150-2-448.

Sanjeev Khanna was supported in part by the National Science Foundation grants CCF-1116961 and IIS-0904314.

References

1. Albert, R., Jeong, H., Barabasi, A.-L.: Error and attack tolerance of complex networks. Nature 406(6794), 378–382 (2000)
2. Alon, N.: A note on network reliability. In: Discrete Probability and Algorithms, pp. 11–14. Springer (1995)

3. Balthrop, J., Forrest, S., Newman, M.E.J., Williamson, M.M.: Technological networks and the spread of computer viruses. Science 304(5670), 527–529 (2004)
4. Bauer, D., Broersma, H., Schmeichel, E.F.: Toughness in graphs - a survey. Graphs and Combinatorics 22(1), 1–35 (2006)
5. Bazgan, C., Santha, M., Tuza, Z.: On the approximation of finding a(nother) Hamiltonian cycle in cubic Hamiltonian graphs. J. Algorithms 31(1), 249–268 (1999)
6. Britton, T., Janson, S., Martin-Löf, A.: Graphs with specified degree distributions, simple epidemics, and local vaccination strategies. Advances in Applied Probability 39(4), 922–948 (2007)
7. Chen, Z., Chen, C., Ji, C.: Understanding localized-scanning worms. In: IPCCC, pp. 186–193 (2007)
8. Cohen, R., Havlin, S., Ben-Avraham, D.: Efficiency immunization strategies for computer networks and populations. Physical Review Letters 91(24), 247901-1–247901-4 (2003)
9. Crucitti, P., Latora, V., Marchiori, M., Rapisarda, A.: Efficiency of scale-free networks: Error and attack tolerance. Physica A 320(642), 622–642 (2003)
10. Easley, D.A., Kleinberg, J.M.: Networks, Crowds, and Markets - Reasoning About a Highly Connected World. Cambridge University Press (2010)
11. Fürer, M., Raghavachari, B.: Approximating the minimum-degree Steiner tree to within one of optimal. J. Algorithms 17(3), 409–423 (1994)
12. Karger, D.R., Motwani, R., Ramkumar, G.D.S.: On approximating the longest path in a graph. Algorithmica 18(1), 82–98 (1997)
13. Kempe, D., Kleinberg, J.M., Tardos, É.: Maximizing the spread of influence through a social network. In: KDD, pp. 137–146 (2003)
14. Komlós, J., Szemerédi, E.: Limit distributions for the existence of Hamiltonian circuits in a random graph. Discrete Mathematics (43), 55–63 (1983)
15. Ozeki, K., Yamashita, T.: Spanning trees: A survey. Graphs and Combinatorics 27(1), 1–26 (2011)
16. Serazzi, G., Zanero, S.: Computer virus propagation models. In: Calzarossa, M.C., Gelenbe, E. (eds.) MASCOTS 2003. LNCS, vol. 2965, pp. 26–50. Springer, Heidelberg (2004)
17. Weaver, N., Paxson, V., Staniford, S., Cunningham, R.: A taxonomy of computer worms. In: WORM, pp. 11–18 (2003)
18. Win, S.: On a connection between the existence of k-trees and the toughness of a graph. Graphs and Combinatorics 5(1), 201–205 (1989)
19. Yan, G., Chen, G., Eidenbenz, S., Li, N.: Malware propagation in online social networks: nature, dynamics, and defense implications. In: ASIACCS, pp. 196–206 (2011)
20. Zou, C.C., Towsley, D.F., Gong, W.: Email worms modeling and defense. In: ICCCN, pp. 409–414 (2004)
21. Zou, C.C., Towsley, D.F., Gong, W.: On the performance of internet worm scanning strategies. Perform. Eval. 63(7), 700–723 (2006)
22. Zou, C.C., Towsley, D.F., Gong, W., Cai, S.: Routing worm: A fast, selective attack worm based on IP address information. In: PADS, pp. 199–206 (2005)

On the Choice of Kernel and Labelled Data in Semi-supervised Learning Methods

Konstantin Avrachenkov[1], Paulo Gonçalves[2], and Marina Sokol[1]

[1] Inria Sophia Antipolis, 2004 Route des Lucioles, Sophia-Antipolis, France
[2] Inria Rhone-Alpes and ENS Lyon, 46 Allée Italie, Lyon, France

Abstract. Semi-supervised learning methods constitute a category of machine learning methods which use labelled points together with unlabelled data to tune the classifier. The main idea of the semi-supervised methods is based on an assumption that the classification function should change smoothly over a similarity graph, which represents relations among data points. This idea can be expressed using kernels on graphs such as graph Laplacian. Different semi-supervised learning methods have different kernels which reflect how the underlying similarity graph influences the classification results. In the present work, we analyse a general family of semi-supervised methods, provide insights about the differences among the methods and give recommendations for the choice of the kernel parameters and labelled points. In particular, it appears that it is preferable to choose a kernel based on the properties of the labelled points. We illustrate our general theoretical conclusions with an analytically tractable characteristic example, clustered preferential attachment model and classification of content in P2P networks.

1 Introduction

The first principal idea of the semi-supervised learning methods is to use few labelled points (points with known classification) together with the unlabelled data to tune the classifier. This drastically reduces the size of the training set. The second principal idea of the semi-supervised learning methods is to use a (weighted) similarity graph. If two data points are connected by an edge, this indicates some similarity of these points. Then, the weight of the edge, if present, reflects the degree of similarity. Later in the paper we show how the similarity graph can be constructed in a specific application. Each class has a classification function defined over all data points which gives a degree of relevance to the class for each data point. The third principal idea of the semi-supervised learning methods is that the classification function should change smoothly over the similarity graph. Intuitively, nodes of the similarity graph that are closer together in some sense are more likely to have the same label. This idea of classification function smoothness can be expressed using graph Laplacian or its modification. In particular, the authors of [14] proposed transductive learning, a semi-supervised learning method based on the Standard Laplacian. The authors of [13] and [15] used the Normalized Laplacian (or diffusion kernel). And the

A. Bonato, M. Mitzenmacher, and P. Prałat (Eds.): WAW 2013, LNCS 8305, pp. 56–67, 2013.
© Springer International Publishing Switzerland 2013

authors of [3] used the Markov kernel. We observe that if one takes the method of [1] for detecting local cuts and takes seeds in [1] as the labelled data and considers sweeps as classification functions, then because the degrees of data points in different sweeps are the same, the resulting method will be equivalent to the semi-supervised method proposed in [3]. Recently in [5], the authors proposed a generalized optimization formulation which gives the above mentioned methods as particular cases. In the present work we provide more insights about the differences among the semi-supervised methods based on random walk theory, and give recommendations on how to choose the kernel and labelled points (of course, when there is some freedom in the choice of labelled points). It appears that the choice of labelled points influences the choice of kernel. In particular, we show that if the labelled points are chosen uniformly at random, the PageRank based method is the best choice for the semi-supervised kernel. On the other hand, if one can choose labelled points with large degrees or we know that labelled points given to us have large degrees, the Standard Laplacian method is the best choice.

The paper is organized as follows: In the next section we briefly describe the graph-based semi-supervised learning methods. We refer readers interested in more details on semi-supervised methods to several excellent surveys [8,16,17]. In Section 3 we provide general theoretical insights about semi-supervised learning methods and suggest how to choose the kernel and labelled points. Then, in Section 4 we illustrate the theoretical conclusions on an analytically tractable characteristic network example, on clustered preferential attachment model and with application to P2P content classification. In particular, for this specific application we show that with the right combination of labelled points and kernel one can achieve 95% precision with as little as 50 points per class for several hundred thousands unlabelled points. Finally, in Section 5 we give conclusions and provide directions for future research.

2 Semi-supervised Learning Methods

Suppose we need to classify N data points into K classes and assume P data points are labelled. That is, we know the class to which each labelled point belongs. Denote by V_k, the set of labelled points in class $k = 1, ..., K$. Thus, $|V_1| + ... + |V_K| = P$.

The graph-based semi-supervised learning approach uses a weighted graph connecting data points. The weight matrix, or similarity matrix, is denoted by W. Here we assume that W is symmetric and the underlying graph is connected. Each element $w_{i,j}$ represents the degree of similarity between data points i and j. Denote by D a diagonal matrix with its (i,i)-element equal to the sum of the i-th row of matrix W: $d_i = \sum_{j=1}^{N} w_{i,j}$. Later in the paper we demonstrate how to construct the similarity matrix for a specific application.

Define an $N \times K$ matrix Y as

$$Y_{ik} = \begin{cases} 1, & \text{if } i \in V_k, \text{ i.e., point } i \text{ is labelled as a class } k \text{ point,} \\ 0, & \text{otherwise.} \end{cases}$$

We refer to each column Y_{*k} of matrix Y as a labeling function. Also define an $N \times K$ matrix F and call its columns F_{*k} classification functions. The general idea of the graph-based semi-supervised learning is to find classification functions so that on the one hand they will be close to the corresponding labeling function and on the other hand they will change smoothly over the graph associated with the similarity matrix. This general idea can be expressed by means of the following optimization formulation [5]:

$$\min_F \{ \sum_{i=1}^N \sum_{j=1}^N w_{ij} \| d_i^{\sigma-1} F_{i*} - d_j^{\sigma-1} F_{j*} \|^2 + \mu \sum_{i=1}^N d_i^{2\sigma-1} \| F_{i*} - Y_{i*} \|^2 \}, \quad (1)$$

where μ is a regularization parameter. In fact, the parameter μ represents a trade-off between the closeness of the classification function to the labeling function and its smoothness.

The first order optimality condition gives explicit expressions for the classification functions

$$F_{*k} = \frac{\mu}{2+\mu} \left(I - \frac{2}{2+\mu} D^{-\sigma} W D^{\sigma-1} \right)^{-1} Y_{*k}, \quad k = 1, ..., K. \quad (2)$$

Once the classification functions are obtained, the points are classified according to the rule

$$F_{ik} > F_{ik'}, \forall k' \neq k \quad \Rightarrow \quad \text{Point } i \text{ is classified into class } k.$$

The ties can be broken in arbitrary fashion. We would like to note that our general scheme allows us to retrieve as particular cases:

- The Standard Laplacian method ($\sigma = 1$), [14]:

$$F_{*k} = \frac{\mu}{2+\mu} \left(I - \frac{2}{2+\mu} D^{-1} W \right)^{-1} Y_{*k},$$

- The Normalized Laplacian method ($\sigma = 1/2$), [13]:

$$F_{*k} = \frac{\mu}{2+\mu} \left(I - \frac{2}{2+\mu} D^{-\frac{1}{2}} W D^{-\frac{1}{2}} \right)^{-1} Y_{*k},$$

- The PageRank based method ($\sigma = 0$), [3]:

$$F_{*k} = \frac{\mu}{2+\mu} \left(I - \frac{2}{2+\mu} W D^{-1} \right)^{-1} Y_{*k}.$$

In the present work we try to answer the questions: which kernel (or which values of σ and μ) one needs to choose? and which points to label if we have some freedom with respect to labelling points? It turns out that these questions are not independent and one has to choose the kernel depending on the information available while labelling the points.

3 General Theoretical Considerations

First, let us transform the expression (2) to a more convenient form.

$$\begin{aligned} F_{*k} &= \frac{\mu}{2+\mu}\left(I - \frac{2}{2+\mu}D^{-\sigma}WD^{\sigma-1}\right)^{-1} Y_{*k} \\ &= \frac{\mu}{2+\mu}\left(D^{-\sigma}\left(I - \frac{2}{2+\mu}WD^{-1}\right)D^{\sigma}\right)^{-1} Y_{*k} \\ &= \frac{\mu}{2+\mu}D^{-\sigma}\left(I - \frac{2}{2+\mu}WD^{-1}\right)^{-1} D^{\sigma}Y_{*k}. \end{aligned}$$

Denoting $\alpha = 2/(2+\mu)$, transposing and using the fact that W is symmetric, we obtain

$$F_{*k}^T = (1-\alpha)Y_{*k}^T D^{\sigma}\left(I - \alpha D^{-1}W\right)^{-1} D^{-\sigma}. \tag{3}$$

Next we apply to the above expression the Blackwell series expansion [7,12]

$$(1-\alpha)\left(I - \alpha D^{-1}W\right)^{-1} = \underline{1}\pi + (1-\alpha)H + o(1-\alpha), \tag{4}$$

where π is the stationary distribution of the standard random walk ($\pi D^{-1}W = \pi$), $\underline{1}$ is a vector of ones of appropriate dimension and $H = (I - D^{-1}W + \underline{1}\pi)^{-1} - \underline{1}\pi$ is the deviation matrix. We note that since the similarity matrix W is symmetric, the random walk governed by the transition matrix $D^{-1}W$ is time-reversible and its stationary distribution is given in the explicit form

$$\pi = (\underline{1}^T D\underline{1})^{-1}\underline{1}^T D. \tag{5}$$

Combining (3), (4) and (5), we can write

$$F_{*k}^T = (\underline{1}^T D\underline{1})^{-1} Y_{*k}^T D^{\sigma}\underline{1}\underline{1}^T D^{1-\sigma} + (1-\alpha)Y_{*k}^T D^{\sigma} H D^{-\sigma} + o(1-\alpha).$$

In particular, we have

$$F_{ik} = \frac{d_i^{1-\sigma}}{\sum_{j=1}^N d_j}\sum_{p\in V_k} d_p^{\sigma} + (1-\alpha)d_i^{-\sigma}\sum_{p\in V_k} d_p^{\sigma}H_{pi} + o(1-\alpha), \tag{6}$$

and, consequently, if $\sum_{p\in V_k} d_p^{\sigma} \neq \sum_{p\in V_{k}'} d_p^{\sigma}$ for some k and k', in the case when the parameter α is close to 1 (equivalently when μ is close to 0), then all points will be classified into the classes with the largest value of $\sum_{p\in V_k} d_p^{\sigma}$. An interesting exception is the case when $\sigma = 0$ and $|V_k| = const(k)$. In such a case, the zero order terms in the Blackwell expansions for the classification functions are the same for all classes and we need to compare the first order terms. Recall [10] that there is a connection between the mean first passage time of the standard random walk from node i to node j, m_{ij}, and the elements of the deviation matrix, namely, $m_{ij} = (\delta_{ij} + H_{jj} - H_{ij})/\pi_j$, where δ_{ij} is the Kronecker delta. If $\sigma = 0$ and $|V_k| = const(k)$, substituting (6) into $F_{ik} - F_{ik'} > 0$ with $H_{pi} = H_{ii} - \pi_i m_{pi}$ for $i \neq p$ results in the condition

$$\sum_{s \in V_{k'}} m_{si} > \sum_{p \in V_k} m_{pi}.$$

This condition has a clear probabilistic interpretation: point i is classified into class k if the sum of mean passage times from the labelled points to point i is smallest for class k over all classes.

In addition to the standard random walk, it will also be helpful to consider a random walk with absorption $\{S_t \in \{1,...,N\},\ t = 0,1,...\}$. At each step with probability α the random walk chooses next node among its neighbours uniformly and with probability $1-\alpha$ goes into the absorbing state. The probabilities of visiting nodes before absorption given the random walk starts at node j, $S_0 = j$, are provided by the distribution

$$\mathrm{ppr}(j) = (1-\alpha)e_j^T \left(I - \alpha D^{-1}W\right)^{-1}, \tag{7}$$

which is the personalized PageRank vector with respect to seed node j [9]. Here e_j denotes the j-th element of the standard basis.

Now we are ready to formulate the first result explaining the classification by the semi-supervised learning methods.

Theorem 1. *Data point i is classified by the generalized semi-supervised learning method (1) into class k, if*

$$\sum_{p \in V_k} d_p^\sigma q_{pi} > \sum_{s \in V_{k'}} d_s^\sigma q_{si}, \quad \forall k' \neq k, \tag{8}$$

where q_{pi} is the probability of reaching state i before absorption if $S_0 = p$.

Proof: Since $Y_{*k}^T = \sum_{p \in V_k} e_p^T$ and $F_{ik} = F_{*k}^T e_i$, from (3) we obtain

$$F_{ik} = \sum_{p \in V_k} d_p^\sigma (1-\alpha) e_p^T \left(I - \alpha D^{-1}W\right)^{-1} e_i d_i^{-\sigma} = \frac{1}{d_i^\sigma} \sum_{p \in V_k} d_p^\sigma \mathrm{ppr}_i(p). \tag{9}$$

It has been shown in [6] that

$$\left(I - \alpha D^{-1}W\right)_{pi}^{-1} = q_{pi} \left(I - \alpha D^{-1}W\right)_{ii}^{-1},$$

where $(\cdot)_{pi}^{-1}$ denotes the (p,i)-element of the inverse matrix. Multiplying the above equation by $(1-\alpha)$ yields

$$\mathrm{ppr}_i(p) = q_{pi} \mathrm{ppr}_i(i). \tag{10}$$

Thus, using relation (10) and equation (9), we conclude that for point i to be classified into class k we need

$$F_{ik} - F_{ik'} = \frac{\mathrm{ppr}_i(i)}{d_i^\sigma} \left(\sum_{p \in V_k} d_p^\sigma q_{pi} - \sum_{s \in V_{k'}} d_s^\sigma q_{si} \right) > 0, \quad \forall k' \neq k,$$

or, equivalently (8). □

Let us discuss the implications of Theorem 1. First, it is very interesting to observe that, using (8), one can decouple the effects from the choice of α and σ. A change in the value of α only influences the factor q_{pi} and a change in the value of σ only affects the factor d_p^σ. Second, the results of Theorem 1 are consistent with the conclusions obtained with the help of the Blackwell expansion. When α goes to one, q_{pi} goes to one and indeed classes with the largest value of $\sum_{p \in V_k} d_p^\sigma$ attract all points. Thus, the case of $\sigma = 0$ and $|V_k| = const(k)$ is especially interesting. In this case there is stability of classification even when α is close to one. Third, if $\sigma = 0$ and $|V_k| = const(k)$, one can expect that smaller classes will attract a larger number of "border points" than larger classes. Suppose that class k is smaller than class k'. Then, it is natural to expect that $q_{pi} > q_{si}$ with $p \in V_k$ and $s \in V_{k'}$. This observation will be confirmed by examples in the next section. This effect, if needed, can be compensated by increasing σ away from zero. And finally, fourth, we have the following rather surprising conclusion.

Corollary 1. *If labelled points have the same degree ($d_p = d$, $p \in V_k$, $k = 1, \ldots, K$), all considered semi-supervised learning methods provide the same classification.*

Now with the help of the following lemma, we can obtain another alternative condition for semi-supervised learning classification.

Lemma 1. *If the graph is undirected ($W^T = W$), then the following relation holds*

$$\mathrm{ppr}_j(i) = \frac{d_j}{d_i} \mathrm{ppr}_i(j). \tag{11}$$

Proof: We can rewrite (7) as follows

$$\mathrm{ppr}(i) = (1-\alpha) e_i^T [D - \alpha W]^{-1} D,$$

and hence,

$$\mathrm{ppr}(i) D^{-1} = (1-\alpha) e_i^T [D - \alpha W]^{-1}.$$

Since matrix W is symmetric, $[D - \alpha W]^{-1}$ is also symmetric and we have

$$[\mathrm{ppr}(i) D^{-1}]_j = (1-\alpha) e_i^T [D - \alpha W]^{-1} e_j = (1-\alpha) e_j^T [D - \alpha W]^{-1} e_i = [\mathrm{ppr}(j) D^{-1}]_i.$$

Thus, $\mathrm{ppr}_j(i)/d_j = \mathrm{ppr}_i(j)/d_i$, which completes the proof. □

Theorem 2. *Data point i is classified by the generalized semi-supervised learning method (1) into class k, if*

$$\sum_{p \in V_k} \frac{\mathrm{ppr}_p(i)}{d_p^{1-\sigma}} > \sum_{s \in V_{k'}} \frac{\mathrm{ppr}_s(i)}{d_s^{1-\sigma}}, \quad \forall k' \neq k. \tag{12}$$

Proof: Follows from equation (9) and Lemma 1. □

We note that in the statement of Theorem 2 the "reversed" PageRank is used instead of the PageRank in (9). In particular, this provides another interesting interpretation of the PageRank based method. If we set $\sigma = 0$ in (12), it appears that we need to compare the reversed PageRanks divided by the degrees of the labelled points. As already mentioned in the Introduction, if one considers the sweeps from [1] as classification functions, then the degrees of the nodes to be classified are cancelled in the sweeps. However, if we now view the PageRank method in terms of the reversed PageRank, the division by the degree of the PageRank values remains essential. This provides another interesting interpretation of sweeps defined in [1].

4 Evaluation

Let us illustrate the theoretical results with the help of a characteristic network example, clustered preferential attachment graph and application to P2P content classification.

Characteristic Network Example: Let us first consider an analytically tractable network example. Despite its simplicity, it clearly demonstrates major properties of graph-based semi-supervised learning methods. There are two classes, A and B with $|A| = N_1$ and $|B| = N_2$. Each class is represented by a star network. The two classes are connected by a link connecting two leaves. The graph of the model is given in Figure 2(a).

The central nodes with indices 1 and $N_1 + N_2$ are the obvious choice for labelled points. In order to determine the classification functions analytically, we need to calculate the matrix $Z = [I - \alpha D^{-1}W]^{-1}$. It is easier to calculate the symmetric matrix $C = [D - \alpha W]^{-1}$. Once the matrix C is calculated, we can immediately retrieve the elements of matrix Z by the formula

$$Z_{ij} = C_{ij} d_j. \tag{13}$$

Thus we need to solve a system of equations $[D - \alpha W]C_{*,j} = e_j$. Since we have chosen the central nodes as labelled points and due to the symmetry of the graph, we actually need to solve only one system for $j = 1$ of six equations

$$\begin{aligned}
(N_1-1)C_{1,1} - (N_1-2)\alpha C_{2,1} - \alpha C_{N_1,1} &= 1\\
C_{2,1} &= \alpha C_{1,1}\\
C_{N_1-1,1} &= \alpha C_{1,1}\\
-\alpha C_{1,1} + 2C_{N_1,1} - \alpha C_{N_1+1,1} &= 0\\
-\alpha C_{N_1,1} + 2C_{N_1+1,1} - \alpha C_{N_1+N_2,1} &= 0\\
C_{N_1+2,1} &= \alpha C_{N_1+N_2,1}\\
-\alpha C_{N_1+1,1} - (N_2-2)\alpha C_{N_1+2,1} + (N_2-1)C_{N_1+N_2,1} &= 0
\end{aligned}$$

Solving the above system, in particular, we obtain

$$C_{N_1,1} = \frac{\alpha(2N_2 - 2 - \alpha^2(2N_2-3))}{R}, \tag{14}$$

$$C_{N_1+1,1} = \frac{\alpha^2(N_2 - 1 - \alpha^2(N_2 - 2))}{R}, \tag{15}$$

with

$$R = (1-\alpha^2)(-2\alpha^4 N_2 - 2\alpha^4 N_1 + 4\alpha^4 + \alpha^4 N_2 N_1 - 9\alpha^2$$
$$+7\alpha^2 N_2 + 7\alpha^2 N_1 - 5N_2\alpha^2 N_1 + 4N_2 N_1 + 4 - 4N_1 - 4N_2).$$

Consider first the PageRank based method ($\sigma = 0$). According to the theoretical consideration, it is very likely that some points will be misclassified into a smaller class. Suppose that $N_1 < N_2$ and consider border points. The point $N_1 + 1$ will be classified into class B by the PageRank based method if and only if

$$\frac{Z_{1,N_1+1}}{Z_{N_1+N_2,N_1+1}} = \frac{C_{1,N_1+1}}{C_{N_1+N_2,N_1+1}} < 1.$$

Using slightly more convenient notation $n_i = N_i - 1, i = 1, 2$, we can rewrite the above condition as follows:

$$\frac{\alpha(n_2 - \alpha^2(n_2 - 1))}{2n_1 - \alpha^2(2n_1 - 1)} < 1,$$

or, equivalently, $(1 - n_2)\alpha^2 + (2n_1 - n_2)\alpha + 2n_1 > 0$. If $2n_1 + 1 > n_2$, the above inequality holds for any $\alpha \in (0, 1)$. And consequently, for any $\alpha \in (0, 1)$ the point $N_1 + 1$ is classified into class B. However, if $2n_1 + 1 < n_2$ (class A is significantly smaller than class B), for $\alpha \in (\bar{\alpha}, 1)$ point $N_1 + 1$ will be erroneously classified into class A. The expression for $\bar{\alpha}$ is given by

$$\bar{\alpha} = \frac{(n_2 - 2n_1) + \sqrt{(2n_1 + n_2)^2 - 8n_1}}{2(n_2 - 1)}.$$

If we fix the value of n_1 and let n_2 go to infinity, we get $\bar{\alpha} \to 0$. Thus, if the sizes of A and B are very different, the point $N_1 + 1$ will be misclassified for nearly all values of the parameter α.

Now we analyse the performance of the Standard Laplacian method ($\sigma = 1$). According to the general theoretical considerations, the Standard Laplacian method has a tendency to classify more points into a larger class. We consider the classification of the point with index N_1 (still assuming $N_1 < N_2$). It will be classified correctly if and only if

$$\frac{Z_{N_1,1}}{Z_{N_1,N_1+N_2}} > 1,$$

or, equivalently,

$$\frac{n_1(2n_2 - \alpha^2(2n_2 - 1))}{n_2\alpha(n_1 - \alpha^2(n_1 - 1))} > 1$$

which results in the following cubic inequality

$$\alpha^3 n_2(n_1 - 1) - \alpha^2 n_1(2n_2 - 1) - \alpha n_2 n_1 + 2n_2 n_1 > 0.$$

Consider a linear scaling $n_2 = Kn_1, K > 1$. Then, the above inequality can be rewritten in the form

$$\alpha^3\left(1-\frac{1}{n_1}\right)-\alpha^2\left(2-\frac{1}{Kn_1}\right)-\alpha+2>0.$$

This inequality can be regarded as a regularly perturbed inequality with respect to $1/n_1$ (see e.g., [2]). If we let n_1 go to infinity, the limiting inequality can be easily factored, i.e.,$(1-\alpha)(1+\alpha)(2-\alpha)>0$. Since the perturbation is regular, when n_1 varies in the vicinity of infinity the roots change slightly. In particular, using the implicit function theorem, we can find that the root near 1 changes as follows:

$$\bar{\bar{\alpha}} = 1 - \frac{K-1}{2K}\frac{1}{n_1} + o\left(\frac{1}{n_1}\right).$$

In particular, this means that if the sizes of classes are large, the Standard Laplacian method performs well for nearly all values of α from the interval $(0,1)$. This is in contrast with the PageRank based method.

We summarize and illustrate various considered cases by means of numerical examples presented in Table 1. Our main conclusion from this characteristic network model is that the PageRank based method is a safe choice as it can misclassify at most one point in this particular example whereas with α close to one the Standard Laplacian method can classify all points in the largest class. On the other hand if parameter α is chosen appropriately, the Standard Laplacian method gives a perfect classification for nearly all values of α, even when classes have many points and very different sizes.

Table 1. Comparison between different methods in terms of classification errors

N_1	N_2	PR	SL
20	100	$v_{N_1+1} \mapsto A$ if $\alpha \geq \bar{\alpha} = 0.3849$	$v_{N_1} \mapsto B$ if $\alpha \geq \bar{\bar{\alpha}} = 0.9803$, $A \mapsto B$ if $\alpha \geq 0.9931$
20	200	$v_{N_1+1} \mapsto A$ if $\alpha \geq \bar{\alpha} = 0.1911$	$v_{N_1} \mapsto B$ if $\alpha \geq \bar{\bar{\alpha}} = 0.9780$, $A \mapsto B$ if $\alpha \geq 0.9923$
200	2000	$v_{N_1+1} \mapsto A$ if $\alpha \geq \bar{\alpha} = 0.1991$	$v_{N_1} \mapsto B$ if $\alpha \geq \bar{\bar{\alpha}} = 0.9978$, $A \mapsto B$ if $\alpha \geq 0.9992$

Clustered Preferential Attachment Model: Let us now consider a synthetic graph generated according to the clustered preferential attachment model. Our model has 5 unbalanced classes (1500 / 240 / 120 / 100 / 50). Once a node is generated, it has two links which it attaches independently with probability 0.98 within its class and with probability 0.02 outside its class. In both cases a link is attached to a node with probability proportional to the number of existing links. First, we test the case of random labelled points. Five labelled points were chosen randomly for each class and results are averaged over 100 realizations. The precision of classification for various values of σ and α is given in Figure 1(a). Then, in each class we have chosen 5 labelled points with maximal degrees. The results of classification are given in Figure 1(b). We obtain conclusions consistent with the characteristic network model. If no information is available for assignment of the labelled points, the PageRank method is a safe choice. If one can choose labelled points with large degrees, it is better to use the Standard Laplacian method. There could be a significant gain in precision (roughly from 70% to 95%). It can be observed that the Standard Laplacian method is not too sensitive to the value of α if we stay well away from $\alpha = 1$.

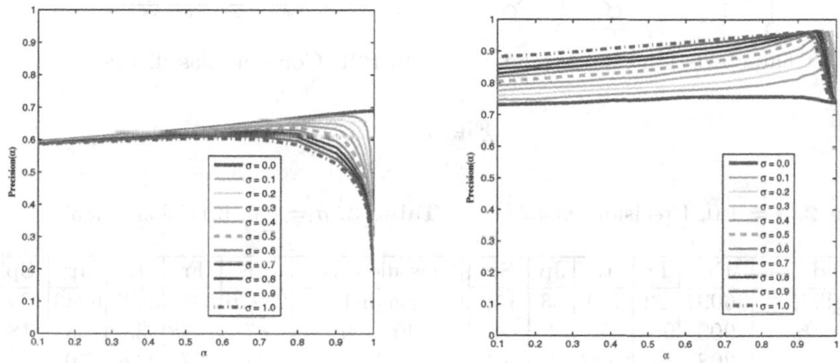

(a) Random Labelled Points (b) Max Degree Labelled Points

Fig. 1. Clustered Preferential Attachment Model: Precision of classification

Application to P2P Content Classification: Finally, we would like to conclude the illustration with an application to P2P content classification. For lack of space, here we just very briefly outline the experiment and the results. An interested reader can find more details about this application in [4]. Using the technology developed in [11] we had an access to all world-wide Torrents managed by BitTorrents protocol. In particular, within one week we could observe 200413 different content files. Each file is a data point and we create an edge between two data points i and j if the same user downloaded two files i and j. By such a construction, graph has 50726946 edges. Consider an example of classification of the content by language (e.g., language of a movie or language of a book). Fortunately, a big portion of the content is tagged, so we can compare with the ground truth for some content. We have chosen to classify the content according to five major languages (English, French, Italian, Japanese, Spanish). For each language we have chosen 50 labelled points with the maximal degree within the ground truth points. Since we do not have ground truth for all the points, it is assumed that choosing random points from the ground truth will not be representative (popular content is more likely to be tagged). The precision of classification for $\sigma = 0.0; 0.5; 1.0$ and various values of α is given in Figure 2(b). The figure is consistent with Figure 1(b). In Tables 2 and 3 we provide cross-validation matrices for the Standard Laplacian and PageRank based methods

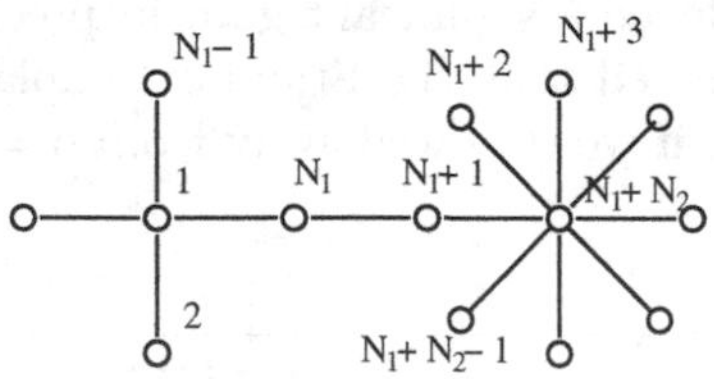

(a) Characteristic network model.

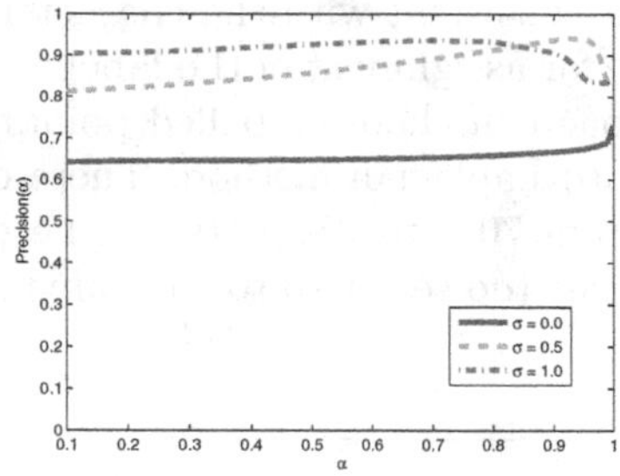

(b) P2P Content Classification.

Fig. 2.

Table 2. $\sigma = 1.0$, Precision 93.43%

Classified as→	En	Fr	It	Jp	Sp
English	36097	22	134	53	159
French	903	909	7	1	4
Italian	308	1	2123	1	17
Japanese	583	7	4	120	6
Spanish	662	1	14	0	1804

Table 3. $\sigma = 0.0$, precision 65.85%

Classified as→	En	Fr	It	Jp	Sp
English	22276	3812	3095	6233	1049
French	87	1618	38	63	18
Italian	24	27	2329	40	30
Japanese	45	43	25	568	39
Spanish	124	78	83	52	2144

with $\alpha = 0.8$. We can observe that as in the previous examples, the PageRank method pulls elements from the largest class to the smaller classes and the Standard Laplacian method does the opposite. Thus, in the case of unbalanced classification, by choosing σ, one admits a trade off between precision and recall for smaller classes.

5 Conclusion and Future Research

Using random walk theory, we provide insights about different graph-based semi-supervised learning methods. We also suggest the following recommendations. If possible, choose labelled points with large degrees. Then, adopt the Standard Laplacian method with α in the upper-middle range of the interval $(0, 1)$. If finding large degree points is not feasible or recall is more important than precision for small classes, choose the PageRank based method. In our near future research we plan to study in more detail the choice of the regularization parameter.

Acknowledgements. This research is funded by Inria Alcatel-Lucent Joint Lab. We also would like to thank P.G. Howlett, J.K. Sreedharan and anonymous reviewers whose comments helped to improve the presentation of the results.

References

1. Andersen, R., Chung, F., Lang, K.: Using pagerank to locally partition a graph. Internet Mathematics 4(1), 35–64 (2007)
2. Avrachenkov, K.: Analytic Perturbation Theory and its Applications, PhD Thesis. University of South Australia, Adelaide, Australia (1999)
3. Avrachenkov, K., Dobrynin, V., Nemirovsky, D., Pham, S.K., Smirnova, E.: Pagerank based clustering of hypertext document collections. In: Proceedings of the 31st Annual International ACM Conference on Research and Development in Information Retrieval, SIGIR 2008, pp. 873–874. ACM (2008)
4. Avrachenkov, K., Gonçalves, P., Legout, A., Sokol, M.: Classification of content and users in bittorrent by semi-supervised learning methods. In: 2012 8th International Wireless Communications and Mobile Computing Conference (IWCMC), Workshop on Traffic Analysis and Classification, pp. 625–630 (2012)
5. Avrachenkov, K., Gonçalves, P., Mishenin, A., Sokol, M.: Generalized optimization framework for graph-based semi-supervised learning. In: Proceedings of SIAM Conference on Data Mining (SDM 2012), 9 pages (2012)
6. Avrachenkov, K., Litvak, N.: The effect of new links on google pagerank. Stochastic Models 22(2) (2006)
7. Blackwell, D.: Discrete dynamic programming. Ann. Math. Statist. 33, 719–726 (1962)
8. Guo, Z., Zhang, Z., Xing, E.P., Faloutsos, C.: Semi-supervised learning based on semiparametric regularization. In: SDM 2008 Proceedings, pp. 132–142 (2008)
9. Haveliwala, T.H.: Topic-sensitive pagerank. In: Proceedings of the 11th International Conference on World Wide Web (WWW 2002), pp. 517–526 (2002)
10. Kemeny, J.G., Snell, J.L.: Finite Markov chains, 1st edn. Springer (1976)
11. Le Blond, S., Legout, A., Lefessant, F., Dabbous, W., Kaafar, M.A.: Spying the world from your laptop: identifying and profiling content providers and big downloaders in bittorrent. In: Proceedings of the 3rd USENIX Conference on Large-Scale Exploits and Emergent Threats: Botnets, Spyware, Worms, and More, LEET 2010, p. 4. USENIX Association, Berkeley (2010)
12. Puterman, M.L.: Markov Decision Processes: Discrete Stochastic Dynamic Programming, 1st edn. John Wiley & Sons, Inc., New York (1994)
13. Zhou, D., Bousquet, O., Navin Lal, T., Weston, J., Schölkopf, B.: Learning with local and global consistency. In: Advances in Neural Information Processing Systems 16, pp. 321–328. MIT Press (2004)
14. Zhou, D., Burges, C.J.C.: Spectral clustering and transductive learning with multiple views. In: Proceedings of the 24th International Conference on Machine Learning, ICML 2007, pp. 1159–1166. ACM (2007)
15. Zhou, D., Schölkopf, B.: A regularization framework for learning from graph data. In: Proceedings of the Workshop on Statistical Relational Learning at Twenty-First International Conference on Machine Learning (ICML 2004), Canada, 6 pages (2004)
16. Zhu, X.: Semi-supervised learning literature survey. Technical report 1530, Department of computer sciences, University of wisconsin, Madison (2005)
17. Zhu, X., Goldberg, A.B.: Introduction to semi-supervised learning. Synthesis Lectures on Artificial Intelligence and Machine Learning 3(1), 1–130 (2009)

A Nearly-Sublinear Method for Approximating a Column of the Matrix Exponential for Matrices from Large, Sparse Networks

Kyle Kloster[1,*] and David F. Gleich[2,*]

[1] Purdue University, Mathematics Department
[2] Purdue University, Computer Science Department
{kkloste,dgleich}@purdue.edu
http://www.cs.purdue.edu/homes/dgleich/codes/nexpokit

Abstract. We consider random-walk transition matrices from large social and information networks. For these matrices, we describe and evaluate a fast method to estimate one column of the matrix exponential. Our method runs in sublinear time on networks where the maximum degree grows doubly logarithmic with respect to the number of nodes. For collaboration networks with over 5 million edges, we find it runs in less than a second on a standard desktop machine.

Keywords: Matrix exponential, Gauss-Southwell, local algorithms.

1 Introduction

The matrix exponential is a standard tool in network analysis. Its uses include node centrality [9,11,10], link-prediction [15], graph kernels [14], and clustering [8]. For the particular problems of node centrality, graph kernels, and clustering, what is most valuable is a coarse estimate of a column of the matrix exponential. In this paper, we consider computing $\exp\{\mathbf{P}\}\mathbf{e}_c$ where $\mathbf{P}$ is the random-walk transition matrix for a directed or undirected graph and $\mathbf{e}_c$ is the cth column of the identity matrix. More precisely, and to establish some notation for the paper, let $\mathbf{G}$ be an $n \times n$ adjacency matrix for a *directed or undirected* graph, let $\mathbf{e} = [1, \cdots, 1]^T$ the vector (of appropriate dimensions) of all 1s, and let $\mathbf{D} = \text{diag}(\mathbf{Ge})$ so that $\mathbf{D}_{ii} = d_i$ is the degree of node i (and $d = \max_i\{d_i\}$ is the maximum degree). We consider $\mathbf{P} = \mathbf{GD}^{-1}$. This case suffices for many of the problems studied in the literature and allows us to compute exponentials of the negative normalized Laplacian $-\hat{\mathbf{L}} = \mathbf{D}^{-1/2}\mathbf{GD}^{-1/2} - \mathbf{I}$ as well. Observe that

$$\begin{aligned}\exp\{\mathbf{D}^{-1/2}\mathbf{GD}^{-1/2} - \mathbf{I}\}\mathbf{e}_c &= e^{-1}\mathbf{D}^{-1/2}\exp\{\mathbf{GD}^{-1}\}\mathbf{D}^{1/2}\mathbf{e}_c \\ &= \sqrt{d_c}e^{-1}\mathbf{D}^{-1/2}\exp\{\mathbf{GD}^{-1}\}\mathbf{e}_c,\end{aligned}$$

so computing a column of either $-\hat{\mathbf{L}}$ or $\mathbf{P}$ allows computation of the other at the cost of scaling the solution vector.

* Supported by NSF CAREER award 1149756-CCF.

A. Bonato, M. Mitzenmacher, and P. Prałat (Eds.): WAW 2013, LNCS 8305, pp. 68–79, 2013.
© Springer International Publishing Switzerland 2013

Computing accurate matrix exponentials has a lengthy and "dubious" history [17]. Let $\mathbf{A}$ be a general $n \times n$ matrix that is large and sparse and let $\mathbf{b}$ be a general vector. A popular method for computing $\exp\{\mathbf{A}\}\mathbf{b}$ involves using an m-step Krylov approximation of the matrix $\mathbf{A}$ yielding $\mathbf{A} \approx \mathbf{V}_m \mathbf{H}_m \mathbf{V}_m^T$. If we use this form, we can approximate $\exp\{\mathbf{A}\}\mathbf{b} \approx \mathbf{V}_m \exp\{\mathbf{H}\}\mathbf{e}_1$. For $m \ll n$, the computation is reduced to the much smaller $\exp\{\mathbf{H}\}$ at the cost of using m matrix-vector products to generate $\mathbf{V}_m$. Such an approximation works well for computing both the entire exponential $\exp\{\mathbf{A}\}$ and its action on a vector: $\exp\{\mathbf{A}\}\mathbf{b}$. This idea underlies the implementation of the Matlab package ExpoKit [21], which has been a standard for computing $\exp\{\mathbf{A}\}\mathbf{b}$ for some time.

While these algorithms are fast and accurate (see references [13], [12], and [2], for the numerical analysis), they depend on matrix-vector products with the matrix $\mathbf{P}$ and orthogonalization steps between successive vectors. When a Krylov method approximates a sparse matrix arising from a graph with a small diameter, then the vectors involved in the matrix-vector products become dense after two or three steps, even if the vector starting the Krylov approximation has only a single non-zero entry. Subsequent matrix-vector products take $O(|E|)$ work where $|E|$ is the number of edges in the graph. For networks with billions of edges, we want an alternative to Krylov-based methods that prioritizes speed and sparsity over accuracy. In particular, we would like an algorithm to estimate a column of the matrix exponential in less work than a single matrix-vector product.

Local methods perform a computation by accessing a small region of a matrix or graph. These are a practical alternative to Krylov methods for solving massive linear systems from network problems that have sparse right hand sides; see, for instance, references [3,7]. Rather than matrix-vector products, these procedures use steps that access only a single row or column of the matrix. We design a local algorithm for computing $\exp\{\mathbf{P}\}\mathbf{e}_c$ by translating the problem of computing the exponential into solving a linear system, and then using a local algorithm.

We present an algorithm that approximates a specified column of $\exp\{\mathbf{P}\}$ for column stochastic $\mathbf{P}$ (Section 4, Figure 1). The algorithm uses the Gauss-Southwell method (Section 2) for solving a linear system to approximate a degree N Taylor polynomial (Section 3). The error after l iterations of the algorithm is bounded by $\frac{1}{N!N} + l^{-1/(2d)}$ as shown in Theorem 2, and the runtime is $O(ld + ld\log(ld))$ (Section 5.3). Given an input error ε, the runtime to produce a solution vector with error less than ε is sublinear in n for graphs with $d \leq O(\log\log n)$. We acknowledge that this doubly logarithmic scaling of the maximum degree is unrealistic for social and information networks where the maximum degree typically scales almost linearly with n. Nevertheless, the existence of a bound suggests that it may be possible to improve or establish a matching lower-bound.

2 Local Computations and the Gauss-Southwell Method

The Gauss-Southwell (GS) iteration is a classic stationary method for solving a linear system related to the Gauss-Seidel and coordinate descent methods [16]. It is especially efficient when the desired solution vector is sparse or localized [7,5] and the goal is a coarse $O(10^{-4})$ approximation. In these cases, GS produces a sparse approximation that is accurate for only the largest entries. The coarse nature of the GS approximation is acceptable because the primary use is to find those nodes having the largest values in link-prediction or clustering problems.

The GS iteration is simple. Select the largest entry in the residual vector and perform a coordinate descent step in that component of the solution. Let $\mathbf{x}^{(l)}$ and $\mathbf{r}^{(l)}$ be the solution and residual iterates for $\mathbf{Ax} = \mathbf{b}$ after l steps. In the $(l+1)$st step, pick q so that $m_l = \mathbf{r}_q^{(l)}$ is the maximum magnitude entry of the residual vector. Next update:

$$\begin{aligned} \mathbf{x}^{(l+1)} &= \mathbf{x}^{(l)} + m_l \mathbf{e}_q \\ \mathbf{r}^{(l+1)} &= \mathbf{r}^{(l)} - m_l \mathbf{A}\mathbf{e}_q. \end{aligned} \tag{1}$$

The iteration consists of a single-entry update to $\mathbf{x}$ and a few updates to the residual vector if $\mathbf{A}$ is sparse. This method converges for diagonally dominant and symmetric positive definite linear systems [16].

Applied to a matrix from a graph, each iteration of GS requires at most $O(d \log n)$ operations, where the $\log n$ term comes from heap updates to maintain the largest residual entry. This procedure underlies Berkhin's bookmark coloring algorithm [5] for PageRank and a related method avoids the heap [3].

3 Exponentials via the Taylor Series Approximation

The GS method is most effective on sparse linear systems. We now design a large, sparse linear system to compute a Taylor polynomial appromation of $\exp\{\mathbf{P}\}\mathbf{e}_c$.

3.1 The Truncated Taylor Series of the Exponential

The Taylor series for the matrix $\exp\{\mathbf{A}\}$ is

$$\exp\{\mathbf{A}\} = \sum_{k=0}^{\infty} \tfrac{1}{k!}\mathbf{A}^k.$$

It converges for any matrix $\mathbf{A}$. Truncating to N terms, we arrive at an algorithm. If $\|\mathbf{A}\|$ is large with mixed sign then the summands $\mathbf{A}^k$ may be large and cancel only in exact arithmetic, resulting in poor accuracy. However, a stochastic matrix $\mathbf{P}$ is non-negative and has $\|\mathbf{P}\|_1 = 1$, so the approximation converges quickly and reliably (Lemma 1). Using an N-degree Taylor approximation to compute the cth column results in a simple iteration. Let $\mathbf{x}_N$ be the N-degree Taylor approximation:

$$\mathbf{x}_N = \sum_{k=0}^{N} \tfrac{1}{k!}\mathbf{P}^k\mathbf{e}_c \approx \exp\{\mathbf{P}\}\mathbf{e}_c.$$

Then

$$\mathbf{x}_N = \sum_{k=0}^{N} \mathbf{v}_k \quad \mathbf{v}_0 = \mathbf{e}_c, \quad \mathbf{v}_1 = \mathbf{P}\mathbf{v}_0, \quad \mathbf{v}_{k+1} = \mathbf{P}\mathbf{v}_k/k \quad \text{for } k = 1, \cdots, N.$$

If $\mathbf{x}_{\text{true}}$ is the actual column of $\exp\{\mathbf{P}\}$ we are trying to compute, note that $\mathbf{x}_N$ converges to $\mathbf{x}_{\text{true}}$ as N tends to infinity. For practical purposes, we want to ensure that $\|\mathbf{x}_N - \mathbf{x}_{\text{true}}\|_1$ is small so that our approximation of $\mathbf{x}_N$ is near $\mathbf{x}_{\text{true}}$. The next Lemma shows that $N = 11$ yields a 1-norm error of 2.3×10^{-9}. This is sufficiently small for our purposes and from now on, we use $N = 11$ throughout.

Lemma 1. *The degree N Taylor approximation satisfies* $\|\mathbf{x}_{true} - \mathbf{x}_N\|_1 \leq \frac{1}{N!N}$.

Proof. The truncation results in a simple error analysis:

$$\|\mathbf{x}_{\text{true}} - \mathbf{x}_N\|_1 = \left\| \sum_{k=N+1}^{\infty} \frac{\mathbf{P}^k}{k!} \mathbf{e}_c \right\|_1 = \sum_{k=N+1}^{\infty} \frac{1}{k!}, \tag{2}$$

which follows because $\mathbf{P}$ and $\mathbf{e}_c$ are nonnegative and $\mathbf{P}$ is column stochastic. By factoring out $\frac{1}{(N+1)!}$ and majorizing $\frac{(N+1)!}{(N+1+k)!} \leq \left(\frac{1}{N+1}\right)^k$ for $k \geq 0$, we finish:

$$\|\mathbf{x}_{\text{true}} - \mathbf{x}_N\|_1 \leq \left(\frac{1}{(N+1)!}\right) \sum_{k=0}^{\infty} \left(\frac{1}{N+1}\right)^k = \frac{1}{(N+1)!} \frac{N+1}{N} \tag{3}$$

after substituting the limit for the convergent geometric series. □

3.2 Forming a Linear System

We now devise a linear system to compute the intermediate terms $\mathbf{v}_k$ and $\mathbf{x}_N$. From the identity $\mathbf{v}_{k+1} = \frac{\mathbf{P}}{k+1} \cdot \mathbf{v}_k$ we see that the $\mathbf{v}_k$ satisfy

$$\begin{bmatrix} \mathbf{I} & & & & \\ -\mathbf{P}/1 & \mathbf{I} & & & \\ & -\mathbf{P}/2 & \ddots & & \\ & & \ddots & \mathbf{I} & \\ & & & -\mathbf{P}/N & \mathbf{I} \end{bmatrix} \begin{bmatrix} \mathbf{v}_0 \\ \mathbf{v}_1 \\ \vdots \\ \vdots \\ \mathbf{v}_N \end{bmatrix} = \begin{bmatrix} \mathbf{e}_c \\ 0 \\ \vdots \\ \vdots \\ 0 \end{bmatrix}. \tag{4}$$

For convenience of notation, let $\mathbf{S}_{N+1}$ be the $(N+1) \times (N+1)$ zero matrix with first subdiagonal $[\frac{1}{1}, \frac{1}{2}, \cdots, \frac{1}{N}]$. Let $\mathbf{v} = [\mathbf{v}_0; \cdots; \mathbf{v}_N]$. Then we can rewrite (4):

$$(\mathbf{I}_{N+1} \otimes \mathbf{I}_n - \mathbf{S}_{N+1} \otimes \mathbf{P})\, \mathbf{v} = \mathbf{e}_1 \otimes \mathbf{e}_c. \tag{5}$$

The left- and right-hand sides of (5) are sparse, making this linear system a candidate for a Gauss-Southwell iteration. Note also that we need never form this large block-wise system explicitly and can work with it implicitly. Each row of the system is uniquely defined by a node index i and a step index k.

We now show that approximating $\mathbf{v}$ will help approximate $\mathbf{x}_N$. Let $\mathbf{M} = \mathbf{I}_{N+1} \otimes \mathbf{I}_n - \mathbf{S}_{N+1} \otimes \mathbf{P}$ from (5). With an approximation $\hat{\mathbf{v}}$ to the solution $\mathbf{v}$, we can approximate $\mathbf{x}_N$ by summing components of $\hat{\mathbf{v}}$: $\hat{\mathbf{x}}_N = \sum_{k=0}^{N} \hat{\mathbf{v}}_k$. Given that our primary purpose is computing $\mathbf{x}_N$, we want to know how accurately $\hat{\mathbf{x}}_N$ approximates $\mathbf{x}_N$. With that in mind, we state the following:

Lemma 2. *If* $\hat{\mathbf{v}} \leq \mathbf{v}$ *component-wise, then* $\|\mathbf{x}_N - \hat{\mathbf{x}}_N\|_1 = \|\mathbf{v} - \hat{\mathbf{v}}\|_1$.

Proof. The vector $\mathbf{x}_N$ is the sum of the block-vectors composing $\mathbf{v} = [\mathbf{v}_0; \cdots; \mathbf{v}_N]$, and similarly $\hat{\mathbf{x}}_N$ is the sum of the block-vectors of $\hat{\mathbf{v}}$. Thus, $\mathbf{e}^T\mathbf{v} = \mathbf{e}^T\mathbf{x}_N$, and $\mathbf{e}^T\hat{\mathbf{v}} = \mathbf{e}^T\hat{\mathbf{x}}_N$. Since $\hat{\mathbf{v}}$ approaches $\mathbf{v}$ from below (by assumption), we have that $\|\mathbf{x}_N - \hat{\mathbf{x}}_N\|_1 = \mathbf{e}^T(\mathbf{x}_N - \hat{\mathbf{x}}_N) = \mathbf{e}^T(\mathbf{v} - \hat{\mathbf{v}}) = \|\mathbf{v} - \hat{\mathbf{v}}\|_1$. □

4 Approximating the Taylor Polynomial with GS

We apply Gauss-Southwell (GS) to $\mathbf{M}\mathbf{v} = \mathbf{e}_1 \otimes \mathbf{e}_c$ staring with $\mathbf{x}^{(0)} = 0$. The block structure of $\mathbf{M}$ makes the solution and residual update simple. Recall that q was the index of the largest entry in the residual and m_l was the value of the entry. Let $\mathbf{e}_q = \mathbf{e}_k \otimes \mathbf{e}_i$ where $\mathbf{e}_k$ is a length $N+1$ vector indicating the step number and $\mathbf{e}_i$ indicates the node number. By substituting this into (1), we find:

$$\mathbf{v}^{(l+1)} = \mathbf{v}^{(l)} + m_l(\mathbf{e}_k \otimes \mathbf{e}_i) \tag{6}$$
$$\mathbf{r}^{(l+1)} = \mathbf{r}^{(l)} - m_l\mathbf{M}(\mathbf{e}_k \otimes \mathbf{e}_i). \tag{7}$$

Note that (6) simply adds $m_l\mathbf{e}_i$ to the block of $\mathbf{v}$ corresponding to $\mathbf{v}_{k-1}$. However, since we intend to add together the $\mathbf{v}_i$ at the end (to produce the full Taylor approximation), in practice we simply add $m_l\mathbf{e}_i$ directly to to our matrix-exponential-column approximation vector $\hat{\mathbf{x}}^{(l)}$. Thus we satisfy the requirements of Lemma 2. When $k < N+1$ the residual update can also be adapted using

$$\mathbf{M}(\mathbf{e}_k \otimes \mathbf{e}_i) = \mathbf{e}_k \otimes \mathbf{e}_i - (\tfrac{1}{k}\mathbf{e}_{k+1} \otimes (\mathbf{P}\mathbf{e}_i)) \tag{8}$$

In the case that $k = N+1$, then $\mathbf{S}_{N+1}\mathbf{e}_{N+1} = 0$, so we have simply $\mathbf{M}(\mathbf{e}_{N+1} \otimes \mathbf{e}_i) = \mathbf{e}_{N+1} \otimes \mathbf{e}_i$. Substituting (8) into the residual update in (7) gives

$$\mathbf{r}^{(l+1)} = \mathbf{r}^{(l)} - m_l\mathbf{e}_k \otimes \mathbf{e}_i + \tfrac{m_l}{k}(\mathbf{e}_{k+1} \otimes \mathbf{P}\mathbf{e}_i). \tag{9}$$

Because the indices k and i are chosen so that the entry in the vector $\mathbf{e}_k \otimes \mathbf{e}_i$ corresponds to the entry of $\mathbf{r}^{(l)}$ that is largest, we have $m_l = (\mathbf{e}_k \otimes \mathbf{e}_i)^T \mathbf{r}^{(l)}$. Thus, $(\mathbf{e}_k \otimes \mathbf{e}_i)^T \mathbf{r}^{(l+1)} = (\mathbf{e}_k \otimes \mathbf{e}_i)^T \mathbf{r}^{(l)} - m_l(\mathbf{e}_k \otimes \mathbf{e}_i)^T \mathbf{e}_k \otimes \mathbf{e}_i = 0$, and this iteration zeros out the largest entry of $\mathbf{r}^{(l)}$ at each step. See Figure 1 for working code.

5 Convergence Results for Gauss-Southwell

Our convergence analysis has two stages. First, we show that the algorithm in Figure 1 produces a residual that converges to zero (Theorem 1). Second, we establish the rate at which the error in the computed solution $\hat{\mathbf{x}}_N$ converges to zero (Theorem 2). From this second bound, we arrive at a sublinear runtime bound in the case of a slowly growing maximum degree.

```
function x = gexpm(P,c,tol)
n = size(P,1); N = 11; sumr=1;
r = zeros(n,N+1); r(c,1) = 1; x = zeros(n,1);      % the residual and solution
while sumr >= tol % use max iteration too
  [ml,q]=max(r(:)); i=mod(q-1,n)+1; k=ceil(q/n); % use a heap in practice for max
  r(q) = 0; x(i) = x(i)+ml; sumr = sumr-ml;% zero the residual, add to solution
  [nset,~,vals] = find(P(:,i)); ml=ml/k;      % look up the neighbors of node i
  for j=1:numel(nset)                             % for all neighbors
    if k==N, x(nset(j)) = x(nset(j)) + vals(j)*ml;  % add to solution
    else, r(nset(j),k+1) = r(nset(j),k+1) + vals(j)*ml;% or add to next residual
          sumr = sumr + vals(j)*ml;
end, end, end                                 % end if, end for, end while
```

Fig. 1. Pseudo-code for our nearly sublinear time algorithm as Matlab code. In practice, the solution vector `x` and residual `r` should be stored as hash-tables, and the entries of the residual as a heap. Note that the command `[nset,~,vals] = find(P(:,i))` returns the neighbors of the ith node (`nset`) along with values of $\mathbf{P}$ for those neighbors.

5.1 Convergence of the Residual

Theorem 1. *Let* $\mathbf{P}$ *be column-stochastic and* $\mathbf{v}^{(0)} = 0$. *(Nonnegativity) The iterates and residuals are nonnegative:* $\mathbf{v}^{(l)} \geq 0$ *and* $\mathbf{r}^{(l)} \geq 0$ *for all* $l \geq 0$. *(Convergence) The residual satisfies the following bound and converges to 0:*

$$\|\mathbf{r}^{(l)}\|_1 \leq \prod_{k=1}^{l}\left(1 - \frac{1}{2dk}\right) \leq l^{\left(-\frac{1}{2d}\right)} \tag{10}$$

Proof. (Nonnegativity) Since $\mathbf{v}^{(0)} = 0$ we have $\mathbf{r}^{(0)} = \mathbf{e}_1 \otimes \mathbf{e}_c - \mathbf{M}\cdot 0 = \mathbf{e}_1 \otimes \mathbf{e}_c \geq 0$, establishing both base cases. Now assume by way of induction that $\mathbf{v}^{(l)} \geq 0$ and $\mathbf{r}^{(l)} \geq 0$. Then the GS update gives $\mathbf{v}^{(l+1)} = \mathbf{v}^{(l)} + m_l \mathbf{e}_k \otimes \mathbf{e}_i$, and since $m_l \geq 0$ (because it is taken to be the largest entry in $\mathbf{r}^{(l)}$, which we have assumed is nonnegative) we have that $\mathbf{v}^{(l+1)} \geq 0$.

From (9) we have $\mathbf{r}^{(l+1)} = \mathbf{r}^{(l)} - m_l \mathbf{e}_k \otimes \mathbf{e}_i + \frac{m_l}{k} \mathbf{e}_{k+1} \otimes \mathbf{P}\mathbf{e}_i$, but we have assumed $\mathbf{P}$ is stochastic, so $\mathbf{e}_{k+1} \otimes \mathbf{P}\mathbf{e}_i \geq 0$. Then, note that $\mathbf{r}^{(l)} - m_l \mathbf{e}_k \otimes \mathbf{e}_i \geq 0$ because by the inductive hypothesis $\mathbf{r}^{(l)} \geq 0$ and subtracting m_l simply zeros out that entry of $\mathbf{r}^{(l)}$. Thus, $\mathbf{r}^{(l+1)} \geq 0$, as desired.

(Convergence) Because the residual is always nonnegative, we can use the identity $\|\mathbf{r}^{(l)}\|_1 = \mathbf{e}^T \mathbf{r}^{(l)}$. Left multiplying by $\mathbf{e}^T$ in (9) yields $\|\mathbf{r}^{(l+1)}\|_1 = \|\mathbf{r}^{(l)}\|_1 - m_l + \frac{m_l}{k}\left(\mathbf{e}^T \mathbf{e}_{k+1} \otimes \mathbf{e}^T \mathbf{P}\mathbf{e}_i\right)$. Since we've assumed $\mathbf{P}$ is column stochastic, $\mathbf{e}^T \mathbf{P}\mathbf{e}_i = 1$, and so this simplifies to $\|\mathbf{r}^{(l+1)}\|_1 = \|\mathbf{r}^{(l)}\|_1 - m_l + \frac{m_l}{k} = \|\mathbf{r}^{(l)}\|_1 - m_l\left(1 - \frac{1}{k}\right)$.

Since m_l is the largest entry in $\mathbf{r}^{(l)}$, we know it must be at least as big as the average value of an entry of $\mathbf{r}^{(l)}$. After l iterations, the residual can have no more than dl nonzero entries, because no more than d nonzeros can be added each iteration. Hence we have $m_l \geq \frac{\|\mathbf{r}^{(l)}\|_1}{dl}$. After the first iteration, we know $k \geq 2$

because the $k = 1$ block of $\mathbf{r}$ (denoted $\mathbf{r}_0$ in the notation of Section 3.2) is empty. If we put these bounds together, we have: $\|\mathbf{r}^{(l+1)}\|_1 \le \|\mathbf{r}^{(l)}\|_1 - \frac{\|\mathbf{r}^{(l)}\|_1}{dl}\left(1 - \frac{1}{2}\right) = \|\mathbf{r}^{(l)}\|_1 \left(1 - \frac{1}{2dl}\right)$. Iterating this inequality yields the bound:

$$\|\mathbf{r}^{(l)}\|_1 \le \prod_{j=1}^{l} \left(1 - \tfrac{1}{2dj}\right) \cdot \|\mathbf{r}^{(0)}\|_1, \tag{11}$$

and since $\mathbf{r}^{(0)} = \mathbf{e}_1 \otimes \mathbf{e}_c$ we have $\|\mathbf{r}^{(0)}\|_1 = 1$, proving the first inequality.

The second inequality of (10) follows from using the facts $(1 + x) \le e^x$ (for $x > -1$) and $\log(l) < \sum_{j=1}^{l} \frac{1}{j}$ to write

$$\prod_{j=1}^{l} \left(1 - \tfrac{1}{2dj}\right) \le \exp\{-\tfrac{1}{2d} \sum_{j=1}^{l} \tfrac{1}{j}\} \le \exp\{-\tfrac{\log l}{2d}\} = l^{(-\frac{1}{2d})}. \qquad \square$$

The inequality $(1 + x) \le e^x$ follows from the Taylor series $e^x = 1 + x + o(x^2)$, and the lowerbound for the partial harmonic sum $\sum_{j=1}^{l} \frac{1}{j}$ follows from the left-hand rule integral approximation $\log(l) = \int_1^l \frac{1}{x} \mathrm{dx} \le \sum_{j=1}^{l} \frac{1}{j}$.

5.2 Convergence of Error

Although the previous theorem establishes that GS will converge, we need a more precise statement about the error to bound the runtime. We will first state such a bound and use it to justify the claim of "nearly" sublinear runtime before formally proving it.

Theorem 2. *In the notation described above, the error of the approximation from l iterations of GS satisfies*

$$\|\hat{\mathbf{x}}^{(l)} - \mathbf{x}_{true}\|_1 \le \tfrac{1}{N!N} + e \cdot l^{-\frac{1}{2d}} \quad \textit{where} \quad e = \exp(1). \tag{12}$$

Nearly Sublinear Runtime Given an input tolerance ε, Theorem 2 shows that the number of iterations l required to produce $\hat{\mathbf{x}}^{(l)}$ satisfying $\|\mathbf{x}_{\text{true}} - \hat{\mathbf{x}}^{(l)}\|_1 < \varepsilon$ depends on d alone (since we have control over N, and so can just choose N such that $\frac{1}{N!N} < \varepsilon/2$). For $\frac{1}{N!N} < \frac{\varepsilon}{2}$ to hold, it suffices to take $N = 2\log(1/\varepsilon)$ because $\frac{1}{N!N} < e^{-N}/2 = \frac{\varepsilon}{2}$ for $N > 5$. So $N = 2\log(1/\varepsilon)$.

Next, we need a value of l for which $e \cdot l^{-\frac{1}{2d}} < \frac{\varepsilon}{2}$ holds. Taking logs and exponentiating yields the bound

$$l > \exp\{2d(1 + \log(2) + \log(1/\varepsilon))\}. \tag{13}$$

If $d = C \log\log n$ for a constant C, then the desired error is guaranteed by

$$l > (\log n)^{2C(1+\log(2)+\log(1/\varepsilon))} \tag{14}$$

which grows sublinearly in n.

Proof of Theorem 2 By the triangle inequality $\|\mathbf{x}_{\text{true}} - \hat{\mathbf{x}}^{(l)}\|_1 \leq \|\mathbf{x}_{\text{true}} - \mathbf{x}_N\|_1 + \|\mathbf{x}_N - \hat{\mathbf{x}}^{(l)}\|_1$, so we can apply Lemma 1 to get $\|\mathbf{x}_{\text{true}} - \hat{\mathbf{x}}^{(l)}\|_1 \leq \frac{1}{N!N} + \|\mathbf{x}_N - \hat{\mathbf{x}}^{(l)}\|_1$, and then Lemma 2 to obtain

$$\|\mathbf{x}_{\text{true}} - \hat{\mathbf{x}}^{(l)}\|_1 \leq \frac{1}{N!N} + \|\mathbf{v} - \hat{\mathbf{v}}^{(l)}\|_1. \tag{15}$$

Theorem 1 gives the residual of the GS solution vector after l iterations, $\hat{\mathbf{v}}^{(l)}$, but we want the *error*, i.e. the difference between $\hat{\mathbf{v}}^{(l)}$ and $\mathbf{v}$. To obtain this quantity, we use a standard relationship between the residual and error specialized to our system: $\|\mathbf{v} - \hat{\mathbf{v}}^{(l)}\|_1 \leq \|\mathbf{M}^{-1}\|_1 \|\mathbf{r}^{(l)}\|_1$. To complete the proof of Theorem 2, it suffices, then, to show that $\|\mathbf{M}^{-1}\|_1 \leq e$ and use Theorem 1. The next lemma establishes this remaining bound. We suspect that the following result is already known in the literature and regret the tedious proof.

Lemma 3. *Matrices* $\mathbf{M}$ *of the form* $\mathbf{M} = \mathbf{I}_{N+1} \otimes \mathbf{I}_n - \mathbf{S}_{N+1} \otimes \mathbf{P}$ *for column-stochastic* $\mathbf{P}$ *satisfy* $\|\mathbf{M}^{-1}\|_1 \leq e$ *.*

Proof. Write $\mathbf{M} = \mathbf{I} - \mathbf{S}_{N+1} \otimes \mathbf{P}$ and note that, since $\mathbf{S}_{N+1}^{N+1} = 0$ we have $(\mathbf{S}_{N+1} \otimes \mathbf{P})^{N+1} = 0$, i.e. $\mathbf{S}_{N+1} \otimes \mathbf{P}$ is nilpotent, so $\mathbf{M} = \mathbf{I} - \mathbf{S}_{N+1} \otimes \mathbf{P}$ has inverse

$$\mathbf{M}^{-1} = \mathbf{I} + (\mathbf{S}_{N+1} \otimes \mathbf{P}) + (\mathbf{S}_{N+1} \otimes \mathbf{P})^2 + \cdots + (\mathbf{S}_{N+1} \otimes \mathbf{P})^N. \tag{16}$$

To upperbound $\|\mathbf{M}^{-1}\|_1$ observe that each term $(\mathbf{S}_{N+1} \otimes \mathbf{P})^j$ is nonnegative, and so $\mathbf{M}^{-1}$ is itself nonnegative. Thus, $\|\mathbf{M}^{-1}\|_1$ is simply the largest column-sum of $\mathbf{M}^{-1}$, i.e. $\max_{k,i}\{\mathbf{e}^T \mathbf{M}^{-1} (\mathbf{e}_k \otimes \mathbf{e}_i)\}$. For convenience of notation, define $t_k = \mathbf{e}^T \mathbf{M}^{-1} (\mathbf{e}_k \otimes \mathbf{e}_i)$ (we will show that this value, t_k, is independent of i). Multiplying (16) by $\mathbf{e}^T$ and $(\mathbf{e}_k \otimes \mathbf{e}_i)$ and distributing produces

$$t_k = \left(\mathbf{e}^T (\mathbf{e}_k \otimes \mathbf{e}_i) + (\mathbf{e}^T \mathbf{S}_{N+1} \mathbf{e}_k) \otimes (\mathbf{e}^T \mathbf{P} \mathbf{e}_i) + \cdots + (\mathbf{e}^T \mathbf{S}_{N+1}^N \mathbf{e}_k) \otimes (\mathbf{e}^T \mathbf{P}^N \mathbf{e}_i)\right) \tag{17}$$

but since $\mathbf{e}^T \mathbf{P}^j \mathbf{e}_i = 1$ for all j, we end up with

$$t_k = \mathbf{e}^T \mathbf{S}_{N+1}^0 \mathbf{e}_k + \mathbf{e}^T \mathbf{S}_{N+1}^1 \mathbf{e}_k + \cdots + \mathbf{e}^T \mathbf{S}_{N+1}^N \mathbf{e}_k. \tag{18}$$

This justifies the notation t_k, since the value is seen to be independent of i here.

We set out to upperbound $\|\mathbf{M}^{-1}\|_1$, and we've now reduced the problem to simply bounding the t_k above. To do this, first note that $\mathbf{S}_{N+1} \mathbf{e}_k = \frac{1}{k} \mathbf{e}_{k+1}$ for all $1 \leq k \leq N$. Repeatedly left multiplying by $\mathbf{S}_{N+1}$ establishes that for $k+j > N+1$ we have $\mathbf{S}_{N+1}^j \mathbf{e}_k = 0$ if $j \geq 1$, but for $k+j \leq N$ we have $\mathbf{S}_{N+1}^j \mathbf{e}_k = \frac{(k-1)!}{(k-1+j)!} \mathbf{e}_{k+j}$ for $j = 0, \cdots, N-k+1$ and $k = 1, \cdots, N$. Using these we rewrite (18):

$$t_k = \sum_{j=0}^{N+1-k} \frac{(k-1)!}{(k-1+j)!}. \tag{19}$$

Observe that $t_1 = \sum_{j=0}^{N} \frac{1}{j!} \leq e$. The inequality $t_{k+1} \leq t_k$ implies that $t_k \leq t_1 < e$, and so to prove Lemma 3 it now suffices to prove this inequality. From (19) we have $t_k = \sum_{j=0}^{N-k} \frac{(k-1)!}{(k-1+j)!} + \frac{(k-1)!}{N!}$ and $t_{k+1} = \sum_{j=0}^{N-k} \frac{k!}{(k+j)!}$. The general terms satisfy $\frac{(k-1)!}{(k-1+j)!} \geq \frac{k!}{(k+j)!}$ because multiplying both sides by $\frac{(k-1+j)!}{(k-1)!}$ yields $1 \geq \frac{k}{k+j}$. Hence $t_k \geq t_{k+1}$, and so the lemma follows. □

5.3 Complexity

Each iteration requires updating the residual for each adjacent node, which in turn requires updating the residual heap. This step involves $O(d \log(ld))$ work to maintain the heap, since there are at most d entries added to $\mathbf{r}$, and each update of the heap requires $O(\log(ld))$ work, where ld is an upperbound on the size of the heap after l iterations. Since each vector add takes at most $O(d)$ operations, the total operation count for l iterations of the algorithm is $O(ld + ld \log(ld))$. If we do not have a sublinear number of steps l, then note that heap updates never take more than $\log(nN)$ work.

6 Experimental Results

We evaluate this method on a desktop computer with an Intel i7-990X, 3.47 GHz CPU and 24 GB of RAM. As described below, we implement two variations of our algorithm in C++ using the Matlab MEX interface. The graphs we use come from a variety of sources and range between 10^3 and 10^7 nodes and edges. All are undirected and connected. They include the dblp and flickr graphs [7], Newman's netscience, condmat-2003, and condmat-2005 graphs [18,19], and Arenas's pgp graph [6]. These results are representative of a larger collection. In the spirit of reproducible research, we make our code available. See the URL in the title.

6.1 Notes on Implementation

We do not yet have a fully efficient implementation of our algorithm. In particular, we maintain full solution vectors and full residual vectors that take $O(n)$ and $O(nN)$ work to initialize and store. In the future, we plan to use a hash table for these vectors. For the current scale of our graphs – $500,000$ nodes – we do not expect this change to make a large difference. We found that the runtime of Gauss-Southwell varied widely due to our use of the heap structure (see the TSGS line in Figure 3). To address this performance, we implemented an idea inspired by Andersen et al.'s replacement of a heap with a queue [3]. Rather than choose the largest element in the residual at each step, we pick an element from a queue that stores all residual elements larger than $\tau/(nN)$. Note that once all elements have been removed from the queue, the norm of the residual will be less than τ. We have not conducted an error analysis to determine how many steps are necessary to satisfy this condition, but empirically we find it performs similarly. For the accuracy results below, we use the method with the queue.

6.2 Accuracy and Runtime

While we know that our method converges as the number of steps runs to infinity, in Figure 2, we study the precision of approximate solutions. Recall that precision is the size of the intersection of the set of indices of the k largest entries from our approximation vector with the set of indices of the k largest entries from the

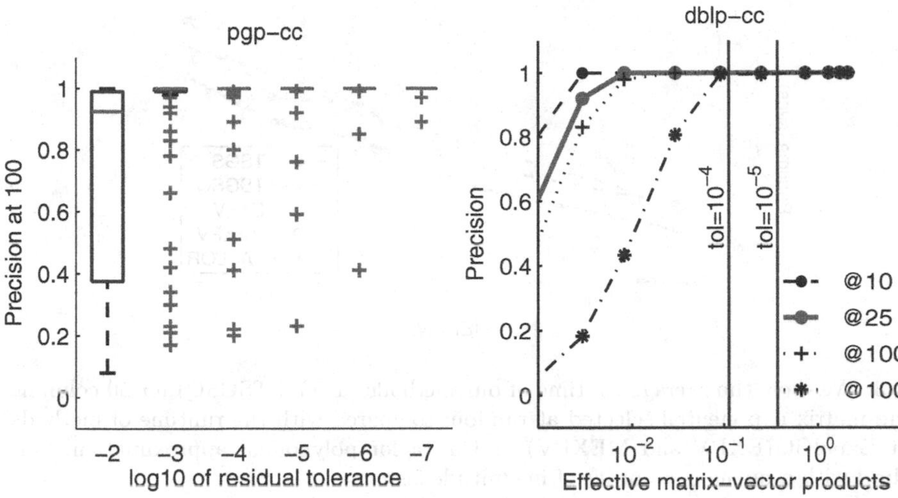

Fig. 2. At left, the precision of the top 100 for the largest entries in a column of the matrix exponential of the pgp graph (without the node's immediate neighbors). We show a boxplot summarizing the results over 50 trials. The "+" marks indicate outliers outside of the 75th percentile. At right, we vary the number of steps of the method and find we get high precision for the top-25 set on the large dblp graph (226,413 nodes) after 1% of the work involved in a matrix vector product. This second plot is indicative of results we see for other choices of the column as well. Solving to tolerance 10^{-5} takes 33% of the work of a matrix-vector product.

true solution, divided by k. In the vector $\exp\{\mathbf{P}\}\mathbf{e}_c$ the entry with index c and the entries corresponding to neighbors of node c are always large in both the true and the approximate solution vectors and so artificially inflate the scores. Because of this, we ignore those entries in both solution vectors and instead look at the indices of the next k largest entries (excluding node c and its neighbors).

We show results for the pgp network (10k nodes) with a boxplot representing 50 random columns. The plot suggests that a residual tolerance of 10^{-4} yields useful results in the majority of cases. We note that this figure represents the worst results we observed and many graphs showed no errors at low tolerances. Next, for a larger graph, we show how the precision in the top-k sets evolves for the dblp graph (226k nodes) as we perform additional steps. Based on both of these results, we suggest a residual tolerance of 10^{-5}.

Next, we compare the runtime of our method using a heap (TSGS) and using a queue (TSGSQ) to the runtime of three other methods solved to tolerance of 10^{-5}. We present the results for all 6 graphs. Each runtime is an average of 50 randomly chosen columns. Both EXPV and MEXPV are from ExpoKit [21] where MEXPV is a special routine for stochastic matrices. The Taylor method simply computes $\lceil 3\log_2(n)\rceil$ terms of the Taylor expansion. These results show that TSGSQ is an order of magnitude faster than the other algorithms and this performance difference persists over graphs spanning four orders of magnitude.

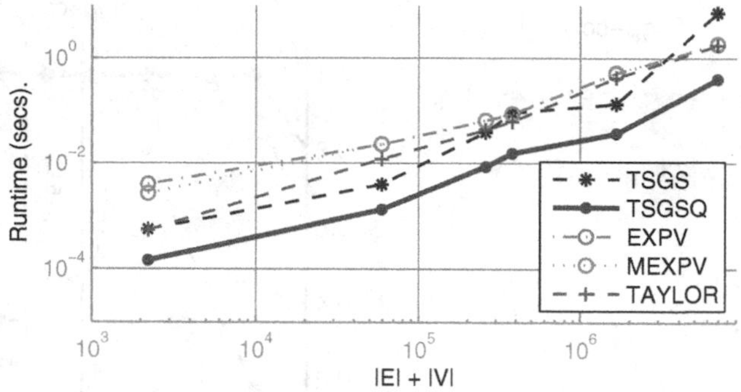

Fig. 3. We show the average runtime of our methods (TSGS, TSGSQ) for 50 columns of the matrix exponential selected at random compared with the runtime of methods from ExpoKit (EXPV and MEXPV) and a Taylor polynomial approximation. Our method with a queue is an order of magnitude faster.

7 Related Work and Discussion

Virtually all recent work on computing $\exp\{\mathbf{A}\}\mathbf{b}$ or even a single element of $\exp\{\mathbf{A}\}$ involves a Krylov or Lanczos approximation [21,4,20,1]. These methods all have runtimes that are $O(|E|)$, or worse, when the matrix comes from a graph. Given our strong assumption about the scaling of the maximum degree, it is possible that these algorithms would also enjoy sublinear runtimes and we are currently studying their analysis. We are also currently searching for additional work that may be related to our current local method to approximate the exponential.

As mentioned before, the doubly logarithmic bound on the maximum degree is unrealistic for social and information networks. We are currently working to improve the bound and believe that using a residual in a weighted ∞-norm, as used by Andersen et al. [3], may succeed. We also note that this method outperforms state of the art Krylov solvers on networks with nearly 10 million nodes and edges. Thus, we believe it to be useful independently of the runtime bound. We are currently working to extend the analysis to functions of scaled stochastic matrices, $\exp\{\alpha\mathbf{P}\}$ with $\alpha < 1$, and the adjacency matrix, $\exp\{\mathbf{A}\}$, for other link prediction methods.

References

1. Afanasjew, M., Eiermann, M., Ernst, O.G., Güttel, S.: Implementation of a restarted Krylov subspace method for the evaluation of matrix functions. Linear Algebra Appl. 429(10), 2293–2314 (2008)
2. Al-Mohy, A.H., Higham, N.J.: Computing the action of the matrix exponential, with an application to exponential integrators. SIAM J. Sci. Comput. 33(2), 488–511 (2011)

3. Andersen, R., Chung, F., Lang, K.: Local graph partitioning using PageRank vectors. In: FOCS 2006 (2006)
4. Benzi, M., Boito, P.: Quadrature rule-based bounds for functions of adjacency matrices. Linear Algebra and its Applications 433(3), 637–652 (2010)
5. Berkhin, P.: Bookmark-coloring algorithm for personalized PageRank computing. Internet Mathematics 3(1), 41–62 (2007)
6. Boguñá, M., Pastor-Satorras, R., Díaz-Guilera, A., Arenas, A.: Models of social networks based on social distance attachment. Phys. Rev. E 70(5), 056122 (2004)
7. Bonchi, F., Esfandiar, P., Gleich, D.F., Greif, C., Lakshmanan, L.V.: Fast matrix computations for pairwise and columnwise commute times and Katz scores. Internet Mathematics 8(1-2), 73–112 (2012)
8. Chung, F.: The heat kernel as the PageRank of a graph. Proceedings of the National Academy of Sciences 104(50), 19735–19740 (2007)
9. Estrada, E.: Characterization of 3d molecular structure. Chemical Physics Letters 319(5-6), 713–718 (2000)
10. Estrada, E., Higham, D.J.: Network properties revealed through matrix functions. SIAM Review 52(4), 696–714 (2010)
11. Farahat, A., LoFaro, T., Miller, J.C., Rae, G., Ward, L.A.: Authority rankings from HITS, PageRank, and SALSA: Existence, uniqueness, and effect of initialization. SIAM Journal on Scientific Computing 27(4), 1181–1201 (2006)
12. Gallopoulos, E., Saad, Y.: Efficient solution of parabolic equations by Krylov approximation methods. SIAM J. Sci. Stat. Comput. 13(5), 1236–1264 (1992)
13. Hochbruck, M., Lubich, C.: On Krylov subspace approximations to the matrix exponential operator. SIAM J. Numer. Anal. 34(5), 1911–1925 (1997)
14. Kondor, R.I., Lafferty, J.D.: Diffusion kernels on graphs and other discrete input spaces. In: ICML 2002, pp. 315–322 (2002)
15. Kunegis, J., Lommatzsch, A.: Learning spectral graph transformations for link prediction. In: Proceedings of the 26th Annual International Conference on Machine Learning, ICML 2009, pp. 561–568. ACM, New York (2009)
16. Luo, Z.Q., Tseng, P.: On the convergence of the coordinate descent method for convex differentiable minimization. J. Optim. Theory Appl. 72(1), 7–35 (1992)
17. Moler, C., Van Loan, C.: Nineteen dubious ways to compute the exponential of a matrix, twenty-five years later. SIAM Review 45(1), 3–49 (2003)
18. Newman, M.E.J.: The structure of scientific collaboration networks. Proceedings of the National Academy of Sciences 98(2), 404–409 (2001)
19. Newman, M.E.J.: Finding community structure in networks using the eigenvectors of matrices. Phys. Rev. E 74(3), 036104 (2006)
20. Orecchia, L., Sachdeva, S., Vishnoi, N.K.: Approximating the exponential, the Lanczos method and an $\tilde{O}(m)$-time spectral algorithm for balanced separator. In: STOC 2012, pp. 1141–1160 (2012)
21. Sidje, R.B.: ExpoKit: a software package for computing matrix exponentials. ACM Trans. Math. Softw. 24, 130–156 (1998)

Evolution of the Media Web*

Damien Lefortier[1], Liudmila Ostroumova[1,2], and Egor Samosvat[1,3]

[1] Yandex, Moscow, Russia
[2] Moscow State University, Moscow, Russia
[3] Moscow Institute of Physics and Technology, Moscow, Russia

Abstract. We present a detailed study of the part of the Web related to media content, i.e., the Media Web. Using publicly available data, we analyze the evolution of incoming and outgoing links from and to media pages. Based on our observations, we propose a new class of models for the appearance of new media content on the Web where different *attractiveness* functions of nodes are possible including ones taken from well-known preferential attachment and fitness models. We analyze these models theoretically and empirically and show which ones realistically predict both the incoming degree distribution and the so-called *recency property* of the Media Web, something that existing models did not capture well. Finally we compare these models by estimating the likelihood of the real-world link graph from our data set given each model and obtain that models we introduce are significantly more accurate than previously proposed ones. One of the most surprising results is that in the Media Web the probability for a post to be cited is determined, most likely, by its quality rather than by its current popularity.

Keywords: Media Web, random graph models, recency.

1 Introduction

Numerous models have been suggested to reflect and predict the growth of the Web [6,9,14]. The most well-known ones are preferential attachment models (see Section 2 for a more thorough discussion about previous work). One of the main drawbacks of these models is that they pay too much attention to old pages and do not realistically explain how links pointing to newly-created pages appear (as we discuss below). In this paper, we are interested in the Media Web, i.e., the highly dynamic part of the Web related to media content where a lot of new pages appear daily. We show that the Media Web has some specific properties and should therefore be analyzed separately. Note that some other parts of the Web have already been studied, for example in [17] a model for the Social Web is suggested.

Most new media pages like news and blog posts are popular only for a short period of time, i.e., such pages are mostly cited and visited for several days after they appeared [15]. We analyze this thoroughly later in the paper and introduce

* The authors are given in alphabetical order.

A. Bonato, M. Mitzenmacher, and P. Prałat (Eds.): WAW 2013, LNCS 8305, pp. 80–92, 2013.
© Springer International Publishing Switzerland 2013

a *recency property*, which reflects the fact that new media pages tend to connect to other media pages of similar age (see Section 3).

In this context, we propose a new class of models for the appearance of new media content on the Web where different *attractiveness* functions of nodes are possible including ones taken from well-known preferential attachment and fitness models, but also new ones accounting for specificities of the Media Web. We analyze these models theoretically and empirically using MemeTracker public data set [1] and show which ones realistically predict both the distribution of incoming degrees and the *recency property* of the Media Web, something that existing models did not capture well. Finally we compare these models by estimating the likelihood of the real-world link graph from this data set given each model and obtain that models we introduce in this paper are significantly more accurate than previously proposed ones. One of the most surprising results is that in the Media Web the probability for a post to be cited is determined, most likely, by its quality rather than by its current popularity.

The contributions of this paper are the following:

- We suggest a new class of models for the appearance of new media content on the Web where different *attractiveness* functions of nodes are possible;
- We analyze these models theoretically and empirically and show which ones realistically depict the behavior of the Media Web;
- We compare these models by estimating the likelihood of the real-world link graph from our data set given each model.

The rest of the paper is organized as follows. In Sections 2 and 3, we discuss related work and experimental results, which motivated this work. In Section 4, based on the results of our experiments, we define our class of models. We analyze theoretically some properties of these models in Section 5, while in Section 6 we validate our models by computing the likelihood of the real-world link graph from our data given each model.

2 Related Work

One of the first attempts to propose a realistic mathematical model of the Web growth was made in [3]. The main idea is to start with the assumption that new pages often link to old popular pages. Barabási and Albert defined a graph construction stochastic process, which is a Markov chain of graphs, governed by the *preferential attachment*. At each step in the process, a new node is added to the graph and is joined to m different nodes already existing in the graph that are chosen with probabilities proportional to their incoming degree (the measure of popularity). This model successfully explained some properties of the Web graph like its small diameter and power law distribution of incoming degrees. Later, many modifications to the Barabási–Albert model have been proposed, e.g., [11,13,12], in order to more accurately depict these but also other properties (see [2,7] for details).

It was noted by Bianconi and Barabási in [5] that in real networks some nodes are gaining new incoming links not only because of their incoming degree, but also because of their own intrinsic properties. For example, new Web pages containing some really popular content can acquire a large number of incoming links in a short period of time and become more popular than older pages. Motivated by this observation, Bianconi and Barabási extended preferential attachment models with pages' inherent quality or *fitness* of nodes. When a new node is added to the graph, it is joined to some already existing nodes that are chosen with probabilities proportional to the product of their fitness and incoming degree. This model was theoretically analyzed in [10].

In the context of our research, the main drawback of these models is that, as said, they pay too much attention to old pages and do not realistically explain how links pointing to newly-created pages appear. Note also that highly dynamic parts of the Web like social networks or weblogs exhibit a specific behavior and should therefore be modeled separately (see Section 3). In [17], the evolution of social networks, or the Social Web, was thoroughly investigated and, based on their results, a model was suggested. In turn, we suggest a model for the Media Web. The main idea is to combine preferential attachment and fitness models with a recency factor. This means that pages are gaining incoming links according to their *attractiveness*, which is determined by the incoming degree of the page, its *inherent popularity* (some page-specific constant) and age (new pages are gaining new links more rapidly).

3 Recency Property of the Media Web

In this section, we present experiments, which motivated us to propose a new model for the Media Web. Our model is based on these experimental results.

3.1 Experimental Setup

We use MemeTracker public data set [1], which covers 9 months of Media Web activity – quite a significant time period. Note that only outgoing links from the content part of the post were extracted (no toolbar, sidebar links). See [16] for details on how this data was collected.

From this data set we kept only links pointing to documents also in the data set, i.e., links with known timestamps both for the source and the destination. We assume that these timestamps correspond to the time when each document was posted on the Web, and we also filtered out links for which the timestamp of the destination is greater than for the source (impossible situation). This can happen because timestamps are noisy and therefore not always reliable. We finally obtained a data set of about 18M links and 6.5M documents that we use in the following experiments.

3.2 Recency Property

Let us define the *recency property* for a graph evolving in time. Denote by $e(T)$ the fraction of edges connecting nodes whose age difference is greater than T.

We analyze the behavior of $e(T)$ and show that media pages tend to connect to pages of similar age. We plotted $e(T)$ for our dataset and noted that $e(T)$ is decreasing exponentially fast (see Figure 1), which is not the case for preferential attachment model as we show later in this paper (Section 5.2).

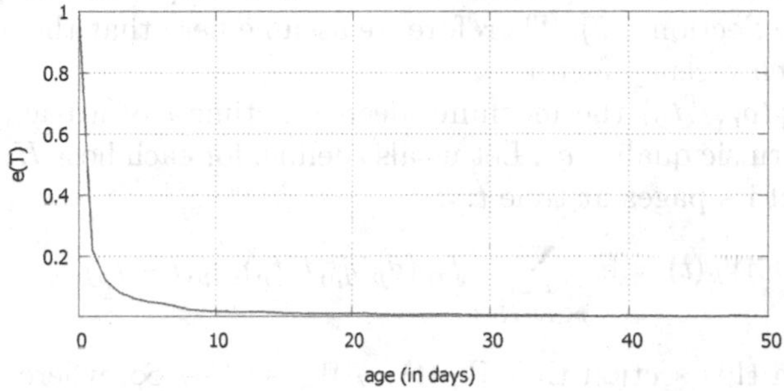

Fig. 1. The recency property

4 Model

Suppose that we have a fixed set of hosts $H_1, \dots, H_n$. Each host H_i has its own rate of new pages appearance λ_i. At the beginning of the process, we have no pages. We assume that new pages appear on a host H_i according to a Poisson process with parameter λ_i. A Poisson process is often used to model a sequence of random events that happen independently with a fixed rate over time.Poisson processes for different hosts are independent.

When a new page p is created on a host i, it has m_p mutually independent outgoing links pointing to already existing media pages. The target page of each link is determined as follows. First, the target host k is chosen with probability ρ_{ik} ($\sum_{k=1}^{n} \rho_{ik} = 1$). Then, the probability of each page r on the host k to be chosen is proportional to the *attractiveness* of r, which is some function of d_r (current incoming degree of r), q_r (intrinsic quality of r), and a_r (current age of r). Different attractiveness functions are possible:

$$f_{\tau_k}(d,q,a) = (1 \text{ or } q) \cdot (1 \text{ or } d) \cdot \left(1 \text{ or } e^{-\frac{a}{\tau_k}}\right).$$

Where τ_k corresponds to the mean lifetime of the decaying attractiveness for media pages on host k.

E.g., $f_{\tau_k}(d,q,a) = d$ leads to preferential attachment, while $f_{\tau_k}(d,q,a) = q \cdot d$ leads to fitness model. In this paper, we study different options and show which ones best depict the behavior of the Media Web.

Let us denote by $\Omega(H_i)$ the set of pages, which belong to a host H_i. We assume that the distributions of q_p and m_p for $p \in \Omega(H_i)$ are the properties of H_i. The only thing we assume about these distributions is that q_p and m_p have finite expectations.

5 Theoretical Analysis

5.1 Distribution of Incoming Degrees

In [5,8,11], models without *recency factor* (i.e., without the factor $e^{-\frac{a}{\tau_k}}$ in the attractiveness function) have been analyzed. On the contrary, in this paper we show that we need the recency factor to reflect some important properties of the Media Web (see Section 5.2). Therefore we assume here that the attractiveness function has such recency factor.

Denote by $d_p(q_p, t, t_p)$ the incoming degree at time t of a page p created at time t_p with intrinsic quality q_p. Let us also define, for each host H_k, the average attractiveness of its pages at time t:

$$W_k(t) = \mathbb{E} \sum_{p \in \Omega(H_k)} f_{\tau_k}(d_p(q_p, t, t_p), q_p, t - t_p)\,. \tag{1}$$

We will show in this section that $W_k(t) \to W_k$ as $t \to \infty$, where W_k are some positive constants.

Let M_k be the average number of outgoing links of pages $p \in \Omega(H_k)$. Then $N_k = \sum_i \lambda_i M_i \rho_{ik}$ is the average rate of new links pointing to host H_k appearance.

Theorem 1. *Let $p \in \Omega(S_k)$ be a page with quality q_p and time of creation t_p.*

(1) If $f_{\tau_k} = q \cdot d \cdot e^{-\frac{a}{\tau_k}}$, *then* $d_p(q_p, t, t_p) = e^{\frac{N_k \tau_k q_p}{W_k}\left(1 - e^{\frac{t_p - t}{\tau_k}}\right)}$,

(2) If $f_{\tau_k} = q \cdot e^{-\frac{a}{\tau_k}}$, *then* $d_p(q_p, t, t_p) = \frac{N_k \tau_k q_p}{W_k}\left(1 - e^{\frac{t_p - t}{\tau_k}}\right)$.

It follows from Theorem 1 that in the first case, in order to have a power law distribution of d_p, we need to have q_p distributed exponentially. In this case, for each host, the parameter of the power law distribution equals $\frac{N_k \tau_k \mu}{W_k}$, where μ is the parameter of exponential distribution. It is interesting to note that this latter parameter cannot affect the parameter of the power law distribution. Indeed, if we multiply μ by some constant, then W_k will also be multiplied by the same constant (see (1)). Therefore, we can change the parameter of the power law distribution only by varying N_k and τ_k. The problem is that the constant W_k depends on N_k and τ_k (see equation (3) in the proof). Hence, it is impossible to find analytical expressions for N_k and τ_k, which give us the desired parameter of the power law distribution.

In the second case, a power law distribution of q_p leads to a power law distribution of d_p with the same constant. Therefore, it is easy to get a realistic distribution of incoming degrees in this case.

In both cases, we cannot avoid the quality factor because if we do not have it in the attractiveness function (i.e., if q_p is constant for all media pages), then the solution does not depend on q_p and we do not have a power law for the distribution of incoming degrees.

To illustrate the results of Theorem 1, we generated graphs according to our model with different functions f_{τ_k}. Obtained results are shown on Figure 2.

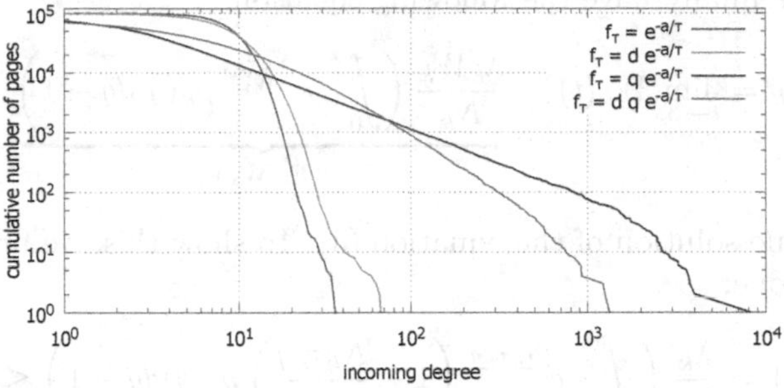

Fig. 2. Distribution of incoming degrees for each model

Proof. In mean-field approximation, we have the following differential equation:

$$\frac{\partial d_p(q_p, t, t_p)}{\partial t} = N_k \frac{f_{\tau_p}(d_p(q_p, t, t_p), q_p, t - t_p)}{W_k(t)},$$

here $p \in \Omega(H_k)$.

In the case $f_{\tau_k}(d, q, a) = q \cdot d \cdot e^{-\frac{a}{\tau_k}}$ we have:

$$\frac{\partial d_p(q_p, t, t_p)}{\partial t} = N_k \frac{q_p \cdot d_p(q_p, t, t_p) \cdot e^{-\frac{t-t_p}{\tau_k}}}{W_k(t)} \tag{2}$$

Later in this section, we show that for each k, $W_k(t)$ tends to some positive constant W_k: $\lim_{t\to\infty} W_k(t) = W_k$.

We thus have the following solution of the equation (2):

$$d_p = e^{\frac{N_k \tau_k q_p}{W_k}\left(1 - e^{\frac{t_p - t}{\tau_k}}\right)} \xrightarrow{t\to\infty} e^{\frac{N_k \tau_k q_p}{W_k}}$$

In case $f_{\tau_k}(d, q, a) = q \cdot e^{-\frac{a}{\tau_k}}$, by similar but even simpler calculations, we obtain:

$$d_p = \frac{N_k \tau_k q_p}{W_k}\left(1 - e^{\frac{t_p - t}{\tau_k}}\right) \xrightarrow{t\to\infty} \frac{N_k \tau_k q_p}{W_k}$$

Let us now check that $\lim_{t\to\infty} W_k(t)$ is indeed a constant. Consider the case $f_{\tau_k}(d, q, a) = q \cdot d \cdot e^{-\frac{a}{\tau_k}}$. Let $\rho_k(q)$ be the probability density function of q_p for $p \in \Omega(H_k)$. Therefore:

$$W_k(t) = \int_0^\infty \left(\int_0^t \lambda_k q \rho_k(q) d(q, t, x) \cdot e^{-\frac{t-x}{\tau_k}} dx \right) dq =$$

$$= \int_0^\infty \left(\int_0^t \lambda_k q \rho_k(q) e^{\frac{N_k \tau_k q}{W_k}\left(1 - e^{\frac{x-t}{\tau_k}}\right)} \cdot e^{\frac{x-t}{\tau_k}} dx \right) dq =$$

$$= \int_0^\infty \frac{\lambda_k W_k}{N_k} \left(e^{\frac{N_k \tau_k q}{S}\left(1 - e^{\frac{-t}{\tau_k}}\right)} - 1 \right) \rho_k(q) dq\,.$$

Thus for W_k we finally have the following equation:

$$W_k = \lim_{t\to\infty} W_k(t) = \underbrace{\frac{\lambda_k W_k}{N_k}\left(\int_0^\infty e^{\frac{N_k\tau_k q}{W_k}}\rho_k(q)dq - 1\right)}_{F_k(W_k)}. \tag{3}$$

There is a unique solution of the equation (3). To show this, we first check that $F_k(x)$ is monotone:

$$F_k'(x) = \frac{\lambda_k}{N_k}\left(\int_0^\infty e^{\frac{N_k\tau_k q}{x}}\left(1 - \frac{N_k\tau_k q}{x}\right)\rho_k(q)dq - 1\right) \le 0\,,$$

since:

$$e^{\frac{N_k\tau_k q}{x}}\left(1 - \frac{N_k\tau_k q}{x}\right) \le 1 \text{ and } \int_0^\infty \rho_k(q)dq = 1.$$

Also $F_k(x) \to \tau_k\lambda_k\mathbb{E}_{p\in\Omega(S_k)}q_p$ as $x \to \infty$ and $F_k(x) \to \infty$ as $x \to 0$. From these observations, it follows that $y = x$ and $y = F_k(x)$ have a unique intersection. In other words, $x = F_k(x)$ has a unique solution.

Similarly, we can show that $\lim_{t\to\infty} W_k(t) = W_k$ for the attractiveness function $f_{\tau_k} = q \cdot e^{-\frac{a}{\tau_k}}$.

5.2 Recency Property

In this section, we show that we need a recency factor $e^{-\frac{a}{\tau_k}}$ in the formula for the attractiveness function f_{τ_k}. We prove that because of the recency factor, the number of edges, which connect nodes with time difference greater than T decreases exponentially in T. We prove the following theorem.

Theorem 2. *For $f_{\tau_k} = q \cdot d \cdot e^{-\frac{a}{\tau_k}}$ or $f_{\tau_k} = q \cdot e^{-\frac{a}{\tau_k}}$ we have*

$$e(T) \sim \sum_k N_k C_k e^{\frac{-T}{\tau_k}}\,,$$

where C_k are some constants.

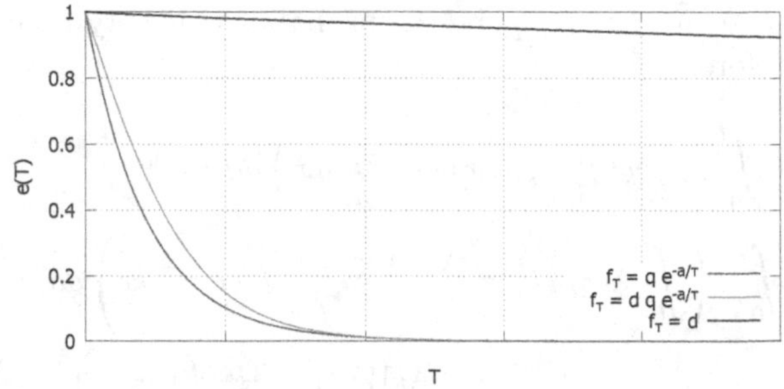

Fig. 3. Recency property in the model

Due to space constraints, we move the proof of Theorem 2 to Appendix. To illustrate the results obtained, we plot $e(T)$ for different attractiveness functions on Figure 3. Note that if we have a recency factor in the attractiveness function, then $e(T)$ approaches its upper bound exponentially fast. In contrast, if the attractiveness function equals d (preferential attachment), then $e(T)$ grows almost linearly with a small rate.

6 Validation

The idea of using Maximum Likelihood in order to compare different graph models and estimate their parameters was suggested in [4]. Since then this method was used for several models (see, e.g., [17,18]). Motivated by these works we also use the idea of Maximum Likelihood in order to compare new models we suggest in this paper with preferential attachment and fitness models.

6.1 Parameters Estimation

In order to do simulations, we first need to estimate all parameters of our models. Note that we are not trying to find the best parameters here. Instead we propose to use simple estimations, which are enough to show the improvements obtained by using our new models.

Host-to-Host Probabilities. We estimated the matrix ρ_{ij} by counting the fraction of edges going from hosts H_i to H_j. Note that 74% of all edges are host internal. We also add host to host probabilities to fitness and preferential attachment models and, as we show later in Section 6.2, this assumption allows to improve these models.

Estimation of τ. In order to estimate τ_k for each host H_k, we consider the histogram of age difference of connected pages. Let x_i ($i \geq 0$) be the number of links which connect pages with age difference greater than i but less than $i+1$ days. If we assume an exponential decay, then for $i < j$ we have $\frac{x_i}{x_j} = e^{\frac{(i-j)T}{\tau_k}}$, i.e., $\tau_k = \frac{(i-j)T}{\log \frac{x_i}{x_j}}$, where T is the time interval of one day. Therefore, we take:

$$\tau_k = \sum_{\substack{0 \leq i < j < 10: \\ x_i \neq 0, x_j \neq 0}} \frac{(i-j)T}{\binom{10}{2} \log \frac{x_i}{x_j}}.$$

We make a cut-off at 10 days because even though the tail of the histogram is heavier than exponential, the most important for us is to have a good estimation when pages are young, i.e. when most incoming links appear.

Estimation of Quality. Given the final incoming degree d of a node, we can use Theorem 1 to find its quality, i.e., we have $q = \frac{Wd}{N_k \tau_k}$ in the case of $f_\tau = qe^{\frac{-a}{T}}$ and $q = \frac{W \ln d}{N_k \tau_k}$ in the case of $f_\tau = dqe^{\frac{-a}{T}}$. Note that the factor $\frac{W}{N_k \tau_k}$ is common for all pages created on host H_k and can be cancelled so we finally used the following estimations: $q = d$ and $q = \ln d$ respectively.

6.2 Likelihood

In order to validate our model, we propose to use the data described in Section 3.1 and estimate the likelihood of the real-world link graph from this data set given each model discussed in this paper. We do this as follows.

We add edges one by one according to their real temporal order and compute their probability given the model under consideration. The sum of logarithms of all obtained probabilities gives us the log-likelihood of our graph. We normalize this sum by the number of edges and obtained results are presented in Table 1.

Table 1. Log-likelihood table: average logarithm of edge probability

d	q	$e^{\frac{-a}{T}}$	dq	$de^{\frac{-a}{T}}$	$qe^{\frac{-a}{T}}$	$dqe^{\frac{-a}{T}}$
-6.11	-5.56	-5.34	-6.08	-5.50	**-5.17**	-5.45

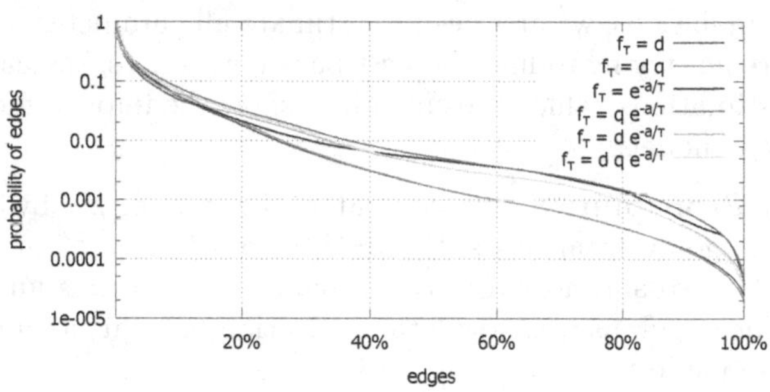

Fig. 4. Distribution of edges' probabilities

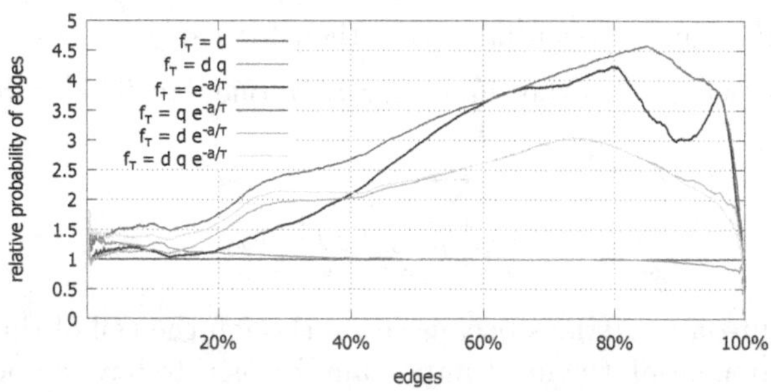

Fig. 5. Distribution of edges' probabilities relative to preferential attachment model

We see that the most likely model here is with $f_T = qe^{\frac{-a}{T}}$. However, since timestamps are noisy and therefore not always reliable (see Section 3.1), these results might not be representative (for example, if the probability of one edge is very small, it can heavily affect the final likelihood). Hence, in addition to this log-likelihood, which is strongly affected by outliers, we also performed the analysis of edges' probabilities, i.e. we try to understand which model is better on a per-edge basis. We believe that such deeper analysis allows to reduce the influence of outliers when validating our models. To the best of our knowledge, this is the first time such analysis is made when using Maximum Likelihood in order to compare different graph models.

Each edge has different probabilities according to different models and there is one model M for which this probability is the largest. In this case, we say that the model M *wins* on this edge (see Table 2). Also, for each pair of models M_1 and M_2, we computed the percentage of edges which have greater probability according to M_1 than according to M_2 (see Table 3). It can be clearly seen from both tables that the recency factor plays a very important role.

Table 2. Winner table: fraction of edges on which model wins all others

d	q	$e^{\frac{-a}{\tau}}$	dq	$de^{\frac{-a}{\tau}}$	$qe^{\frac{-a}{\tau}}$	$dqe^{\frac{-a}{\tau}}$
0.03	0.07	0.28	0.07	0.07	**0.30**	0.16

Table 3. Competition table: the value in (a,b) is the fraction of edges where a wins b

	d	q	$e^{\frac{-a}{\tau}}$	dq	$de^{\frac{-a}{\tau}}$	$qe^{\frac{-a}{\tau}}$	$dqe^{\frac{-a}{\tau}}$
d	-	0.22	0.30	0.43	0.18	0.22	0.19
q	0.78	-	0.38	0.76	0.41	0.23	0.40
$e^{\frac{-a}{\tau}}$	0.70	0.62	-	0.69	0.54	0.40	0.53
dq	0.57	0.24	0.31	-	0.24	0.23	0.17
$de^{\frac{-a}{\tau}}$	0.82	0.59	0.44	0.76	-	0.39	0.43
$qe^{\frac{-a}{\tau}}$	0.78	0.77	0.60	0.77	0.61	-	0.62
$dqe^{\frac{-a}{\tau}}$	0.81	0.60	0.47	0.83	0.57	0.38	-

Then, for each model, we sorted edges' probabilities in decreasing order on Figure 4. Furthermore, in order to more clearly visualize the differences between models, we normalized the probability of each edge in all models by dividing it by the corresponding probability in the sorted order of the preferential attachment model (see Figure 5). One can see that the model with $f_\tau = qe^{\frac{-a}{\tau}}$ again shows the best result in our tests. This means that in the Media Web the probability for a post to be cited is determined, most likely, by its quality rather than by its current popularity (i.e., incoming degree). Finally, the importance of host-to-host probabilities ρ_{ij} can be illustrated by Figure 6.

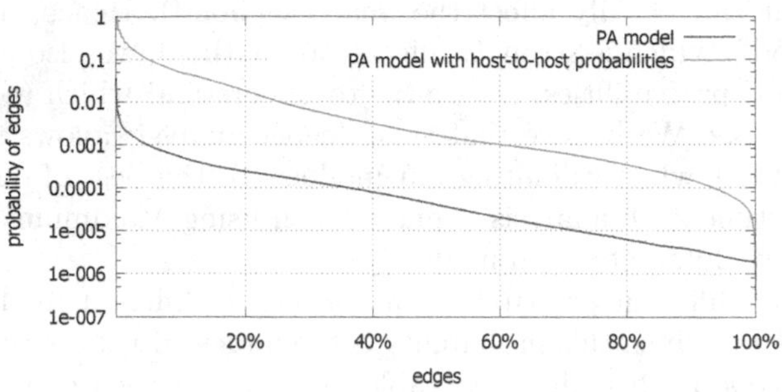

Fig. 6. The influence of host-to-host probabilities, e.g., on preferential attachment model (PA)

7 Conclusion

In this paper, we presented a detailed study of the Media Web. We proposed a new class of models for the appearance of new media content on the Web where different attractiveness functions of nodes are possible including ones taken from well-known preferential attachment and fitness models, but also new ones accounting for specificities of this part of the Web. Our new models are based on the observation that media pages tend to connect with other media pages of similar age.

We analyzed these models theoretically and empirically using publicly available data and show which ones realistically predict both the distribution of incoming degrees and the so-called *recency property* of the Media Web, something that existing models did not capture well.

Finally we compared these models by estimating the likelihood of the real-world link graph from our data set given each model and obtained that new models we introduce, with a recency factor, are significantly more accurate than previously proposed ones. One of the most surprising results is that in the Media Web the probability for a post to be cited is determined, most likely, by its quality rather than by its current popularity.

References

1. `http://www.memetracker.org/data.html`
2. Albert, R., Barabási, A.L.: Statistical mechanics of complex networks. Reviews of Modern Physics 74, 47–97 (2002)
3. Barabási, A.L., Albert, R.: Emergence of scaling in random network. Science 286(5439), 509–512 (1999)
4. Bezáková, I., Kalai, A., Santhanam, R.: Graph model selection using maximum likelihood. In: Proceedings of the 23rd International Conference on Machine Learning, ICML, pp. 105–112 (2006)
5. Bianconi, G., Barabási, A.L.: Bose-Einstein condensation in complex networks. Physical Review Letters 86(24), 5632–5635 (2001)

6. Boccaletti, S., Latora, V., Moreno, Y., Chavez, M., Hwang, D.U.: Complex networks: structure and dynamics. Physics Reports 424(45), 175–308 (2006)
7. Bollobás, B.: Mathematical results on scale-free random graphs. In: Handbook of Graphs and Networks, pp. 1–34 (2003)
8. Bollobás, B., Riordan, O., Spencer, J., Tusnády, G.: The degree sequence of a scale-free random graph process. Random Structures and Algorithms 18(3), 279–290 (2001)
9. Bonato, A.: A Survey of models of the web graph. In: López-Ortiz, A., Hamel, A.M. (eds.) CAAN 2004. LNCS, vol. 3405, pp. 159–172. Springer, Heidelberg (2005)
10. Borgs, C., Chayes, J., Daskalakis, C., Roch, S.: First to market is not everything: an analysis of preferential attachment with fitness. In: Proceedings of the Thirty-Ninth Annual ACM Symposium on Theory of Computing, pp. 135–144 (2007)
11. Buckley, P.G., Osthus, D.: Popularity based random graph models leading to a scale-free degree sequence. Discrete Mathematics 282(1-3), 53–68 (2004)
12. Cooper, C., Frieze, A.: A general model of web graphs. Random Structures and Algorithms 22(3), 311–335 (2003)
13. Holme, P., Kim, B.: Growing scale-free networks with tunable clustering. Physical Review E 65(2) (2002)
14. Kumar, R., Raghavan, P., Rajagopalan, S., Sivakumar, D., Tomkins, A., Upfal, E.: Web as a graph. In: Proceedings of the Nineteenth ACM SIGMOD-SIGACT-SIGART Symposium on Principles of Database Systems, pp. 1–10 (2000)
15. Lefortier, D., Ostroumova, L., Samosvat, E., Serdyukov, P.: Timely crawling of high-quality ephemeral new content. arXiv preprint arXiv:1307.6080 (2013)
16. Leskovec, J., Backstrom, L., Kleinberg, J.: Meme-tracking and the dynamics of the news cycle, pp. 497–506 (2009)
17. Leskovec, J., Backstrom, L., Kumar, R., Tomkins, A.: Microscopic evolution of social networks. In: Proceedings of the 14th ACM SIGKDD International Conference on Knowledge Discovery and Data Mining, pp. 462–470 (2008)
18. Leskovec, J., Chakrabarti, D., Kleinberg, J., Faloutsos, C., Ghahramani, Z.: Kronecker Graphs: An Approach to Modeling Networks. The Journal of Machine Learning Research 11, 985–1042 (2010)

Appendix: Proof of Theorem 2

To analyze the behavior of $e(T)$, we need to estimate the average attractiveness of all media pages created in the last T seconds at time t at a host k:

$$\mathcal{W}_k(T,t) = \mathbb{E} \sum_{\substack{p \in \Omega(H_k) \\ |t-t_p|<T}} f_{\tau_k}(d_p(q_p, t, t_p), q_p, t - t_p).$$

We will show that if $t > T$, then this function does not depend on t.

We can analyze the function $\mathcal{W}_k(T,t)$ using the technique we used in Section 5.1. Consider the case $f_{\tau_k}(d,q,a) = q \cdot d \cdot e^{-\frac{a}{\tau_k}}$:

$$\mathcal{W}_k(T,t) =$$

$$= \int_0^\infty \left(\int_{t-T}^t \lambda_k q \rho_k(q) d(q,t,x) \cdot e^{-\frac{t-x}{\tau_k}} dx \right) dq =$$

$$= \int_0^\infty \left(\int_{t-T}^t \lambda_k q \rho_k(q) e^{\frac{N_k \tau_k q}{W_k}\left(1 - e^{\frac{x-t}{\tau_k}}\right)} \cdot e^{\frac{x-t}{\tau_k}} dx \right) dq =$$

$$= \frac{\lambda_k W_k}{N_k} \int_0^\infty \left(e^{\frac{N_k \tau_k q}{W_k}\left(1 - e^{\frac{-T}{\tau_k}}\right)} - 1 \right) \rho_k(q) dq \,.$$

We proved that $\mathcal{W}_k(T,t)$ does not depend on t and will use the notation $\mathcal{W}_k(T) = \mathcal{W}_k(T,t)$ from now on. Also

$$W_k - \mathcal{W}_k(T) =$$

$$= \frac{\lambda_k W_k}{N_k} \int_0^\infty \left(1 - e^{-\frac{N_k \tau_k q}{W_k} e^{\frac{-T}{\tau_k}}} \right) e^{\frac{N_k \tau_k q}{W_k}} \rho_k(q) dq \sim$$

$$\sim \frac{\lambda_k W_k}{N_k} e^{\frac{-T}{\tau_k}} \int_0^\infty \frac{N_k \tau_k q}{W_k} e^{\frac{N \tau_k q}{W}} \rho_k(q) dq \sim C_k e^{\frac{-T}{\tau_k}},$$

where the constants C_k do not depend on T.

Note that the portion of links which point to the host H_k and have the age difference less than T is $\frac{W_k - \mathcal{W}_k(T)}{W_k}$. Thus, using N_k, which is the average rate of new links pointing to host H_k appearance (see Section 5.1), we can write the following equation for $e(T)$:

$$e(T) = \sum_k N_k \frac{W_k - \mathcal{W}_k(T)}{W_k} \sim \sum_k \frac{N_k C_k}{W_k} e^{\frac{-T}{\tau_k}}$$

The same analysis can be made for the case $f_{\tau_k}(d,q,a) = q \cdot e^{-\frac{a}{\tau_k}}$. In this case we get:

$$\mathcal{W}_k(T) = \int_0^\infty \left(\int_{t-T}^t \lambda_k q \rho_k(q) \cdot e^{-\frac{t-x}{\tau_k}} dx \right) dq = \lambda_k \tau_k \left(1 - e^{-\frac{T}{\tau_k}} \right) \mathbb{E}_{p \in \Omega(H_k)} q_p \,,$$

and further reasonings are the same.

Random Intersection Graph Process

Mindaugas Bloznelis[1,*] and Michał Karoński[2]

[1] Faculty of Mathematics and Informatics,
Vilnius University, 03225 Vilnius, Lithuania
mindaugas.bloznelis@mif.vu.lt
[2] Faculty of Mathematics and Computer Science,
Adam Mickiewicz University, 60769 Poznań, Poland

Abstract. We introduce a random intersection graph process aimed at modeling sparse evolving affiliation networks. We establish the asymptotic degree distribution and provide explicit asymptotic formulas for assortativity and clustering coefficients showing how these edge dependence characteristics vary over time.

Keywords: random graph process, random intersection graph, degree distribution, power law, clustering, assortativity.

1 Introduction

Given non-negative weights $x = \{x_i\}_{i \geq 1}$ and $y = \{y_j\}_{j \geq 1}$, and a non-decreasing positive sequence $\{\tau(t)\}_{t \geq 1}$, satisfying $\lim_{t \to +\infty} \tau(t) = +\infty$, let $H_{x,y}$ be the random bipartite graph with bipartition $V = \{v_1, v_2, \dots\}$ and $W = \{w_1, w_2, \dots\}$, where edges $\{w_i, v_j\}$ are inserted independently and with probabilities

$$p_{ij} = \min\Big\{1, \frac{x_i y_j}{\sqrt{ij}}\Big\} \mathbb{I}_{\{a\tau(j) \leq i \leq b\tau(j)\}}. \tag{1}$$

Here $b > a > 0$ are fixed numbers. $H_{x,y}$ defines the graph $G_{x,y}$ on the vertex set V such that any $u, v \in V$ are declared adjacent (denoted $u \sim v$) whenever they have a common neighbor in $H_{x,y}$.

Consider, for example, a library where items $w_1, w_2, \dots$ are acquired one after another, and where users $v_1, v_2, \dots, v_j, \dots$ are registered at times $j = 1, 2, \dots$. User v_j picks at random items from the "contemporary literature collection" $[a\tau(j), b\tau(j)] = \{w_i : a\tau(j) \leq i \leq b\tau(j)\}$ relevant to its arrival time j. An item w_i is picked with probability proportional to the activity y_j of the user v_j and the attractiveness x_i of the item w_i, cf. (1). The realized bipartite graph $H_{x,y}$ represents "library" records, whereas realized $G_{x,y}$ represents adjacency relations between users in the resulting affiliation network.

Theoretical analysis of such a network becomes simpler if we impose some regularity conditions on the weight sequences x and y. A convenient assumption is that x and y are realized values of iid sequences $X = \{X_i\}_{i \geq 1}$ and $Y = \{Y_j\}_{j \geq 1}$. In this

* Corresponding author.

A. Bonato, M. Mitzenmacher, and P. Prałat (Eds.): WAW 2013, LNCS 8305, pp. 93–105, 2013.
© Springer International Publishing Switzerland 2013

way we obtain random graphs $H_{X,Y}$ and $G_{X,Y}$. The parameters of the model are the probability distributions of X_1, Y_1, function τ (the speed of item acquisition), and cut-offs $a < b$, which together with $\tau(j)$ determine the interval $[a\tau(j), b\tau(j)]$, which we also interpret as the lifespan of a user v_j.

Random graph $G_{X,Y}$ is aimed at modeling sparse affiliation networks that evolve in time, such as, e.g., the actor network, where two actors are declared adjacent whenever they have played in the same movie, a collaboration network where two authors are declared adjacent if they have co-authored a paper. Such networks display a natural bipartite structure: actors are linked to films, authors are linked to papers. An underlying bipartite structure seems to be present in many social networks: members of a network become acquainted because they share some common interests, [11]. Furthermore, the bipartite structure is helpful in explaining an important feature of social networks that they have a non-trivial clustering (network transitivity), and a positive correlation between degrees of adjacent vertices (network assortativity), [16], [19].

We note that each vertex of the graph $G_{X,Y}$ can be identified with the random subset of W, consisting of items selected by that vertex, and two vertices are adjacent in $G_{X,Y}$ whenever their subsets intersect. Graphs describing such adjacency relations between members of a *finite* family $V_n = \{v_1, \ldots, v_n\}$ of random subsets of a given *finite* set $W_m = \{w_1, \ldots, w_m\}$ have been introduced in [13], see also [9]. They are called random intersection graphs. We remark that random intersection graphs reproduce empirically observed clustering properties of the actor network with remarkable accuracy ([2], [5]). Unfortunately, they do not account for the evolving nature of the network (actors acting in 2013 are unlikely to be adjacent to those that acted in 1913) and, therefore, can not explain how various characteristic of an evolving network varies over time. This drawback of the static model has motivated our interest in evolving random intersection graph $G_{X,Y}$. A related empirical study showing how various characteristics of an evolving collaboration network had changed over a hundred year period is presented in [14], see also [15].

The random graph $G_{X,Y}$ can be considered as a random process evolving in time, where the vertex v_t that arrived at time t can only establish adjacency relations with *contemporaries* v_s such that the intervals $[a\tau(t), b\tau(t)]$ and $[a\tau(s), b\tau(s)]$ (lifespans of v_t and v_s) intersect. We show that the random intersection graph process admits an asymptotic power law distribution of the degree $d(v_t)$ of a vertex v_t as $t \to +\infty$. Moreover, we give an explicit description of the asymptotic degree distribution. Furthermore, we show that $G_{X,Y}$ admits a non-trivial *clustering* and *assortativity* coefficients and calculate their first order asymptotics.

The intuition behind these results is explained as follows. We first observe that choosing inhomogeneous weight sequences x and y one may expect to obtain an inhomogeneous degree sequence of the graph $G_{x,y}$: vertices with larger weights attract larger numbers of neighbors. Consequently, in the case where the probability distributions of X_1 and Y_1 have heavy tails, we obtain a heavy tailed asymptotic degree distribution. Secondly, we observe that if the set $W(t)$ of items selected by a user v_t is (stochastically) bounded and the lifespans of two neighbors of v_t, say v_s and v_u, intersect, then with a non-vanishing probability v_s and v_u share an item from $W(t)$. Consequently, the conditional probability $\alpha_{t|su} = \mathbf{P}(v_s \sim v_u | v_s \sim v_t, v_t \sim v_u)$, called the clustering coefficient, is positive and bounded away from zero. In particular, the underlying bipartite graph structure serves as a clustering mechanism. A similar argument applies to the assortativity coefficient (Pearson's correlation coefficient between degrees of adjacent vertices)

$$r_{s,t} = \frac{\mathbf{E}_{st}d(v_s)d(v_t) - \mathbf{E}_{st}d(v_s)\mathbf{E}_{st}d_v(t)}{\sqrt{\mathbf{Var}_{st}d(v_s)\mathbf{Var}_{st}d(v_t)}}. \tag{2}$$

Here an item shared by adjacent vertices $v_s \sim v_t$ attracts a number of common neighbors of v_s and v_t. This makes the correlation coefficient positive and bounded away from zero. Here $\mathbf{E}_{st}$ denotes the conditional expectation given the event $v_s \sim v_t$ and $\mathbf{Var}_{st}d(v_s) = \mathbf{E}_{st}d^2(v_s) - (\mathbf{E}_{st}d_v(s))^2$.

The rest of the paper is organized as follows. Results are stated in Sect. 2. Proofs are given in Sect. 3.

2 Results

Degree. For $\tau(t)$ growing linearly in t we obtain a compound probability distribution in the limit. We remark that in this case $G_{X,Y}$ admits non-trivial asymptotic clustering coefficients. For $\tau(t)$ growing faster than linearly in t, we obtain a mixed Poisson asymptotic degree distribution. In this case clustering coefficients vanish. We denote $a_k = \mathbf{E}X_1^k$ and $b_k = \mathbf{E}Y_1^k$.

Theorem 1. *Let $b > a > 0$. Let $\tau(t) = t$. Suppose that $\mathbf{E}X_1^2 < \infty$ and $\mathbf{E}Y_1 < \infty$. For $t \to +\infty$ the random variable $d(v_t)$ converges in distribution to the random variable*

$$d_* = \sum_{j=1}^{\Lambda_1} \varkappa_j, \tag{3}$$

where $\varkappa_1, \varkappa_2, \ldots$ are independent and identically distributed random variables independent of the random variable Λ_1. They are distributed as follows. For $r = 0, 1, 2, \ldots$, we have

$$\mathbf{P}(\varkappa_1 = r) = \frac{r+1}{\mathbf{E}\Lambda_2}\mathbf{P}(\Lambda_2 = r+1), \qquad \mathbf{P}(\Lambda_i = r) = \mathbf{E}\, e^{-\lambda_i}\frac{\lambda_i^r}{r!}, \qquad i = 1, 2. \tag{4}$$

Here $\lambda_1 = 2(b^{1/2} - a^{1/2})a_1 Y_1$ and $\lambda_2 = 2(a^{-1/2} - b^{-1/2})b_1 X_1$.

We remark that the second moment condition $\mathbf{E}X_1^2 < \infty$ of Theorem 1 seems to be redundant and could perhaps be waived.

Theorem 2. *Let $b > a > 0$ and $\nu > 1$. Let $\tau(t) = t^\nu$, $t = 1, 2 \ldots$. Suppose that $\mathbf{E}X_1^2 < \infty$ and $\mathbf{E}Y_1 < \infty$. For $t \to +\infty$ the random variable $d(v_t)$ converges in distribution to the random variable Λ_3 having the probability distribution*

$$\mathbf{P}(\Lambda_3 = r) = \mathbf{E}\, e^{-\lambda_3}\frac{\lambda_3^r}{r!}, \qquad r = 0, 1, 2, \ldots. \tag{5}$$

Here $\lambda_3 = \gamma a_2 b_1 Y_1$ and $\gamma = 4\nu(b^{1/2\nu} - a^{1/2\nu})(a^{-1/2\nu} - b^{-1/2\nu})$.

Remark 1. The probability distributions Λ_i, $i = 1, 2, 3$, are Poisson mixtures. One way to sample from the distribution of Λ_i is to generate random variable λ_i and then, given λ_i, to generate Poisson random variable with the parameter λ_i. In Theorem 1 we obtain a power law asymptotic degree distribution provided that the heavier of the tails $t \to \mathbf{P}(\varkappa_1 > t)$ and $t \to \mathbf{P}(\Lambda_1 > t)$ has a power law, see [8]. In Theorem 2 we obtain a power law asymptotic degree distribution provided that Y_1 has a power law.

Remark 2. The result of Theorem 2 extends to a more general class of increasing non-negative functions τ. In particular, assuming that

$$\lim_{t\to+\infty} \frac{t}{\tau(t)} = 0, \qquad \sup_{t>1} \frac{\tau^{-1}(2t)}{\tau^{-1}(t)} < \infty, \tag{6}$$

and that there exists a finite limit

$$\gamma^* = \lim_{t\to+\infty} t^{-1/2} \sum_{a\tau(t)\le i\le b\tau(i)} i^{-1} \sum_{j:\, a\tau(j)\le i\le b\tau(j)} j^{-1/2},$$

we obtain the convergence in distribution of $d(v_t)$ to Λ_3 defined by (5) with $\lambda_3 = \gamma^* a_2 b_1 Y_1$. Here τ^{-1} denotes the inverse of τ (i.e., $\tau(\tau^{-1}(t)) = t$).

Remark 3. The function $\tau(t) = t \ln t$, which grows slower than any power t^ν, $\nu > 1$, satisfies conditions of Remark 2 with $\gamma^* = 4(a^{-1/2} - b^{-1/2})(b^{1/2} - a^{1/2})$. Furthermore, the functions $\tau_1(t) = e^{\ln^2 t}$ and $\tau_2(t) = e^t$, that grow faster than any power t^ν, satisfy conditions of Remark 2 with $\gamma^* = 0$ and now the asymptotic degree distribution is degenerate.

Clustering. Our next result, Theorem 3, provides explicit asymptotic formulas for clustering coefficients. We note that for $s < t < u$ the conditional probabilities $\alpha_{s|tu}$, $\alpha_{t|su}$ and $\alpha_{u|st}$ are all different and, given $0 < a < b$, they mainly depend on the ratios s/t, s/u and t/u. Denote $p_\Delta := p_\Delta(s, t, u) = \mathbf{P}(v_s \sim v_t, v_s \sim v_u, v_t \sim v_u)$ the probability that v_s, v_t, v_u make up a triangle.

Theorem 3. *Let $b > a > 0$. Let $\tau(t) = t$. Suppose that $\mathbf{E}X_1^3 < \infty$ and $\mathbf{E}Y_1^2 < \infty$. Assume that $s, t, u \to +\infty$ so that $s < t < u$ and $\lceil au \rceil \le \lfloor bs \rfloor$. We have*

$$p_\Delta = \frac{a_3 b_1^3}{\sqrt{stu}} \left(\frac{2}{\sqrt{au}} - \frac{2}{\sqrt{bs}} \right) + o(t^{-2}), \tag{7}$$

$$\alpha_{t|su} = \frac{p_\Delta}{p_\Delta + a_2^2 b_1^2 b_2 t^{-1} (su)^{-1/2} \delta_{t|su}} + o(1), \tag{8}$$

$$\alpha_{s|tu} = \frac{p_\Delta}{p_\Delta + a_2^2 b_1^2 b_2 s^{-1} (tu)^{-1/2} \delta_{s|tu}} + o(1), \tag{9}$$

$$\alpha_{u|st} = \frac{p_\Delta}{p_\Delta + a_2^2 b_1^2 b_2 u^{-1} (st)^{-1/2} \delta_{u|st}} + o(1). \tag{10}$$

Here we denote $a_i = \mathbf{E}X_1^i$, $i = 2, 3$, *and* $b_j = \mathbf{E}Y_1^j$, $j = 1, 2$, *and*

$$\delta_{t|su} = \ln(u/t)\ln(t/s) + \ln(u/t)\ln(bs/au) + \ln(t/s)\ln(bs/au) + \ln^2(bs/au),$$
$$\delta_{s|tu} = \ln(u/t)\ln(bs/au) + \ln^2(bs/au),$$
$$\delta_{u|st} = \ln(t/s)\ln(bs/au) + \ln^2(bs/au).$$

We remark that the condition $\lceil au \rceil \le \lfloor bs \rfloor$ of Theorem 3 excludes the trivial case where $p_\Delta \equiv 0$. Indeed, for $s < u$, the converse inequality $\lceil au \rceil > \lfloor bs \rfloor$ means that the lifetimes of v_s and v_u do not intersect and, therefore, we have $\mathbf{P}(v_s \sim v_u) \equiv 0$. In addition, the inequality $\lceil au \rceil \le \lfloor bs \rfloor$ implies that positive numbers $\delta_{t|su}, \delta_{s|tu}, \delta_{u|st}$ are bounded from above by a constant (only depending on a and b).

Assortativity. Let us consider the sequence of random variables $\{d(v_t)\}_{t\ge 1}$. We assume that $\tau(t) = t$. From Theorem 1 we know about the possible limiting distributions for $d(v_t)$. Moreover, from the fact that $G_{X,Y}$ is sparse we can conclude that, for any given k, the random variables $d(v_t), d(v_{t+1}), \dots, d(v_{t+k})$ are asymptotically independent as $t \to +\infty$. An interesting question is about the statistical dependence between $d(v_s)$ and $d(v_t)$ if we know, in addition, that vertices v_s and v_t are adjacent in $G_{X,Y}$. We assume that $s < t$ and let $s, t \to +\infty$ so that $bs - at \to +\infty$. Note that the latter condition ensures that the shared lifetime of v_s and v_t tends to infinity as $s, t \to +\infty$. In this case we obtain that conditional moments

$$\begin{aligned} \mathbf{E}_{st}d(v_s) &= \mathbf{E}_{st}d(v_t) + o(1) = \delta_1 + o(1), \\ \mathbf{E}_{st}d^2(v_s) &= \mathbf{E}_{st}d^2(v_t) + o(1) = \delta_2 + o(1), \\ \mathbf{E}_{st}d(v_s)d(v_t) &= \delta_2 - \Delta + o(1) \end{aligned} \tag{11}$$

are asymptotically constant. Here $\Delta = h_1^{-1}(2h_3 + 2h_5 + 4(h_6 - h_7))$ and

$$\delta_1 = 1 + h_1^{-1}(h_2 + 2h_3), \qquad \delta_2 = 1 + h_1^{-1}(3h_2 + 6h_3 + h_4 + 6h_5 + 4h_6).$$

Furthermore, we denote

$$\begin{aligned} &h_1 = a_2 b_1^2, \qquad h_2 = a_3 b_1^3 \tilde{\gamma}, \qquad h_3 = a_2^2 b_1^2 b_2 \tilde{\gamma}(\sqrt{b} - \sqrt{a}), \\ &h_4 = a_4 b_1^4 \tilde{\gamma}^2, \qquad h_5 = a_2 a_3 b_1^3 b_2 \tilde{\gamma}^2(\sqrt{b} - \sqrt{a}), \\ &h_6 = a_2^3 b_1^3 b_3 \tilde{\gamma}^2(\sqrt{b} - \sqrt{a})^2, \qquad h_7 = a_2^3 b_1^2 b_2^2 \tilde{\gamma}^2(\sqrt{b} - \sqrt{a})^2, \end{aligned} \tag{12}$$

and $\tilde{\gamma} = 2(a^{-1/2} - b^{-1/2})$. A sketch of the proof of (11) is given in Sect. 3.

From (2) and (11) we obtain that the assortativity coefficient

$$r_{st} = 1 - \frac{\Delta}{\delta_2 - \delta_1^2} + o(1) \tag{13}$$

is asymptotically constant.

Finally, we mention that the degree distribution of the typical vertex of *finite* random intersection graphs have been studied by several authors, see, e.g., [2], [7], [12], [17], the clustering properties have been studied in [2], [4] [7], [10]. Assortativity coefficient has been evaluated in [4].

Future Work. An interesting problem were to study the component structure of $G_{X,Y}$ and that of the subgraph induced by the users contemporary to some v_t for large t. Furthermore, in the construction of $G_{X,Y}$ one can replace the deterministic cut-offs $a < b$ in (1) by random cut-offs $A_j \leq B_j$ (so that the lifespan $[A_j\tau(j), B_j\tau(j)]$ of a user v_j were random). Similarly, abrupt cut-offs can be replaced by some smooth cut-off functions. Another interesting question were about incorporation of the preferred attachment principle, e.g., by increasing the weight (attractiveness) of an item proportionally to the number of users that have chosen this item, similarly to the preferred attachment model [1], where the area of the influence sphere depends on the in-degree of a node.

3 Proofs

Here we give a sketch of the proofs of Theorems 1 and 2. Details can be found in the extended version of the paper [6]. In the proof of Theorems 1, 2 we apply the approach used in [3].

We start with some notation. The intervals

$$T_t = \{k : a\tau(t) \leq k \leq b\tau(t)\}, \qquad T_k^* = \{j : a\tau(j) \leq k \leq b\tau(j)\} \qquad (14)$$

can be interpreted as lifetimes of the user v_t and item w_k, respectively. The event "edge $\{w_k, v_j\}$ is present in $H_{X,Y}$" is denoted $w_k \to v_j$. Introduce random variables

$$\mathbb{I}_{kj} = \mathbb{I}_{\{w_k \to v_j\}}, \qquad u_k = \sum_{j \in T_k^* \setminus \{t\}} \mathbb{I}_{kj}, \qquad L = \sum_{k \in T_t} \mathbb{I}_{kt} u_k, \qquad \lambda_{kt} = X_k Y_t / \sqrt{kt}.$$

Proof (of Theorem 1). We first approximate the random variable $d(v_t)$ by L, using the simple bound $\mathbf{P}(d(v_t) \neq L) = o(1)$. Then we approximate L by $L_1 = \sum_{k \in T_t} \eta_k \xi_k$, where η_k, ξ_k, $k \in T_t$, are conditionally independent, given X, Y, Poisson random variables with the conditional mean values

$$\mathbf{E}(\eta_k | X, Y) = \lambda_{kt}, \qquad \mathbf{E}(\xi_k | X, Y) = 2(a^{-1/2} - b^{-1/2}) X_k b_1, \qquad k \in T_t. \quad (15)$$

We note that L_1 is obtained from L, by replacing $\mathbb{I}_{kt}$ and u_k by the random variables η_k and ξ_k, respectively. In order to show that the error of this replacement is negligible we apply the standard Poisson approximation error bound (see, e.g., [18]) conditionally, given X and Y.

Finally, we prove that L_1 converges in distribution to d_*. Let $Y_\star$ be a random variable with the same distribution as Y_1 and independent of X, Y. Given $X, Y, Y_\star$, we generate independent Poisson random variables $\eta_k^\star$, $k \in T_t$ with (conditional) mean values $\mathbf{E}(\eta_k^\star | X, Y, Y_\star) = \lambda_{k\star}$, where $\lambda_{k\star} = X_k Y_\star (kt)^{-1/2}$. We assume that, given $X, Y, Y_\star$, the family of random variables $\{\eta_k^\star,\ k \in T_t\}$ is conditionally independent of $\{\xi_k,\ k \in T_t\}$. Define $L_\star = \sum_{k \in T_t} \eta_k^\star \xi_k$, and let $d_\star$ be defined in the same way as d_*, but with λ_1 replaced by $\lambda_\star = 2(b^{1/2} - a^{1/2}) a_1 Y_\star$. We note that L_1 has the same distribution as $L_\star$. Similarly, d_* has the same distribution as $d_\star$. Therefore, it suffices to show that $L_\star$ converges in distribution to

$d_\star$ and for this purpose we prove the convergence of Fourier-Stieltjes transforms $\mathbf{E}e^{izL_\star} \to \mathbf{E}e^{izd_\star}$. Here i denotes the imaginary unit, z is a real number. Denote $\Delta^\star(z) = e^{izL_\star} - e^{izd_\star}$ We show that for any real z and any realized value $Y_\star$

$$\limsup_t |\mathbf{E}(\Delta^\star(z)|Y_\star)| = o(1). \tag{16}$$

This bound together with the simple inequality $|\Delta^\star(z)| \le 2$ yields $\mathbf{E}\Delta^\star(z)=o(1)$, by Lebesgue's dominated convergence theorem. Finally, the identity $\mathbf{E}\Delta^\star(z) = \mathbf{E}e^{izL_\star} - \mathbf{E}e^{izd_\star}$ implies the desired convergence $\mathbf{E}e^{izL_\star} \to \mathbf{E}e^{izd_\star}$ as $t \to +\infty$.

In order to prove (16) we observe that, given $Y_\star$, the conditional distribution of $d_\star$ is the compound Poisson distribution with the characteristic function $f(z) = e^{\lambda_\star(f_\varkappa(z)-1)}$, where $f_\varkappa(z) = \mathbf{E}e^{iz\varkappa_1}$. Similarly, given $X, Y, Y_\star$ and $\{\xi_k,\, k \in T_t\}$, the conditional distribution of $L_\star$ is the compound Poisson distribution with the characteristic function $\bar{f}(z) = e^{\bar{\lambda}(\bar{f}_\varkappa(z)-1)}$, where

$$\bar{f}_\varkappa(z) = \sum_{r\ge 0} e^{izr}\bar{p}_r, \qquad \bar{p}_r = \bar{\lambda}^{-1}\sum_{k\in T_t}\lambda_{k\star}\mathbb{I}_{\{\xi_k=r\}}, \qquad \bar{\lambda} = \sum_{k\in T_t}\lambda_{k\star}.$$

Finally, we obtain (16) from the convergence $\bar{\lambda} \to \lambda_\star$ and $\bar{f}_\varkappa(z) \to f_\varkappa(z)$. $\square$

Proof (of Theorem 2). Here we assume that $\tau(t) = t^\nu$, $\nu > 1$. Denote

$$\zeta = \sum_{k\in T_t}\lambda_{kt}\zeta_k, \qquad \zeta_k = \beta_k b_1 X_k, \qquad \beta_k = 2\left(a^{-(2\nu)^{-1}} - b^{-(2\nu)^{-1}}\right)k^{(1-\nu)/(2\nu)},$$

and let ξ_k, $k \in T_t$ be conditionally independent (given X, Y) Poisson random variables with the conditional mean values $\mathbf{E}(\xi_k|X,Y) = \zeta_k$.

In the first step of the proof we proceed similarly as in the proof of Theorem 1. We approximate the random variable $d(v_t)$ by L and then we approximate the distribution of L by that of the random variable $L' = \sum_{k\in T_t}\mathbb{I}_k\xi_k$. Here it is assumed that the random variables $\mathbb{I}_k, \xi_k$, $k \in T_t$ are conditionally independent, given X, Y. Next we observe that $\xi_k = o_P(1)$ for $k \in T_t$ as $t \to +\infty$ and replace ξ_k by the random indicators $\tilde{\mathbb{I}}_k$, $k \in T_t$, with success probabilities

$$\mathbf{P}(\tilde{\mathbb{I}}_k = 1|X,Y) = 1 - \mathbf{P}(\tilde{\mathbb{I}}_k = 0|X,Y) = \zeta_k.$$

It remains to show the asymptotic distribution of $L'' = \sum_{k\in T_k}\mathbb{I}_k\tilde{\mathbb{I}}_k$. Conditionally, given X, Y, we approximate the sum L'' of (conditionally) independent indicators by the Poisson random variable having the mean value $\mathbf{E}(L''|X,Y) = \zeta$. Using the standard Poisson approximation error bounds [18] we show the valid approximation of L'' by the random variable L''', having the Poisson distribution with the random intensity parameter ζ. That is, conditionally, given X, Y, the random variable L''' has the Poisson distribution with the mean value $\mathbf{E}(L'''|X,Y) = \zeta$. Finally we show that L''' converges in distribution to Λ_3. For this purpose we prove the convergence of the Fourier-Stieltjes transforms

$$\mathbf{E}e^{izL'''} \to \mathbf{E}e^{iz\Lambda_3}. \tag{17}$$

We write

$$\mathbf{E}e^{izL'''} = \mathbf{E}\mathbf{E}(e^{izL'''}|X,Y) = \mathbf{E}e^{\zeta(e^{izt}-1)} \quad \text{and} \quad \mathbf{E}e^{iz\Lambda_3} = \mathbf{E}e^{Y_t b_1 a_2 \gamma(e^{iz}-1)}$$

and derive (17) from the convergence $Y_t b_1 a_2 \gamma - \zeta \to 0$ in probability.

We remark that in this abbreviated version of the proof we have skipped the truncation argument which helps to get the result under the minimal moment conditions, see [6]. □

Before the proof of Theorem 3 we state an auxiliary lemma.

Lemma 1. *Denote* $\mathbf{I}_i^x = \mathbb{I}_{\{X_i > i^{1/2}\}}$ *and* $\mathbf{I}_j^y = \mathbb{I}_{\{Y_j > j^{1/2}\}}$. *We have*

$$\lambda_{ij}(1 - \mathbf{I}_i^x - \mathbf{I}_j^y) \le \min\{1, \lambda_{ij}\} \le \lambda_{ij}.$$

Proof (of Lemma 1). The inequality $\mathbb{I}_{\{\lambda_{ij}>1\}} \le \mathbf{I}_i^x + \mathbf{I}_j^y$ implies

$$\lambda_{ij}(1 - \mathbf{I}_i^x - \mathbf{I}_j^y) \le \lambda_{ij} - (\lambda_{ij} - 1)\mathbb{I}_{\{\lambda_{ij}>1\}} = \min\{1, \lambda_{ij}\}.$$

Proof (of Theorem 3). We only prove (7) and (8). The proof of (9), (10) is similar to that of (8).

Before the proof we introduce some notation. Denote

$$T_{st} = T_s \cap T_t, \quad T_{tu} = T_t \cap T_u, \quad T_{stu} = T_s \cap T_t \cap T_u, \quad T = T_s \cup T_t \cup T_u.$$

An item w_i is called witness of the edge $v_j \sim v_k$ whenever $\mathbb{I}_{ij}\mathbb{I}_{ik} = 1$. In this case we say that w_i realizes the edge $v_j \sim v_k$. Let $\Delta_1 = \{\exists i : \mathbb{I}_{is}\mathbb{I}_{it}\mathbb{I}_{iu} = 1\}$ denote the event that all edges of the triangle v_s, v_t, v_u are realized by a common witness. Let Δ_2 denote the event that all edges are realized by different witnesses,

$$\Delta_2 = \{\exists \text{ distinct } i, j, k \text{ such that } \mathbb{I}_{is}\mathbb{I}_{it} = 1, \ \mathbb{I}_{js}\mathbb{I}_{ju} = 1, \ \mathbb{I}_{kt}\mathbb{I}_{ku} = 1\}.$$

Let $\Delta = \{v_s \sim v_t, \ v_s \sim v_u, \ v_t \sim v_u\}$ denote the event that vertices v_s, v_t, v_u make up a triangle. Introduce events $\mathcal{H}_t = \{v_s \sim v_t, v_t \sim v_u\}$ and $\mathcal{K}_t = \{\exists i \neq j : \mathbb{I}_{it}\mathbb{I}_{is}\mathbb{I}_{jt}\mathbb{I}_{ju} = 1\}$, and random variables

$$S = \sum_{au \le k \le bs} \mathbb{I}_{ks}\mathbb{I}_{kt}\mathbb{I}_{ku}, \quad Q = \sum_{au \le i < j \le bs} \mathbb{I}_{is}\mathbb{I}_{it}\mathbb{I}_{iu}\mathbb{I}_{js}\mathbb{I}_{jt}\mathbb{I}_{ju},$$

$$S_t = \sum_{(i,j)\in I} \mathbb{I}_{it}\mathbb{I}_{is}\mathbb{I}_{jt}\mathbb{I}_{ju}, \quad Q_t = \sum_{(i,j)\in I} \ \sum_{(k,r)\in I, (k,r)\neq(i,j)} \mathbb{I}_{is}\mathbb{I}_{it}\mathbb{I}_{jt}\mathbb{I}_{ju}\mathbb{I}_{kt}\mathbb{I}_{ks}\mathbb{I}_{rt}\mathbb{I}_{ru}.$$

Here I denote the set of all ordered pairs $(i, j) \in T \times T$ such that $i \neq j$. We remark that a pair (i, j) corresponds to the pair (w_i, w_j) of attributes that might be witnesses of edges $v_s \sim v_t$ and $v_t \sim v_u$ respectively.

We note that $t/s, u/t, u/s \in [1, b/a]$ for $0 < s < t < u$ satisfying $\lceil au \rceil \le \lfloor bs \rfloor$. Hence the variables $s, t, u \to +\infty$ are of the same order of magnitude.

Let us prove (7). We observe that $\Delta_1 \subset \Delta \subset \Delta_1 \cup \Delta_2$. Hence

$$\mathbf{P}(\Delta_1) \le \mathbf{P}(\Delta) \le \mathbf{P}(\Delta_1) + \mathbf{P}(\Delta_2). \tag{18}$$

Next, by inclusion exclusion, we write $S - Q \leq \mathbb{I}_{\Delta_1} \leq S$ and estimate

$$\mathbf{E}S - \mathbf{E}Q \leq \mathbf{P}(\Delta_1) \leq \mathbf{E}S. \tag{19}$$

Now, combining (18) and (19) with the relations

$$\mathbf{E}S = \sum_{au \leq k \leq bs} \frac{\mathbf{E}X_k^3 Y_s Y_t Y_u}{k^{3/2}\sqrt{stu}} + o(t^{-2}) = \frac{a_3 b_1^3}{\sqrt{stu}} \left(\frac{2}{\sqrt{au}} - \frac{2}{\sqrt{bs}} \right) + o(t^{-2}), \tag{20}$$

$$\mathbf{E}Q \leq \sum_{au \leq i < j \leq bs} \mathbf{E}\lambda_{is}\lambda_{it}\lambda_{iu}\lambda_{js}\lambda_{jt}\lambda_{ju} \leq \frac{a_3^2 b_2^3}{stu} \sum_{au \leq i < j \leq bs} \frac{1}{i^{3/2} j^{3/2}} = O(t^{-4}),$$

$$\mathbf{P}(\Delta_2) \leq \mathbf{E} \sum_{i,j,k \in T,\, i \neq j \neq k} \mathbb{I}_{is}\mathbb{I}_{it}\mathbb{I}_{js}\mathbb{I}_{ju}\mathbb{I}_{kt}\mathbb{I}_{ku} \leq \frac{a_2^3 b_2^3}{stu} \left(\sum_{i \in T} i^{-1} \right)^3 = O(t^{-3}) \tag{21}$$

we obtain (7), since $p_\Delta = \mathbf{P}(\Delta)$. In the first step of (20) we applied Lemma 1, and in the last step of (21) we used $\sum_{i \in T} i^{-1} \leq c$.

Let us prove (8). We observe that (8) follows from (7) and the relation

$$\mathbf{P}(\mathcal{H}_t) = \mathbf{P}(\Delta) + a_2^2 b_1^2 b_2 \frac{1}{t\sqrt{su}} \delta_{t|su} + o(t^{-2}). \tag{22}$$

It remains to show (22). We note that the identity $\mathcal{H}_t = \Delta_1 \cup \mathcal{K}_t$ implies

$$\mathbf{P}(\mathcal{H}_t) = \mathbf{P}(\Delta_1) + \mathbf{P}(\mathcal{K}_t) - \mathbf{P}(\Delta_1 \cap \mathcal{K}_t). \tag{23}$$

Next, by inclusion exclusion, we write $S_t - Q_t \leq \mathbb{I}_{\mathcal{K}_t} \leq S_t$ and obtain

$$\mathbf{E}S_t - \mathbf{E}S_t(1 - \mathbb{I}_{\mathcal{D}_\varepsilon}) - \mathbf{E}Q_t \mathbb{I}_{\mathcal{D}_\varepsilon} \leq \mathbf{E}\mathbb{I}_{\mathcal{K}_t}\mathbb{I}_{\mathcal{D}_\varepsilon} \leq \mathbf{P}(\mathcal{K}_t) \leq \mathbf{E}S_t, \tag{24}$$

for any $\varepsilon \in (0,1)$. Here $\mathcal{D}_\varepsilon$ denotes the event $\{Y_t \leq \varepsilon t\}$. In the remaining part of the proof we show that

$$\mathbf{E}S_t = a_2^2 b_1^2 b_2 \frac{1}{t\sqrt{su}} \delta_{t|s,u} + o(t^{-2}), \tag{25}$$

$$\mathbf{P}(\Delta_1 \cap \mathcal{K}_t) = O(t^{-3}), \tag{26}$$

and that there exists $c^* > 0$ which does not depend on s, t, u and ε such that, for any $\varepsilon \in (0,1)$,

$$\mathbf{E}Q_t \mathbb{I}_{\mathcal{D}_\varepsilon} \leq c^* \varepsilon t^{-2} + O(t^{-3}), \qquad \mathbf{E}S_t(1 - \mathbb{I}_{\mathcal{D}_\varepsilon}) = o(t^{-2}). \tag{27}$$

We observe that (22) follows from (23), (26) and the approximations $\mathbf{P}(\Delta) \approx \mathbf{P}(\Delta_1)$ (see (18), (21)), and $\mathbf{P}(\mathcal{K}_t) \approx a_2^2 b_1^2 b_2 \frac{1}{t\sqrt{su}} \delta_{t|s,u}$ (see (24), (25), (27)).

Let us prove (25). Since the product $\bar{p}_{ij} := p_{is} p_{it} p_{jt} p_{ju}$ is non zero whenever $i \in T_{st}$ and $j \in T_{tu}$, we have

$$\mathbf{E}S_t = \mathbf{E} \sum_{(i,j) \in I} \bar{p}_{ij} = \mathbf{E} \sum_{(i,j):\, i \in T_{st},\, j \in T_{tu},\, i \neq j} \bar{p}_{ij}. \tag{28}$$

We split the set $\{(i,j) : i \in T_{st}, j \in T_{tu}, i \neq j\} = \mathbb{T}_1 \cup \cdots \cup \mathbb{T}_4$, where

$$\mathbb{T}_1 = (T_{st} \setminus T_u) \times T_{tu}, \qquad \mathbb{T}_2 = T_{stu} \times (T_{tu} \setminus T_s),$$
$$\mathbb{T}_3 = \{(i,j) : i,j \in T_{stu}, i < j\}, \quad \mathbb{T}_4 = \{(i,j) : i,j \in T_{stu}, j < i\},$$

and write (28) in the form

$$\mathbf{E}S_t = \mathbf{E}S_{t1} + \ldots + \mathbf{E}S_{t4}, \qquad S_{tk} := \sum_{(i,j)\in\mathbb{T}_k} \bar{p}_{ij}. \tag{29}$$

Now (25) follows from (29) and the relations, for $1 \le k \le 4$,

$$\mathbf{E}S_{tk} = \mathbf{E} \sum_{(i,j)\in\mathbb{T}_k} \lambda_{is}\lambda_{it}\lambda_{jt}\lambda_{ju} + o(t^{-2}) = a_2^2 b_1^2 b_2 \frac{1}{t\sqrt{su}} \sum_{(i,j)\in\mathbb{T}_k} \frac{1}{ij} + o(t^{-2}) \tag{30}$$

and

$$\sum_{1\le k\le 4} \sum_{(i,j)\in\mathbb{T}_k} \frac{1}{ij} = \delta_{t|su} + O(t^{-1}).$$

In the first step of (30) we used Lemma 1.

Let us prove the first bound of (27). We split the collection of vectors(i,j,k,r)

$$\mathbb{Q} = \big\{(i,j,k,r) \in T^4 \text{ such that } i \neq j, k \neq r \text{ and } (i,j) \neq (k,r)\big\}$$

into five non intersecting pieces $\mathbb{Q} = \mathbb{Q}_1 \cup \cdots \cup \mathbb{Q}_5$, where

$$\mathbb{Q}_1 = \big\{(i,j,k,r) : i = k\big\} \cap \mathbb{Q}, \qquad \mathbb{Q}_2 = \big\{(i,j,k,r) : i = r\big\} \cap \mathbb{Q},$$
$$\mathbb{Q}_3 = \big\{(i,j,k,r) : j = k\big\} \cap \mathbb{Q}, \qquad \mathbb{Q}_4 = \big\{(i,j,k,r) : j = r\big\} \cap \mathbb{Q},$$

and $\mathbb{Q}_5 = \big\{(i,j,k,r) : \text{all } i,j,k,r \text{ are distinct }\big\} \cap \mathbb{Q}$, and write

$$Q_t = \sum_{1\le z\le 5} Q_{tz}, \qquad Q_{tz} = \sum_{(i,j,k,r)\in\mathbb{Q}_z} \mathbb{I}_{is}\mathbb{I}_{it}\mathbb{I}_{jt}\mathbb{I}_{ju}\mathbb{I}_{ks}\mathbb{I}_{kt}\mathbb{I}_{rt}\mathbb{I}_{ru}.$$

Denote $\tilde{\mathbb{Q}} = \{(i,j,r) \in T^3 : \text{all } i,j,r \text{ are distinct}\}$. Observing that the typical summand of the sum Q_{t1} is $\mathbb{I}_{is}\mathbb{I}_{it}\mathbb{I}_{jt}\mathbb{I}_{ju}\mathbb{I}_{rt}\mathbb{I}_{ru}$ (since $i = k$), we write

$$\begin{aligned}
\mathbf{E}Q_{t1}\mathbb{I}_{\mathcal{D}_\varepsilon} &\le \mathbf{E} \sum_{(i,j,r)\in\tilde{\mathbb{Q}}} \lambda_{is}\lambda_{it}\lambda_{jt}\lambda_{ju}\lambda_{rt}\lambda_{ru}\mathbb{I}_{\mathcal{D}_\varepsilon} \\
&\le \frac{a_2^3}{s^{1/2}t^{3/2}u} \mathbf{E}Y_s Y_t^3 Y_u^2 \mathbb{I}_{\mathcal{D}_\varepsilon} \left(\sum_{i\in T} \frac{1}{i}\right)^3 \\
&\le c^3 \varepsilon \frac{a_2^3}{s^{1/2}t^{1/2}u} \mathbf{E}Y_s Y_t^2 Y_u^2 \\
&\le c' \varepsilon t^{-2}.
\end{aligned}$$

Here we used inequalities $Y_t t^{-1}\mathbb{I}_{\mathcal{D}_\varepsilon} \le \varepsilon$ and $\sum_{i\in T}\frac{1}{i} \le c$. Similarly, we prove the inequality $\mathbf{E}Q_{t4}\mathbb{I}_{\mathcal{D}_\varepsilon} \le c'\varepsilon t^{-2}$. Furthermore, observing that the typical summand of the sum Q_{t2} is $\mathbb{I}_{is}\mathbb{I}_{it}\mathbb{I}_{iu}\mathbb{I}_{jt}\mathbb{I}_{ju}\mathbb{I}_{ks}\mathbb{I}_{kt}$ (since $i = r$), we write

$$\begin{aligned}
\mathbf{E}Q_{t2}\mathbb{I}_{\mathcal{D}_\varepsilon} &\le \mathbf{E}\sum_{(i,j,k)\in\tilde{\mathbb{Q}}} \lambda_{is}\lambda_{it}\lambda_{iu}\lambda_{jt}\lambda_{ju}\lambda_{ks}\lambda_{kt}\mathbb{I}_{\mathcal{D}_\varepsilon}\\
&\le \frac{a_3 a_2^2}{st^{3/2}u}\mathbf{E}Y_s^2Y_t^3Y_u^2\mathbb{I}_{\mathcal{D}_\varepsilon}\left(\sum_{i\in T}\frac{1}{i}\right)^2\left(\sum_{i\in T}\frac{1}{i^{3/2}}\right)\\
&\le c^3\frac{a_3a_2^2}{stu}\mathbf{E}Y_s^2Y_t^2Y_u^2.
\end{aligned}$$

In the last step we used inequalities $Y_t t^{-1}\mathbb{I}_{\mathcal{D}_\varepsilon} \le 1$ and $\sum_{i\in T}\frac{1}{i^{3/2}} \le ct^{-1/2}$. Hence, $\mathbf{E}Q_{t2}\mathbb{I}_{\mathcal{D}_\varepsilon} = O(t^{-3})$. Similarly, we prove the bound $\mathbf{E}Q_{t3}\mathbb{I}_{\mathcal{D}_\varepsilon} = O(t^{-3})$. Finally, we estimate

$$\begin{aligned}
\mathbf{E}Q_{t5}\mathbb{I}_{\mathcal{D}_\varepsilon} &\le \mathbf{E}\sum_{(i,j,k,r)\in\mathbb{Q}_5} \lambda_{is}\lambda_{is}\lambda_{jt}\lambda_{ju}\lambda_{ks}\lambda_{kt}\lambda_{rt}\lambda_{ru}\mathbb{I}_{\mathcal{D}_\varepsilon}\\
&\le \frac{a_2^4}{st^2u}\mathbf{E}Y_s^2Y_t^4Y_u^2\mathbb{I}_{\mathcal{D}_\varepsilon}\left(\sum_{i\in T}\frac{1}{i}\right)^4\\
&\le c'\varepsilon^2 t^{-2}.
\end{aligned}$$

In the last step we used the inequality $Y_t^2 t^{-2}\mathbb{I}_{\mathcal{D}_\varepsilon} \le \varepsilon^2$. Collecting these upper bounds for $\mathbf{E}Q_{tz}\mathbb{I}_{\mathcal{D}_\varepsilon}$, $1 \le z \le 5$, we obtain the first bound of (27).

The second bound of (27) follows from $\mathbf{E}Y_t^2\mathbb{I}_{\{Y_t\ge\varepsilon t\}} = o(1)$ and inequalities

$$\mathbf{E}S_t(1-\mathbb{I}_{\mathcal{D}_\varepsilon}) \le \mathbf{E}\sum_{i,j\in T,\, i\ne j}\lambda_{is}\lambda_{it}\lambda_{jt}\lambda_{ju}(1-\mathbb{I}_{\mathcal{D}_\varepsilon}) \le \frac{a_2^2b_1^2}{st}\mathbf{E}Y_t^2\mathbb{I}_{\{Y_t\ge\varepsilon t\}}\left(\sum_{i\in T} i^{-1}\right)^2.$$

Let us prove (26). The inequalities $\mathbb{I}_{\mathcal{K}_t} \le S_t$, $\mathbb{I}_{\Delta_1} \le S$ and $S \le \tilde{S}$, where $\tilde{S} = \sum_{k\in T}\mathbb{I}_k^*$ and $\mathbb{I}_k^* = \mathbb{I}_{ks}\mathbb{I}_{kt}\mathbb{I}_{ku}$, imply $\mathbf{P}(\Delta_1 \cap \mathcal{K}_t) = \mathbf{E}\mathbb{I}_{\Delta_1}\mathbb{I}_{\mathcal{K}_t} \le \mathbf{E}S_t\tilde{S}$. We shall show that $\mathbf{E}S_t\tilde{S} = O(t^{-3})$. To this aim we split $S_t\tilde{S} = \tilde{S}_1 + \tilde{S}_2$, where

$$\tilde{S}_1 = \sum_{i\in T}\sum_{j\in T\setminus\{i\}}\mathbb{I}_{is}\mathbb{I}_{it}\mathbb{I}_{jt}\mathbb{I}_{ju}(\mathbb{I}_i^* + \mathbb{I}_j^*), \qquad \tilde{S}_2 = \sum_{(i,j,k)\in\tilde{\mathbb{Q}}}\mathbb{I}_{is}\mathbb{I}_{it}\mathbb{I}_{jt}\mathbb{I}_{ju}\mathbb{I}_k^*,$$

and estimate

$$\begin{aligned}
\mathbf{E}\tilde{S}_1 &\le \mathbf{E}\sum_{i\in T}\sum_{j\in T\setminus\{i\}}\lambda_{is}\lambda_{it}\lambda_{jt}\lambda_{ju}(\lambda_{iu} + \lambda_{js}) = O(t^{-3}),\\
\mathbf{E}\tilde{S}_2 &\le \mathbf{E}\tilde{S}_2' \le \mathbf{E}\sum_{(i,j,k)\in\tilde{\mathbb{Q}}}\lambda_{is}\lambda_{it}\lambda_{jt}\lambda_{ju}\lambda_{ks}\lambda_{ku} = O(t^{-3}). \qquad (31)
\end{aligned}$$

Here $\tilde{S}_2'$ is defined in the same way as $\tilde{S}_2$, but with $\mathbb{I}_k^*$ replaced by $\mathbb{I}_k' = \mathbb{I}_{ks}\mathbb{I}_{ku}$.

□

Proof (of (11)). We only sketch the proof. For $s < t$ such that $\lceil at \rceil \leq \lfloor bs \rfloor$,

$$e_{st} = \sum_{i \in T_s \cap T_t} \mathbb{I}_{is}\mathbb{I}_{it} \quad \text{and} \quad q_{st} = \sum_{\{i,j\} \subset T_s \cap T_t} \mathbb{I}_{is}\mathbb{I}_{it}\mathbb{I}_{js}\mathbb{I}_{jt}$$

count witnesses and pairs of witnesses of the edge $v_s \sim v_t$, respectively ($w_i \in W$ is called a witness of the edge $v_s \sim v_t$ whenever w_i is a common neighbor of v_s and v_t in $H_{X,Y}$). We write, by inclusion-exclusion,

$$e_{st} - q_{st} \leq \mathbb{I}_{\{v_s \sim v_t\}} \leq e_{st}$$

and note that the quadratic term q_{st} is negligibly small. Hence, we approximate

$$\mathbb{I}_{\{v_s \sim v_t\}} = e_{st}(1 + o_P(1)), \qquad \mathbf{P}(v_s \sim v_t) = (1 + o(1))\mathbf{E}e_{st}. \tag{32}$$

Given t and $i, j \in T_t$, we denote $T_{it}^* = T_i^* \setminus \{t\}$ and introduce random variables

$$u_{it} = \sum_{k \in T_{it}^*} \mathbb{I}_{ik}, \quad z_{ijt} = \sum_{k \in T_{it}^* \cap T_{jt}^*} \mathbb{I}_{ik}\mathbb{I}_{jk}, \quad L_t = \sum_{i \in T_t} \mathbb{I}_{it}u_{it}, \quad Q_t = \sum_{\{i,j\} \subset T_t} \mathbb{I}_{it}\mathbb{I}_{jt}z_{ijt}.$$

We remark that L_t counts pairs $(v_s \sim v_t; w_i)$, where w_i is a witness of the edge $v_s \sim v_t$ in $G_{X,Y}$, for some $v_s \in W \setminus \{v_t\}$. In particular, we have $d(v_t) \leq L_t$. Similarly, Q_t counts all triples $(v_s \sim v_t; w_i, w_j)$, where w_i and w_j are distinct witnesses of an edge $v_s \sim v_t$. Note that a neighbor v_s of v_t, which has k witnesses of the edge $\{v_s \sim v_t\}$, contributes 1 to the number $d(v_t)$ of neighbors of v_t. It contributes k to the sum L_t, and it contributes $\binom{k}{2}$ to the sum Q_t. Hence,

$$L_t - Q_t \leq d(v_t) \leq L_t.$$

Here the quadratic term Q_t is negligibly small and we approximate $d(v_t) = L_t(1 + o_P(1))$. Combining this approximation with (32) we obtain

$$\mathbf{E}_{st}d(v_s)d(v_t) = (\mathbf{E}e_{st})^{-1}\mathbf{E}e_{st}L_sL_t + o(1), \tag{33}$$

$$\mathbf{E}_{st}d^r(v_u) = (\mathbf{E}e_{st})^{-1}\mathbf{E}e_{st}L_u^r + o(1), \tag{34}$$

for $r = 1, 2$ and $u = s, t$. Next we evaluate expectations in the right-hand sides of (33), (34). A straightforward but tedious calculation shows that

$$\begin{aligned}
\mathbf{E}e_{st} &= \Theta(1 + o(1))h_1, \\
\mathbf{E}e_{st}L_s &= \mathbf{E}e_{st}L_t + o(\Theta) = \Theta(1 + o(1))(h_1 + h_2 + 2h_3), \\
\mathbf{E}e_{st}L_s^2 &= \mathbf{E}e_{st}L_t^2 + o(\Theta) = \Theta(1 + o(1))(h_1 + 3h_2 + 6h_3 + h_4 + 6h_5 + 4h_6), \\
\mathbf{E}e_{st}L_sL_t &= \Theta(1 + o(1))(h_1 + 3h_2 + 4h_3 + h_4 + 4h_5 + 4h_7).
\end{aligned}$$

Here we denote $\Theta = (st)^{-1/2}\ln(bs/at)$. We recall that h_i are defined in (12) above. Now (11) follows from (33), (34). □

Acknowledgement. M. Bloznelis acknowledges support from the Research Council of Lithuania (grant MIP-053/2011).

References

1. Aiello, W., Bonato, A., Cooper, C., Janssen, J., Prałat, P.: A spatial web graph model with local influence regions. In: Bonato, A., Chung, F.R.K. (eds.) WAW 2007. LNCS, vol. 4863, pp. 96–107. Springer, Heidelberg (2007)
2. Bloznelis, M.: Degree and clustering coefficient in sparse random intersection graphs. The Annals of Applied Probability 23, 1254–1289 (2013)
3. Bloznelis, M., Damarackas, J.: Degree distribution of an inhomogeneous random intersection graph. The Electronic Journal of Combinatorics 20(3), R3 (2013)
4. Bloznelis, M., Jaworski, J., Kurauskas, V.: Assortativity and clustering of sparse random intersection graphs. Electronic Journal of Probability 18, R38 (2013)
5. Bloznelis, M., Kurauskas, V.: Clustering function: a measure of social influence, http://arxiv.org/abs/1207.4941
6. Bloznelis, M., Karoński, M.: Random intersection graph process, http://arxiv.org/abs/1301.5579
7. Deijfen, M., Kets, W.: Random intersection graphs with tunable degree distribution and clustering. Probab. Engrg. Inform. Sci. 23, 661–674 (2009)
8. Foss, S., Korshunov, D., Zachary, S.: An Introduction to Heavy-Tailed and Subexponential Distributions. ACM, New York (2011)
9. Godehardt, E., Jaworski, J.: Two models of random intersection graphs and their applications. Electronic Notes in Discrete Mathematics 10, 129–132 (2001)
10. Godehardt, E., Jaworski, J., Rybarczyk, K.: Clustering coefficients of random intersection graphs. Studies in Classification, Data Analysis and Knowledge Organization, pp. 243–253. Springer, Heidelberg (2012)
11. Guillaume, J.L., Latapy, M.: Bipartite structure of all complex networks. Inform. Process. Lett. 90, 215–221 (2004)
12. Jaworski, J., Karoński, M., Stark, D.: The degree of a typical vertex in generalized random intersection graph models. Discrete Mathematics 306, 2152–2165 (2006)
13. Karoński, M., Scheinerman, E.R., Singer-Cohen, K.B.: On random intersection graphs: The subgraph problem. Combinatorics, Probability and Computing 8, 131–159 (1999)
14. Martin, T., Ball, B., Karrer, B., Newman, M.E.J.: Coauthorship and citation patterns in the Physical Review. Phys. Rev. E 88, 012814 (2013)
15. Newman, M.E.J.: Clustering and preferential attachment in growing networks. Physical Review E 64, 025102 (2001)
16. Newman, M.E.J., Watts, D.J., Strogatz, S.H.: Random graph models of social networks. Proc. Natl. Acad. Sci. USA 99(suppl. 1), 2566–2572 (2002)
17. Stark, D.: The vertex degree distribution of random intersection graphs. Random Structures and Algorithms 24, 249–258 (2004)
18. Steele, J.M.: Le Cam's inequality and Poisson approximations. The American Mathematical Monthly 101, 48–54 (1994)
19. Watts, D.J., Strogatz, S.H.: Collective dynamics of "small-world" networks. Nature 393, 440–442 (1998)

Alpha Current Flow Betweenness Centrality*

Konstantin Avrachenkov[1], Nelly Litvak[2], Vasily Medyanikov[3], and Marina Sokol[1]

[1] Inria Sophia Antipolis, 2004 Route des Lucioles, Sophia-Antipolis, France
[2] University of Twente, P.O. Box 217, 7500AE, Enschede, The Netherlands
[3] St. Petersburg State University, 7-9, Universitetskaya nab., St. Petersburg, Russia

Abstract. A class of centrality measures called betweenness centralities reflects degree of participation of edges or nodes in communication between different parts of the network. The original shortest-path betweenness centrality is based on counting shortest paths which go through a node or an edge. One of shortcomings of the shortest-path betweenness centrality is that it ignores the paths that might be one or two hops longer than the shortest paths, while the edges on such paths can be important for communication processes in the network. To rectify this shortcoming a current flow betweenness centrality has been proposed. Similarly to the shortest-path betweenness, it has prohibitive complexity for large size networks. In the present work we propose two regularizations of the current flow betweenness centrality, α-current flow betweenness and truncated α-current flow betweenness, which can be computed fast and correlate well with the original current flow betweenness.

1 Introduction

A class of centrality measures called betweenness centralities reflects degree of participation of edges or nodes in communication between different parts of the network. The first notion of betweenness centrality was introduced by Freeman [8]. Let $s, t \in V$ be a pair of nodes in an undirected graph $G = (V, E)$. (In the present work we restrict our consideration to undirected graphs.) We denote $|V| = n$, $|E| = m$, and let d_v be the degree of node v. Let $\sigma_{s,t}$ be the number of shortest paths connecting nodes s and t and denote by $\sigma_{s,t}(e)$ the number of shortest paths connecting nodes s and t passing through edge e. Then betweenness centrality of edge e is calculated as follows:

$$C_{\mathrm{B}}(e) = \frac{1}{n(n-1)} \sum_{s,t \in V} \frac{\sigma_{s,t}(e)}{\sigma_{s,t}}. \tag{1}$$

Computational complexity of the best known algorithm for computing the betweenness in (1) is $\mathcal{O}(mn)$ [4]. This limits its applicability for large graphs.

* This research is partially funded by Inria Alcatel-Lucent Joint Lab, by the European Commission within the framework of the CONGAS project FP7-ICT-2011-8-317672, see `www.congas-project.eu`, and by the EU-FET Open grant NADINE (288956).

A. Bonato, M. Mitzenmacher, and P. Prałat (Eds.): WAW 2013, LNCS 8305, pp. 106–117, 2013.
© Springer International Publishing Switzerland 2013

One of shortcomings of the betweenness centrality in (1) is that it takes into accounts only the shortest paths, ignoring the paths that might be one or two steps longer, while the edges on such paths can be important for communication processes in the network. In order to take such paths into account, Newman [12] and Brandes and Fleischer [5] introduced the current flow betweenness centrality (CF-betweenness). In [12,5] the graph is regarded as an electrical network with edges being unit resistances. The CF-betweenness of an edge is the amount of current that flows through it, averaged over all source-destination pairs, when one unit of current is induced at the source, and the destination (sink) is connected to the ground. This exploits the well known relation between electrical networks and reversible Markov chains, see e.g., [1,7].

The computational difficulty of Betweenness and the CF-betweenness is that the computations must be done over the set of all source-destination pairs. The best previously known computational complexity for the CF-betweenness is $\mathcal{O}(I(n-1) + mn\log n)$ where $I(n-1)$ is the complexity of the inversion of matrix of dimension $n-1$.

In the present work we introduce new betweenness centrality measures: α-current flow betweenness (α-CF betweenness) and its truncated version. The main purpose of these new measures is to bring down the high cost of the CF-betweenness computation. Our proposed measures are very close in performance to the CF-betweenness, but they are comparable to the PageRank algorithm [6] in their modest computational complexity. Our goal is to provide and analyze efficient algorithms for computing α-CF betweenness and truncated α-CF betweenness, and to compare the α-CF betweenness to other centrality measures.

2 Alpha Current Flow Betweenness

We view the graph G as an electrical network where each edge has resistance α^{-1}, and each node v is connected to ground node $n+1$ by an edge with resistance $(1-\alpha)^{-1}d_v^{-1}$. This is in the spirit of PageRank. Indeed, the current (probability flow) is inversely proportional to the resistance. Thus, the fraction α of the current from node v flows to the network, while fraction $(1-\alpha)$ of the current is directed to the sink. Since the graph is undirected, we use a convention that (v,w) and (w,v) represent the same arc in E, but depending on the chosen direction the current along this arc is considered to be positive or negative.

Assume that a unit of current is supplied to a source node $s \in V$, and there is a destination node $t \in V$ connected to the ground. Let $\varphi_v^{(s,t)}$ denote the absolute potential of node $v \in V$, if s is the source and t is the destination. Assume without loss of generality that $s = 1$ and $t = n$ ($\varphi_n^{(1,n)} = \varphi_{n+1}^{(1,n)} = 0$, i.e., we set the ground potential to zero). The vector of absolute potentials of the other nodes $\varphi^{(1,n)} = [\varphi_1^{(1,n)}, ..., \varphi_{n-1}^{(1,n)}]^T$ is a solution of the following system of equations (Kirchhoff's current law):

$$[\tilde{D} - \alpha\tilde{A}]\varphi^{(1,n)} = \tilde{b}, \tag{2}$$

where $\tilde{D}$ and $\tilde{A}$ are the degree and adjacency matrices of the graph without node n and $\tilde{b} = [1, 0, ..., 0]^T$, see e.g., [5].

In the following theorem we demonstrate that we do not need to solve a separate linear system for each source-destination pair, it suffies to invert the coefficient matrix $[D - \alpha A]$.

Theorem 1. *The voltage drop along the edge* (v, w) *is given by*

$$\varphi_v^{(s,t)} - \varphi_w^{(s,t)} = (c_{s,v} - c_{s,w}) + \frac{c_{s,t}}{c_{t,t}}(c_{t,w} - c_{t,v}), \tag{3}$$

where $(c_{v,w})_{v,w \in V}$, *are the elements of the matrix* $C = [D - \alpha A]^{-1}$.

Proof: Assume again without loss of generality that $s = 1$ and $t = n$. The matrix $[D - \alpha A]$ can be written in the following block structure

$$D - \alpha A = \begin{bmatrix} \tilde{D} - \alpha \tilde{A} & -\alpha \tilde{a} \\ -\alpha \tilde{a}^T & d_n \end{bmatrix}, \quad \text{with} \quad \tilde{a} = [a_{1,n}, a_{2,n}, \ldots, a_{n-1,n}]^T.$$

Then, divide accordingly the elements of the inverse matrix

$$C = [D - \alpha A]^{-1} = \begin{bmatrix} \tilde{C} & \tilde{c} \\ \tilde{c}^T & c_{n,n} \end{bmatrix}.$$

Writing the relation $[D - \alpha A]C = I$ in the block form yields

$$[\tilde{D} - \alpha \tilde{A}]\tilde{C} - \alpha \tilde{a}\tilde{c}^T = I, \tag{4}$$

$$[\tilde{D} - \alpha \tilde{A}]\tilde{c} - \alpha \tilde{a}\tilde{c}_{n,n} = 0. \tag{5}$$

Premultiplying equation (4) by $[\tilde{D} - \alpha \tilde{A}]^{-1}$, we obtain

$$[\tilde{D} - \alpha \tilde{A}]^{-1} = \tilde{C} - \alpha [\tilde{D} - \alpha \tilde{A}]^{-1} \tilde{a}\tilde{c}^T. \tag{6}$$

And premultiplying (5) by $[\tilde{D} - \alpha \tilde{A}]^{-1}$, we obtain

$$\alpha [\tilde{D} - \alpha \tilde{A}]^{-1} \tilde{a} = \frac{1}{c_{n,n}} \tilde{c}. \tag{7}$$

Combining both equations (6) and (7) gives

$$[\tilde{D} - \alpha \tilde{A}]^{-1} = \tilde{C} - \frac{1}{c_{n,n}} \tilde{c}\tilde{c}^T,$$

and hence $\varphi^{(1,n)} = [\tilde{D} - \alpha \tilde{A}]^{-1} \tilde{b} = \tilde{C}_{\cdot,1} - \frac{c_{1,n}}{c_{n,n}} \tilde{c}$. Thus, we can write

$$\varphi_v^{(1,n)} - \varphi_w^{(1,n)} = (c_{v,1} - c_{w,1}) + \frac{c_{1,n}}{c_{n,n}}(c_{w,n} - c_{v,n})$$

The above expression is symmetric and can be rewritten for any source-target pair (s,t). That is,

$$\varphi_v^{(s,t)} - \varphi_w^{(s,t)} = (c_{v,s} - c_{w,s}) + \frac{c_{s,t}}{c_{t,t}}(c_{w,t} - c_{v,t}).$$

Furthermore, since matrix C is symmetric for undirected graphs, we can rewrite the above equation as

$$\varphi_v^{(s,t)} - \varphi_w^{(s,t)} = (c_{s,v} - c_{s,w}) + \frac{c_{s,t}}{c_{t,t}}(c_{t,w} - c_{t,v}),$$

which completes the proof. □

The potentials $\varphi_v^{(s,t)}$, $v, s, t \in V$, have a clear probabilistic interpretation. Take again $s = 1$ and $t = n$. Then from (2) we readily obtain

$$\varphi_v^{(1,n)} = \mathbf{e}_v^T[\tilde{D} - \alpha\tilde{A}]^{-1}\tilde{b} = \mathbf{e}_v^T[I - \alpha\tilde{P}]^{-1}\tilde{D}^{-1}\tilde{b}, \tag{8}$$

where $\mathbf{e}_v$ is a v-th standard basis column vector, and $\tilde{P}$ is the transition probability matrix for a simple random walk on G with absorption in n. Compare this to the well-known expression for the Personalized PageRank vector $\pi(v) = (\pi_1(v), \ldots, \pi_n(v))$ with teleportation preference concentrated in v and damping factor α: $\pi(v) = (1-\alpha)\mathbf{e}_v^T[I - \alpha P]^{-1}$. Note that the vector $\tilde{\pi}(v) = (1-\alpha)\mathbf{e}_v^T[I - \alpha\tilde{P}]^{-1}$ is very similar to $\pi(v)$, except it nullifies the contribution of node n. Now, recall that $\tilde{b} = (1, 0, \ldots, 0)^T$ to obtain

$$\varphi_v^{(1,n)} = (1-\alpha)^{-1}\tilde{\pi}_1(v)d_1^{-1}.$$

Furthermore, let $\mathbf{1}$ be a column vector of ones. Recall that the PageRank vector with uniform teleportation can be written as $\pi = \frac{1-\alpha}{n}\mathbf{1}^T[I - \alpha P]^{-1}$, and define a similar vector $\tilde{\pi} = \frac{1-\alpha}{n}\mathbf{1}^T[I - \alpha\tilde{P}]^{-1}$. Then

$$\sum_{v \in V} \varphi_v^{(1,n)} = n(1-\alpha)^{-1}\tilde{\pi}_1 d_1^{-1}.$$

It is well-known (see e.g., [9] and references therein) and is also confirmed by our experiments that the PageRank of a node in an undirected graph is strongly correlated to the degree of the node. Thus, with any choice of the source, the sum of the potentials is of similar magnitude, except for the cases when the destination node has a large contribution into the PageRank mass of the source.

Finally, we note that the source node has the highest potential, and from [1, Chapter 3, Section 3] we find

$$\begin{aligned}\varphi_1^{(1,n)} &= [P(\text{random walk returns to node 1 before absorption})]^{-1} \\ &= E(\#\text{ returns to node 1 before absorption}).\end{aligned}$$

Now we are ready to define α-CF betweenness. The current $I_e^{(s,t)}$ through edge $e = (v,w)$ is equal to $\alpha(\varphi_v^{(s,t)} - \varphi_w^{(s,t)})$. Let

$$x_e^{(s,t)} = |\varphi_v^{(s,t)} - \varphi_w^{(s,t)}|, \quad (v,w) \in E,$$

be the difference of potentials, that determines the absolute value of the current on the edge. The α-CF betweenness of edge e is defined by

$$x_e^\alpha = \frac{1}{n(n-1)} \sum_{s,t \in V, s \neq t} x_e^{(s,t)}, \quad e \in E. \tag{9}$$

Further, for each node $v \in V$ its α-CF betweenness is defined as the sum of the α-CF betweenness scores of its adjacent edges:

$$\alpha\text{-CF betweenness}(v) = \sum_{(v,w) \in E} x_{(v,w)}^\alpha, \quad v \in V. \tag{10}$$

With this definition, the node is central if a relatively large amount of current flows from this node to the network. This is in accordance to the original CF-betweenness of [12,5], except we introduced the additional sink ground node $n+1$. This mitigates the computational issues because the original CF-betweenness requires the inversion of the ill-conditioned matrix $[\tilde{D} - \tilde{A}]$, while for computing α-CF betweenness we need to invert the matrix $[D - \alpha A]$, which is a well posed problem, and has many efficient solutions (e.g., power iteration and Monte Carlo methods). In fact, as we shall show below, we need to obtain just a few rows of the inverse matrix $[D - \alpha A]^{-1}$. In the rest of the paper we will discuss the computation and the properties of the α-CF betweenness.

3 Computation of α-CF Betweenness

Due to the presence of the auxiliary node $n+1$, the value of $x_e^{(s,t)}$ on the right-hand side of (9) can be computed efficiently with high precision for any source-destination pair. However, the summation over all $n(n-1)$ pairs is a problem of prohibitive computational complexity even for graphs of modest size. The solution is to perform the computations for sufficiently many source-destination pairs. Since all source-destination pairs contribute equally in (9) we choose to sample them uniformly at random. This results in the next algorithm for computing the α-CF betweenness.

Algorithm 1.

1. Select a set of pairs of nodes $(s_i, t_i), i = 1, ..., N$, uniformly at random;
2. For each s_i and t_i, $i = 1, ..., N$ compute the rows $c_{s_i,\cdot}$, $c_{t_i,\cdot}$ (this can be done either by power iteration or by Monte Carlo algorithm);
3. For each edge $e = (v, w)$ and each pair (s_i, t_i), use (3) to compute

$$x_e^{(s_i,t_i)} = |\varphi_v^{(s_i,t_i)} - \varphi_w^{(s_i,t_i)}|.$$

4. Average over source-destination pairs

$$\bar{x}_e^\alpha = \frac{1}{N} \sum_{i=1}^{N} x_e^{(s_i,t_i)}.$$

Since we chose the pairs (s_i, t_i) uniformly at random then for every edge e, $\bar{x}_e^\alpha$ is just a sample average where all values are between zero and one. Then using the standard approach for the analysis of the series of independent random variables we have the following result.

Theorem 2. *Algorithm 1 approximates the alpha current flow betweenness in* $O(m\log(n)\varepsilon^{-2}\log(\varepsilon)/\log(\alpha))$ *time to within an absolute error of* ε *with arbitrarily high fixed probability.*

Proof: In addition to the proof of Theorem 3 in [5] we just need to note that we can compute Personalized PageRank with precision ε in $O(\log(\varepsilon)/\log(\alpha))$ power iterations. □

4 Truncated α-CF Betweenness

In the experiments we noticed that the values $x_e^{(s,t)}$ have a high variance, which results in poor precision when evaluating x_e^α. A closer analysis revealed that the edges adjacent to the source s receive large values of $x_e^{(s,t)}$, especially when $e = (v, s)$, where v has degree 1, so (v, s) is its only edge, and s has a large degree. This can be explained using the random walk interpretation. Consider a PageRank-type random walk on G. At each node, with probability α, the random walk traverses a randomly chosen edge of this node, and with probability $1-\alpha$ it jumps to the sink, node $n+1$. Denote by T_B the number of steps of the random walk needed to hit set B. It follows from Proposition 10 of [1, Chapter 3] that $\varphi_v^{(s,t)}/\varphi_s^{(s,t)} = P_v(T_{\{s\}} < T_{\{t,n+1\}})$, where $P_v(\cdot)$ is a conditional probability given that the random walk starts at v. Hence, if s is the only neighbor of v then $\varphi_v^{(s,t)}/\varphi_s^{(s,t)} = \alpha$, the probability of no absorption before reaching s. Thus, $|\varphi_s^{(s,t)} - \varphi_v^{(s,t)}| = (1-\alpha)\varphi_s^{(s,t)}$, which can be large if e.g. $\alpha = 0.8$ because $\varphi_s^{(s,t)}$ is the largest potential in the network. In contrast, the original CF-betweenness corresponds to $\alpha = 1$, implying that the current in (v, s) is zero.

This prompts us to introduce the truncated version of α-CF betweenness where for each edge (v, w) we only take into account the scores $x_{(v,w)}^{(s,t)}$ if $v, w \neq s$. In Fig. 1 we present log-linear plots of the empirical complementary distribution function of $x_{(v,w)}^{(s,t)}$ over all pairs (s, t) (solid line), and its truncated version (dashed line). The plots are given for two edges in the Dolphin social network described in Section 5 below. Nodes 1 and 36 are central in the network, so the high α-CF betweenness of (1,36) is expected. Node 60 has degree 1, so edge (32,60) gains an unwanted high betweenness in the non-truncated version.

Since the truncated α-CF betweenness gives lower scores to the edges connected to nodes of degree 1, one can expect that it has a higher correlation with CF-betweenness, especially for middle-range values of α. This is confirmed below in Fig. 2. Moreover, the truncated version removes outliers, and does not have large spread in values, thus standard statistical procedures, based on the Central Limit Theorem can be applied. Also, because of the smaller variance, Algorithm 1 achieves a desired precision with a smaller sample of source-destination pairs.

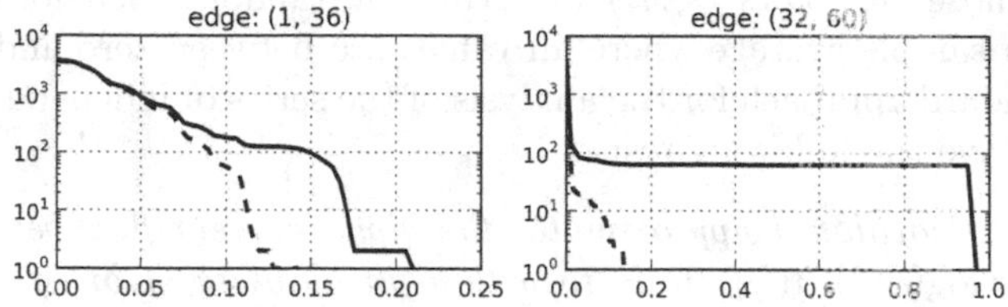

Fig. 1. The number of pairs s, t with $x_{(v,w)}^{(s,t)} > x$ over all pairs (s, t) (solid line) and only pairs with $v, w \neq s$ (dashed line)

5 Datasets

We consider the four graphs described below.

Dolphin Social Network. This small graph represents a social network of frequent associations between 62 dolphins in a community living off Doubtful Sound, New Zealand [11].

Graph of VKontakte Social Network. We have collected data from a popular Russian social network VKontakte. We were considering subgraph representing one of the connected components of people who stated that they were studying at Applied Mathematics - Control Processes Faculty at the St. Petersburg State University in different years. We ran the breadth-first search (BFS) algorithm starting at one specific node of the network and then anonymized the obtained users' data leaving only information about connections between people. Collected network consists of 2092 individuals out of total 8859 denoted the specified faculty in the Education field.

Watts-Strogatz Model. As an artificial example, we used a random graph generated by the Watts-Strogatz model. We have chosen this model as it combines high clustering and short average path length, thus different centrality measures give very different results on this graph. For other random models considered (Erdös-Rényi and Barabási-Albert) all measures are highly correlated and behave very similarly to each other.

Enron Graph. Enron email communication network is a well known test dataset. It covers all the email communication within a dataset of around half million emails between Enron's employees. The nodes are e-mail addresses, and an edge appears if an e-mail message was sent from one e-mail address to another. Although this graph is small compared to, say, web or Twitter samples, it is already prohibitively large for computing the CF-betweenness in its original form.

6 Numerical Results for α-CF Betweenness

To begin with, we compare the two versions of α-CF betweenness (truncated and without truncation) to the CF-betweenness scores defined as in [12,5].

Table 1. Datasets characteristics

	$\|V\|$	$\|E\|$	$\langle deg(v)\rangle$	$diam(G)$	$C_{\text{clustering}}$	$\langle d(u,v)\rangle$
Dolphin social network	62	159	5.13	8	0.259	3.357
VKontakte AMCP social graph	2092	14816	14.16	14	0.338	4.598
Watts-Strogatz ($n = 1000, k = 12, p = 0.150$)	1000	6000	12.00	6	0.422	3.713
Enron	36692	183831	10.02	11	0.4970	≈ 4.8

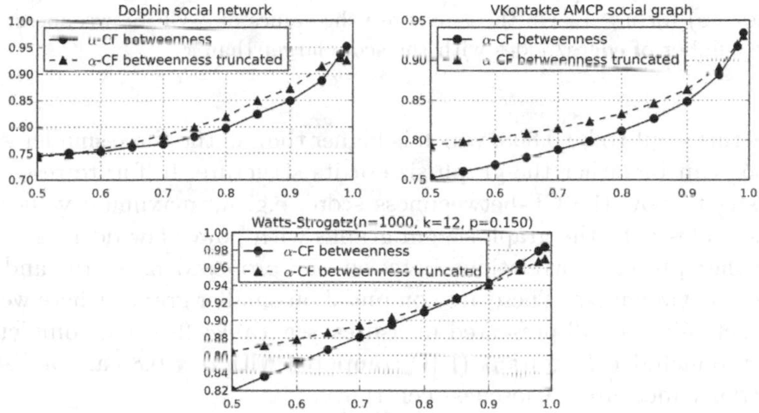

Fig. 2. Correlations between α-CF betweenness and truncated α-CF betweenness with CF-betweenness as a function of α

Fig. 2 presents the results for the three smaller graphs, in which the latter measure could be computed. As a correlation measure we use the Kendall tau rank correlation. We observe that the truncated version is better correlated with the CF-betweenness when α is not very close to one. As explained above, this is because the high probability of absorption in the auxiliary node $n + 1$ results in a relatively high current in the edges connected to the source, which is not necessarily the case if absorption is only possible in the destination node.

Next, we demonstrate that we can compute α-CF betweenness in the Enron graph, where the computation of CF-betweenness is infeasible (at least, with our means). We have evaluated α-CF betweenness, non-truncated and truncated, with $\alpha = 0.98$. We have run Algorithm 1 using $N = 20 \cdot 10^6$ source-destination pairs. In Fig. 3 we plot the complementary distribution function in log-linear scale, of the score $x_e^{0.98}$ across the edges.

Note that distribution over edges (the left plot in Fig. 3) does not have a large spread of values, except one outlier edge that connects two most important hubs. Since the weights of the edges are comparable, it is to be expected that in this graph the nodes of large degrees are also the ones with highest betweenness. Indeed, the Kendall's tau correlation between α-CF betweenness and degree of

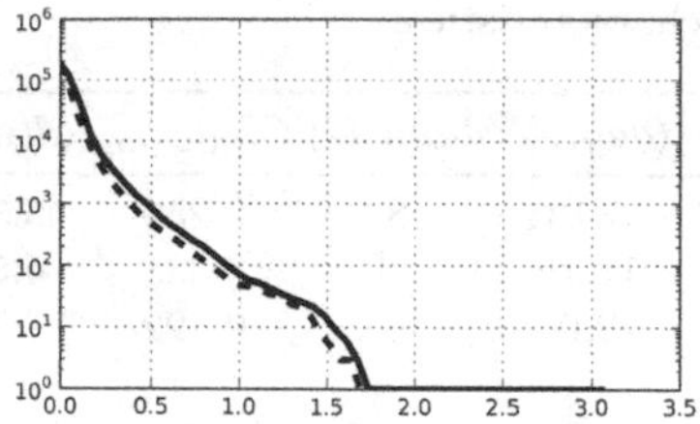

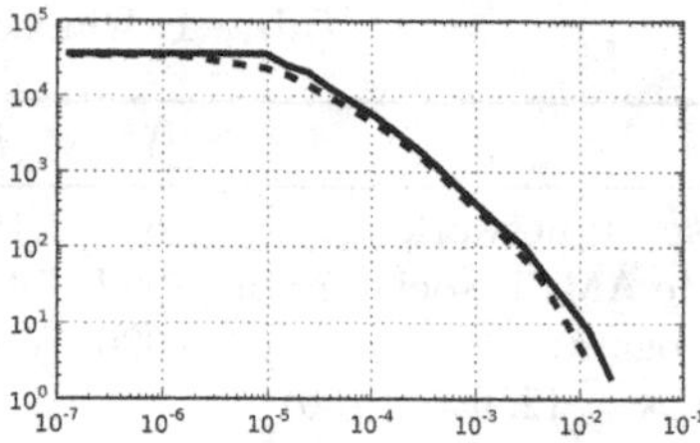

Fig. 3. Distribution of α-CF betweenness scores in the Enron graph, truncated (dashed line) and not truncated (solid line). Left: $10^4 \cdot x_e^{0.98}$ for edges $e \in E$. Right: α-CF betweenness (v) for $v \in V$. On the x-axis are the values of α-CF betweenness, on the y-axis the number of edges/nodes with the score larger than x.

the nodes turns out to be 0.808, which is higher than in the three smaller graphs. The reason can be either the graph size or its structure. In future research we will investigate how the CF-betweenness score, e.g. its maximum value across the edges, scales with the graph size in graphs with power law degrees.

We further present correlations between our proposed measures and other measures of betweenness. These are computed on smaller graphs where we could obtain exact values of all presented measures, see Tables 2–4. For completeness, we have also included PageRank (PR) computed with $\alpha = 0.85$ and a distance-base centrality measure – Closeness centrality:

$$C_C(v) = \frac{n-1}{\sum_{w \in V, w \neq v} d(v,w)},$$

where $d(v, w)$ is the graph distance between v and w. Betweenness (Between.) is computed as in (1).

Table 2. Kendall tau for centrality measures in Dolphin social network

	Degree	PR	Closeness	Between.	CF	αCF(0.8)	αCF-tr(0.8)	αCF(0.98)
Degree	1.000	0.930	0.548	0.665	0.737	0.864	0.855	0.769
PageRank	0.930	1.000	0.458	0.658	0.733	0.872	0.827	0.757
Closeness	0.548	0.458	1.000	0.578	0.575	0.515	0.573	0.591
Between.	0.665	0.658	0.578	1.000	0.829	0.749	0.759	0.828
CF	0.737	0.733	0.575	0.829	1.000	0.798	0.820	0.939
αCF(0.8)	0.864	0.872	0.515	0.749	0.798	1.000	0.925	0.838
αCF-tr(0.8)	0.855	0.827	0.573	0.759	0.820	0.925	1.000	0.876
αCF(0.98)	0.769	0.757	0.591	0.828	0.939	0.838	0.876	1.000

Note that α-CF betweenness is strongly correlated with CF-betweenness. The Closeness Centrality does not agree well with the CF-betweenness, even the PageRank and the degrees have a higher correlations with the CF-betweenness

Table 3. Kendall tau for centrality measures in the social graph VKontakte AMCP

	Degree	PR	Closeness	Between.	CF	αCF(0.8)	αCF-tr(0.8)	αCF(0.98)
Degree	1.000	0.655	0.679	0.521	0.545	0.659	0.668	0.599
PageRank	0.655	1.000	0.375	0.662	0.717	0.833	0.811	0.766
Closeness	0.679	0.375	1.000	0.382	0.356	0.424	0.445	0.395
Between.	0.521	0.662	0.382	1.000	0.761	0.760	0.749	0.778
CF	0.545	0.717	0.356	0.761	1.000	0.812	0.833	0.917
αCF(0.8)	0.659	0.833	0.424	0.760	0.812	1.000	0.938	0.878
αCF-tr(0.8)	0.668	0.811	0.445	0.749	0.833	0.938	1.000	0.903
αCF(0.98)	0.599	0.766	0.395	0.778	0.917	0.878	0.903	1.000

Table 4. Kendall tau for centrality measures in the Watts-Strogatz graph (n=1000, k=12, p=0.150)

	Degree	PR	Closeness	Between.	CF	αCF(0.8)	αCF-tr(0.8)	αCF(0.98)
Degree	1.000	0.891	0.462	0.526	0.610	0.643	0.581	0.612
PageRank	0.891	1.000	0.415	0.485	0.565	0.610	0.546	0.567
Closeness	0.462	0.415	1.000	0.655	0.613	0.647	0.666	0.628
Between.	0.526	0.485	0.655	1.000	0.853	0.819	0.852	0.857
CF	0.610	0.565	0.613	0.853	1.000	0.910	0.914	0.979
αCF(0.8)	0.643	0.610	0.647	0.819	0.910	1.000	0.935	0.923
αCF-tr(0.8)	0.581	0.546	0.666	0.852	0.914	0.935	1.000	0.930
αCF(0.98)	0.612	0.567	0.628	0.857	0.979	0.923	0.930	1.000

in real graphs. Recent paper [2] suggests more measures based on distance, and efficient computation methods for such measures is presented in [3]. In future it will be interesting to compare these new measures to α-CF betweenness.

7 Centrality Measures and Network Vulnerability

We now consider how well the CF-betweenness and α-CF betweenness can indicate the nodes responsible for maintaining connectivity of a network. We follow the methodology of [10]. As measures of connectivity we choose the average inverse distance

$$< d^{-1} >= \frac{1}{n(n-1)} \sum_{u,v \in V, u \neq v} \frac{1}{d(u,v)}$$

and the size of the largest connected component. In the experiment, we remove the top nodes one by one, according to different betweenness measures, and observe how the connectivity of the network changes. In Fig. 4 the results are presented for the inversed average distance.

The results for the social graph VKontakte are especially interesting, because this network turns out to be less vulnerable to the removal of nodes with large degree than nodes with large betweenness and its modifications (CF-betweenness,

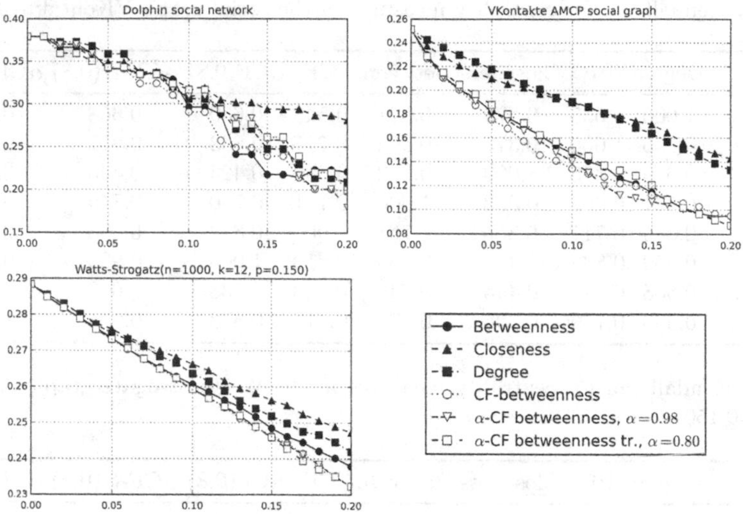

Fig. 4. Inverse average distance as a function of the fraction of removed top-nodes according to different betweenness centrality measures

α-CF betweenness, and truncated α-CF betweenness). On the small Dolphin social network there is no much difference in vulnerability with respect to different centrality measures. Finally, on the artificial Watts-Strogatz graph the CF-betweenness and our proposed two versions of α-CF betweenness find the nodes that are most essential for the network connectivity.

In Fig. 5 we plot the size of the largest connected components against the fraction of removed top-nodes. We do not present the plot for the Watts-Strogatz graph because it remains entirely connected, so the size of its largest connected component equals to the number of remaining nodes irrespectively of which nodes are removed first. For the two real graphs, the CF-betweenness is most efficient in reducing the size of the giant component. On the Dolphin graph, α-CF betweenness performs closely to CF-betweenness, except the interval when 13-18% of nodes are removed. On the graph VKontakte, α-CF betweenness and its truncated version perform comparably to the CF-betweenness. Again, on this graph, degree and Closeness centrality fail to reveal the nodes responsible for good network connectivity. The α-CF betweenness with $\alpha = 0.98$ appears to be a better indicator for vulnerability than the truncated α-CF betweenness with $\alpha = 0.8$. The latter however also gives good results, and can be computed easier on large graphs due to the faster convergence of the power iteration algorithm.

We conclude that both α-CF betweenness and truncated α-CF betweenness provide an adequate measure for the role of a node in network's connectivity. Furthermore, their computational costs are lower than for known measures of betweenness, and the computations can be done in parallel easily. Thus, α-CF betweenness can be applied in large graphs, for which computation of other measures of betweenness is merely infeasible.

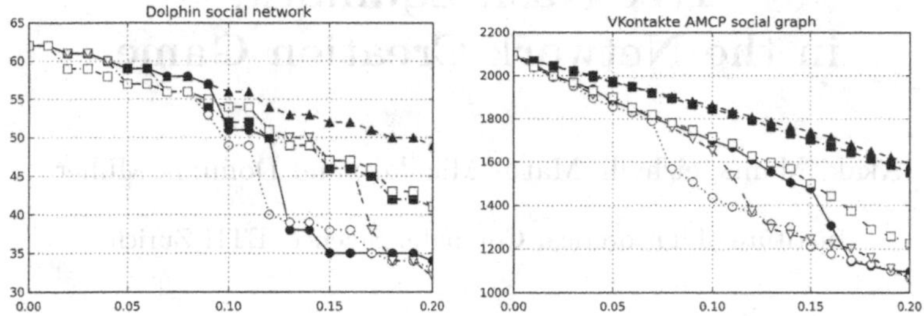

Fig. 5. The size of the largest connected component as a function of the fraction of removed top-nodes according to different betweenness centrality measures. (same legend as in Fig. 4).

References

1. Aldous, D., Fill, J.: Reversible Markov chains and random walks on graphs (1999)
2. Boldi, P., Vigna, S.: Axioms for centrality. arXiv:1308.2140
3. Boldi, P., Vigna, S.: In-core computation of geometric centralities with hyperball: A hundred billion nodes and beyond. arXiv:1308.2144
4. Brandes, U.: A faster algorithm for betweenness centrality. Journal of Mathematical Sociology 25(1994), 163–177 (2001)
5. Brandes, U., Fleischer, D.: Centrality measures based on current flow. In: Proceedings of the 22nd Annual Conference on Theoretical Aspects of Computer Science, pp. 533–544 (2005)
6. Brin, S., Page, L.: The anatomy of a large-scale hypertextual Web search engine. Computer Networks and {ISDN} Systems 30(1-7), 107–117 (1998)
7. Doyle, P.G., Snell, J.L.: Random walks and electric networks. Mathematical Association of America (1984)
8. Freeman, L.C.: A set of measures of centrality based on betweenness. Sociometry (1977)
9. Grolmusz, V.: A note on the pagerank of undirected graphs. arXiv preprint arXiv:1205.1960 (2012)
10. Holme, P., Kim, B.J., Yoon, C.N., Han, S.K.: Attack vulnerability of complex networks. Physical Review. E, Statistical, Nonlinear, and Soft Matter Physics 65(5, pt. 2), 056109 (2002)
11. Lusseau, D., Schneider, K., Boisseau, O.J., Haase, P., Slooten, E., Dawson, S.M.: The bottlenose dolphin community of Doubtful Sound features a large proportion of long-lasting associations. Behavioral Ecology and Sociobiology 54(4), 396–405 (2003)
12. Newman, M.E.J.: A measure of betweenness centrality based on random walks. Social Networks, 1–15 (2005)

Tree Nash Equilibria in the Network Creation Game

Akaki Mamageishvili, Matúš Mihalák, and Dominik Müller

Institute of Theoretical Computer Science, ETH Zurich

Abstract. In the network creation game with n vertices, every vertex (a player) buys a set of adjacent edges, each at a fixed amount $\alpha > 0$. It has been conjectured that for $\alpha \geq n$, every Nash equilibrium is a tree, and has been confirmed for every $\alpha \geq 273 \cdot n$. We improve upon this bound and show that this is true for every $\alpha \geq 65 \cdot n$. To show this, we provide new and improved results on the local structure of Nash equilibria. Technically, we show that if there is a cycle in a Nash equilibrium, then $\alpha < 65 \cdot n$. Proving this, we only consider relatively simple strategy changes of the players involved in the cycle. We further show that this simple approach cannot be used to show the desired upper bound $\alpha < n$ (for which a cycle may exist), but conjecture that a slightly worse bound $\alpha < 1.3 \cdot n$ can be achieved with this approach. Towards this conjecture, we show that if a Nash equilibrium has a cycle of length at most 10, then indeed $\alpha < 1.3 \cdot n$. We further provide experimental evidence suggesting that when the girth of a Nash equilibrium is increasing, the upper bound on α obtained by the simple strategy changes is not increasing. To the end, we investigate the approach for a coalitional variant of Nash equilibrium, where coalitions of two players cannot collectively improve, and show that if $\alpha \geq 41 \cdot n$, then every such Nash equilibrium is a tree.

1 Introduction

Network creation game has been introduced by Fabrikant et al. [8] as a formal model to study the effects of strategic decisions of economically motivated agents in decentralized networks such as the Internet. In such networks, local decisions including those about infrastructure are decided by autonomous systems. Autonomous systems follow their own interest, and as a result, their decisions may be sub-optimal for the whole society. Network creation games allow to formally study the structure of networks created in such a manner, and to compare them with potentially optimal networks (optimal with respect to the whole society).

In the network creation game, there are n players $V = \{1, \ldots, n\}$, each representing a vertex of an undirected graph. The strategy s_i of a player i is to create (or buy) a set of adjacent edges, each at a fixed amount $\alpha > 0$. The played strategies $s = (s_1, \ldots, s_n)$ collectively define an edge-set E_s, and thus a graph $G_s = (V, E_s)$. The goal of every player is to minimize its cost c_i, which is the

A. Bonato, M. Mitzenmacher, and P. Prałat (Eds.): WAW 2013, LNCS 8305, pp. 118–129, 2013.
© Springer International Publishing Switzerland 2013

amount paid for the edges (creation cost), plus the total distances of the player to every other node of the resulting network G (usage cost), i.e.,

$$c_i(s) = \alpha \cdot |s_i| + \sum_{j=1}^{n} \text{dist}(i,j),$$

where $\text{dist}(i,j)$ denotes the distance between i and j in the resulting network G.

A strategy vector $s = (s_1, \ldots, s_n)$ is a *Nash equilibrium* if no player i can change the set s_i of created edges to another set s_i' and improve its cost c_i. Abusing the definition, the resulting graph G_s itself is called a Nash equilibrium, too, and we define its (social) cost $c(G)$ to be the cost $c(s)$, i.e., the cost of the corresponding strategy vector s. The *social cost* $c(s)$ of strategy vector s is the sum of the individual costs, i.e., $c(s) = \sum_{i=1}^{n} c_i(s)$. It is a trivial observation to see that in any Nash equilibrium G_s, no edge is bought more than once. From now on, we only consider such strategy vectors, and observe then that

$$c(s) := \sum_{i=1}^{n} c_i(s) = \alpha \cdot |E_s| + \sum_{i=1}^{n}\sum_{j=1}^{n} d(i,j).$$

A graph $G = (V, E)$ can be created by many strategy vectors s (precisely in $2^{|E|}$ many ways, because every edge in E can be bought by exactly one of its endpoints), but each of such realizations has the same social cost. Graph $G^* = (V, E)$ is an *optimum* graph, if it minimizes the *social cost* $c(s)$ (for any strategy vector s for which $G_s = G$).

Let $\mathcal{N}$ denote the set of all Nash equilibria of a network creation game on n vertices and edge-price α. The *price of anarchy* (PoA) of the network creation game is the ratio

$$\text{PoA} = \max_{s \in \mathcal{N}} \frac{c(G_s)}{c(G^*)}.$$

Price of anarchy expresses the (worst-case) loss of the quality of a network that the society could achieve.

In a series of papers [8,1,6,9] it has been shown that the price of anarchy of the network creation game is $O(1)$, i.e., a constant independent of both n and α, for every value $\alpha > 0$ with the exception of the range $n^{1-\varepsilon} < \alpha < 273 \cdot n$, where $\varepsilon = \Omega(\frac{1}{\log n})$. For the value of α with $n^{1-\varepsilon} < \alpha < 273 \cdot n$, an upper bound of $2^{\sqrt[5]{\log n}}$ on the price of anarchy is known (while no Nash equilibrium with considerably large social cost is known). It is conjectured, however, that the price of anarchy is constant also in this range of α. It remains a major open problem to confirm or disprove this conjecture. It is certainly of interest to note that there are several variants of the network creation game (see, e.g., [2,4,7,3]), but in none of these, with the exception of [5], the price of anarchy could be shown to be constant.

Understanding the structure of Nash equilibria has proven to be important in bounding the price of anarchy. Fabrikant et al. [8] showed that the social cost of any tree G in Nash equilibrium is upper-bounded by $O(1) \cdot c(G^*)$. Therefore, the price of anarchy is $O(1)$ for all values of α for which every Nash equilibrium is a tree. It has been shown that every Nash equilibrium is a tree for all values of α greater than n^2, $12n \log n$, and $273n$, respectively, in [8],[1], and [9]. It has been conjectured that every Nash equilibrium is a tree for every $\alpha \geq n$. Since for $\alpha = n/2$, non-tree Nash equilibria are known, this *tree conjecture* is asymptotically tight.

In this paper, we make steps in the direction of resolving the tree conjecture. We first tighten the tree conjecture and provide a construction of a non-tree Nash equilibrium for every $\alpha = n - 3$ (thus, showing that, asymptotically, one cannot hope to show that every Nash equilibrium is a tree for some value $\alpha < n$). We then apply a "linear-programming-like" approach to show that for $\alpha \geq 65n$, every Nash equilibrium is a tree. To show this, we obtain new structural results on Nash equilibria and combine them with the previous approach of [9]. Towards the end, we make further steps towards the conjecture. We show that if $\alpha \geq n$, then there is no non-tree Nash equilibrium containing exactly one cycle. We then apply the "linear-programming-like" approach again to show that the girth of every non-tree Nash equilibrium (for any $\alpha \geq n$) is at least 6. Using the same ideas, we show that if a non-tree Nash equilibrium has girth at most 10, then $\alpha \leq 1.3n$. By further experimental results, we conjecture that this holds for any girth, i.e., that non-tree Nash equilibria can appear only for $\alpha \leq 1.3n$.

2 Preliminaries

In the following, we will often denote the considered Nash equilibrium graph $G_s = (V, E_s)$ of a network creation game with $\alpha > 0$ simply as $G = (V, E)$. Even though the graph G_s is undirected, we will often direct the edges to express the identity of the player which bought the edge in s; An edge $\{u, v\}$ directed from u to v denotes the fact that u bought/created the edge in s.

Every non-tree G contains a cycle. Let c be the length of a shortest cycle C in G, and let $a_0, a_1, \ldots, a_{c-1}$ be the players that form one such shortest cycle, and where $\{a_i, a_{i+1}\} \in E$ for every $i = 0, 1, \ldots, c-1$ (where indices on vertices of the cycle are in the whole paper to be understood modulo c). Observe the crucial property of a shortest cycle C: the distance between a_i and a_j in the graph G is equal to the distance between a_i and a_j on the cycle C.

We will consider the players on the cycle C and their strategy-changes that involve only the c edges of the cycle. For each strategy-change s'_{a_i} of player a_i, we obtain an inequality $c_i(s) \leq c_i(s_1, \ldots, s'_{a_i}, \ldots, s_n)$ stating simply the fact that in a Nash equilibrium s, player a_i cannot improve by changing its strategy. We will often express such an inequality in the form of "SAVINGS" $\leq$ "INCREASE", where "SAVINGS" denotes the parts of $c_i(s)$ that decreased their value in $c_i(s')$, and "INCREASE" denotes the parts of $c_i(s)$ that increased their value in $c_i(s')$. For example, assume that a_i buys the edge $e = \{a_i, a_{i+1}\}$ (i.e., $e \in s_i$), and let us consider the strategy change where a_i *deletes* the edge e (i.e., $s'_i = s_i \setminus \{e\}$). Recall that $c_i(s) = \alpha \cdot |s_i| + \sum_j d(i, j)$. Then, in such a strategy change, the "SAVINGS" are clearly on the edge-creation side, i.e., the player a_i saves α for

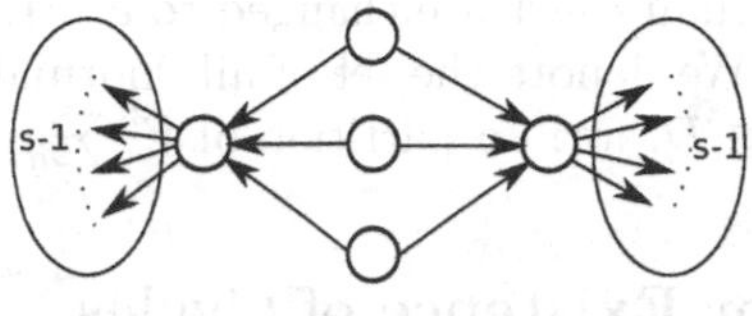

Fig. 1. Non-tree Nash equilibrium for $n = 2s + 3$ players and $\alpha = n - 3$. An edge directed from a node u to a node v denotes that u buys the edge.

not paying for the edge e. At the same time, some distances of player i may have increased – the distance to a vertex v increases, if in G_s every shortest path from a_i to v uses the deleted edge e. But the distance to v could have increased by at most $c-2$ (as before, a_i needed to go to vertex a_{i+1} but now the vertex a_{i+1} can be reached "around" the cycle). Because of the Nash equilibrium property of s, we have "SAVINGS" $\leq$ "INCREASE", which implies $\alpha \leq (c-2)(n-1)$ (as the distance to at most $n-1$ vertices could have increased).

In the following, we will use slightly more involved forms of the just described inequalities. For that reason, we will partition the vertices according to their distances to the vertices from the cycle. Let us fix a vertex $v \in V$. Let $G \setminus C$ be the graph G without the c edges of the cycle C. Let us denote the distances of v to the vertices $a_0, a_1, \ldots, a_{c-1}$ in $G \setminus C$ by the vector $d(v) = (d_0, d_1, \ldots, d_{c-1})$, respectively, where $d_i = \infty$ if a_i and v are disconnected in $G \setminus C$. We call d_i the *outer distance* of v to a_i in the Nash equilibrium G, and d the *vector of outer distances* of v in G. We now partition the vertices of V by this vector of outer distances. We will coarsen the partition in the following way. Observe that $d_s(a_i, v)$ in G_s is now equal to $\min_j(d_s(a_i, a_j) + d_j)$, because there always is a shortest path from a_i to v that first uses a part of the cycle C (until vertex a_j), leaves C and never comes back to C. Therefore, $\min_j d_j \leq d_s(a_i, v) \leq (c-1) + \min_j d_j)$. Moreover, for any strategy change s_i' of player a_i which leaves a_i connected by an edge to a vertex of C, we still have $\min_j d_j \leq d_{s'}(a_i, v) \leq (c-1) + \min_j d_j$ (because there is a path from a_i to the vertex a_j of smallest entry d_j using the edge and the remaining of the cycle). Because we are interested in the changes of the distances from a_i, i.e., in the value of $\Delta := d_{s'}(a_i, v) - d_s(a_i, v)$, we can normalize the vector $d(v)$ by subtracting $\min_j d_j$ from each of the elements $d_0, d_1, \ldots, d_{c-1}$ (which does not change the value of Δ). Observe that after the normalization, there is an entry d_i equal to zero. We will "normalize" the entries further more. Since we are interested in the value Δ, we can handle all entries $d_j \geq c-1$ in the same way: they do not have any influence on Δ at all (no shortest path from vertex a_i, $i \neq j$, will ever use a_j to reach vertex v). We will therefore further modify the vector d by substituting every entry $d_j \geq c-1$ with the value $c-1$.

This gives partition of all vertices into groups V_d, where each group has associated vector of "normalized" outer distances $d = (d_0, \cdots, d_{c-1})$, one of the distances is necessarily equal to 0 and all the distances are upper bounded by $c-1$. Vertices which have vector of outer distances d' containing numbers greater than $c-1$ are associated with the group having a vector d'' obtained from d'

where all entries greater than $c-1$ are changed to $c-1$. In this way, there are $t = c^c - (c-1)^c$ groups. We denote the set of all "normalized" distance vectors by D. Trivially, as V_d, $d \in D$, form a partition of V, $\sum_{d\in D} |V_d| = n$.

3 Bounds on α for Existence of Cycles

We first give in Fig. 1 a construction of a non-tree Nash equilibrium graph for $n = 2s+3$ vertices, and $\alpha = 2s = n-3$, for any integer s. This thus shows that the conjecture "for $\alpha \geq n$, all Nash equilibria are trees" cannot be improved to "for $\alpha \geq (1-\varepsilon)n$, all Nash equilibria are trees". We now proceed and give a lower bound on the length of a shortest cycle in any Nash equilibrium.

Theorem 1. *The length c of a shortest cycle C in any Nash equilibrium is at least $\frac{2\alpha}{n}+2$.*

Proof. We distinguish two cases. First, assume that there is a player, which buys both its adjacent edges on the cycle C. Without loss of generality assume that this player is a_0. Consider the strategy change where a_0 deletes both these edges $\{a_0, a_1\}$ and $\{a_0, a_{c-1}\}$ and buys an edge towards player a_i on the cycle, $i = 2, \ldots, c-2$. The player cannot improve by such a change, and therefore "SAVINGS" $\leq$ "INCREASE". Here, the player saves at least α (by buying one edge less). Let us denote the increase of distances of player a_0 to the players of the group V_d by $c_{i,d}$. Then we get that $\alpha \leq \sum_{d\in D} \delta_{i,d}|V_d|$. Summing up all the $c-3$ inequalities, one for every i, we get $(c-3)\alpha \leq \sum_{i=2}^{c-2}\sum_d \delta_{i,d}|V_d|$.

We now show that for every d, the coefficient $\sum_i \delta_{i,d}$ at $|V_d|$ is at most $(c-2)(c-3)/2$. Consider arbitrary $d = (d_0, d_1, \ldots, d_{c-1})$ of the outer distances of the vertices in V_d. Clearly, the strategy change of a_0 increases its distances to V_d iff every shortest path from a_0 to V_d goes through the deleted edges. Thus, we can assume (for the worst-case) that $d_0 = c-1$. Assume that one shortest path (in G_s) leaves the cycle at a_e, $e \in \{1, \ldots, c-2\}$. In the new graph $G_{s'}$, player a_0 can always use the new edge $\{a_0, a_i\}$ and then go to a_e on the remainder of the cycle C. Thus, the increase of distances $\delta_{i,d}$ is at most $(1+|i-e|)-1 = |i-e|$. In total, we obtain $\sum_{i=2}^{c-2}\delta_{i,d} \leq \sum_i |i-e| \leq \sum_i (i-1) = (c-3)(c-2)/2$, as claimed. Now, since $\sum_{d\in D}|V_d| = n$, we finally get that $\alpha \leq \frac{(c-2)}{2}n$, which gives the claimed $c \geq \frac{2\alpha}{n}+2$.

Consider now the second case where no player buys two of its adjacent edges in C, i.e., every player buys exactly one edge. Without loss of generality assume that every player a_i buys the edge $\{a_i, a_{i+1}\}$. For each player i, we consider the strategy change of deleting the edge $\{a_i, a_{i+1}\}$. Similarly to the previous case, we obtain $\alpha \leq \sum_{d\in D}\delta_{i,d}|V_d|$. Summing for every i, we get $c\alpha \leq \sum_{i=0}^{c-1}\sum_d \delta_{i,d}|V_d|$. We show this time that $\sum_{i=0}^{c-1}\delta_{i,d}$, the coefficient at $|V_d|$, is upper bounded by $1+2+\cdots+(c-2) = (c-2)(c-1)/2$. Consider an arbitrary $d = (d_0, \ldots, d_{c-1}) \in D$, and assume without loss of generality that $d_0 = 0$. For every player a_i, $\delta_{i,d}$ is at most $i-1$, because the worst-case increase in a distance of player a_i to vertices V_d happens when all shortest paths from a_i used the deleted edge $\{a_i, a_{i+1}\}$.

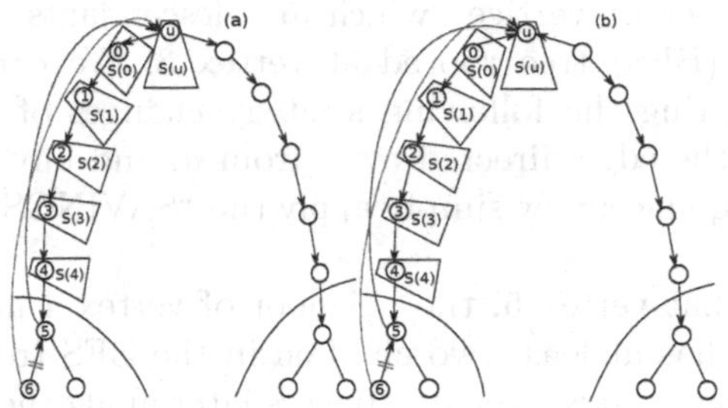

Fig. 2. The 5-Neighborhood $N_5(u)$ of vertex u

But because after the deletion, there is an alternative path from a_i to V_d using a_0, the increase is at most $i-1$. Thus, summing over all i, the total increase in distances to V_d is at most $0+1+2+\cdots+(c-2)=(c-2)(c-1)/2$ as claimed. Plugging this into our inequality, $c\alpha \leq \sum_i \sum_d \delta_{i,d}|V_d|$ and using the fact that $\sum_d |V_d| = n$, we obtain that $c > \frac{2\alpha}{n} + 2$. □

Let H be a non-trivial biconnected component of a non-tree Nash equilibrium, i.e., an induced subgraph of H of at least three vertices containing no bridge. For any vertex $v \in H$, let $S(v)$ be the set of vertices which do not belong to H, and which have v as the closest vertex among all vertices in H. For any vertex $u \in H$, we define $deg_H(u)$ to be the degree of vertex u in the graph induced by H. Furthermore, we define $N_k(u)$ to be the k-th neighborhood of u in H, i.e., $N_k(u) := \{w \in H \mid d(u,w) \leq k\}$. The following lemma has been shown in [9]. We will use it to prove the subsequent lemma.

Lemma 1 ([9]). *If $u, v \in V(H)$ are two vertices in H with $d(u,v) \geq 3$ such that u buys the edge to its adjacent vertex x in a shortest $u-v$-path and v buys the edge to its adjacent vertex y in that path, then $deg_H(x) \geq 3$ or $deg_H(y) \geq 3$.*

Lemma 2. *If H is a biconnected component of G, then for any vertex u, its neighborhood $N_5(u)$ in H contains a vertex v with $deg_H(v) \geq 3$.*

Proof. Assume that this is not true. Then the 5-neighborhood $N_5(u)$ of vertex u is formed by two disjoint paths. (The case that the 5-neighborhood forms a cycle is excluded by Proposition 1 stating that no Nash equilibrium for $\alpha > n$ contains exactly one cycle). We consider two cases. First, we will assume that at least one of the two paths starting at u is directed away from u (see Fig. 2(a)). In the second case, in each of the two paths, there has to be a vertex which buys an edge towards u. It follows from Lemma 1 that these two vertices are the two neighbors of u in $N_5(u)$ (see Fig. 2(b)).

In the first case, there is a sequence of five edges directed away from u, with the naming like in Fig. 2(a)). Let $s_u := |S(u)|$, $s_i = |S(i)|$ for $0 \leq i \leq 4$. Then,

$$s_0 \geq s_1+s_2+s_3+s_4, s_1 \geq s_2+s_3+s_4, s_2 \geq s_3+s_4, s_3 \geq s_4, s_4 \geq k, \quad (1)$$

where k is the number of vertices which are descendants of vertex 5 in the breadth-first-search (BFS) tree rooted at vertex 3. We can obtain these inequalities by considering the following strategy changes of the players u and i, $0 \leq i \leq 3$: delete the edge directed away from u, and buy a new edge to the next vertex in the sequence; now simply apply the "SAVINGS" $\leq$ "INCREASE" principle.

We first assume that vertex 5, the neighbor of vertex 4 in H, has degree at least 3 in H (i.e., it has at least two children in the BFS tree rooted at vertex 3). The case when the degree-3 vertex appears later along the path is easier and will be discussed later. We now distinguish two cases. First, we assume that one of the children of vertex 5 in the considered BFS tree buys an edge to vertex 5. Let us call it vertex 6. The other case is when vertex 5 buys all the edges to its children.

Consider the following strategy change: vertex 6 deletes an edge towards vertex 5 and buys new edge towards vertex u. This decreases its distance cost at least to vertices in $S(0)$ by 4, and to vertices in $S(1)$ by 2, whilst increases distances to vertices in the set of descendants of 5 in the BFS tree rooted at 3 by at most 6, to the vertices in $S(4)$ by 4 and to the vertices in $S(3)$ by two. By this strategy change distance from vertex 6 to any other vertex is not increased, because vertex u is located deeper than vertex 6 in the BFS tree rooted at vertex 3. But then according to the chain of inequalities (1) we get $4s_0 + 2s_1 > 6k + 4s_4 + 2s_3$, and thus the player 6 can improve, a contradiction.

In the case where vertex 5 buys all edges towards its children, consider the following strategy change of vertex 5: delete all the edges to its children (in the considered BFS tree) and buy one edge to vertex u. By this, the "SAVINGS" are at least α. Furthermore, since H is biconnected, the graph remains connected. Distances from vertex 5 are increased only to vertices in the set K – the set of the vertices which are descendants of vertex 5 in the BFS tree rooted at vertex 3. This "INCREASE" is at most $2 \cdot diam(H)$, where $diam(H)$ is the diameter of H. By the "SAVINGS" $\leq$ "INCREASE" principle, we get that $\alpha \leq 2 \cdot diam(h)k$. At the same time, $\alpha \geq (rad(H) - 1)s_0$, where $rad(H)$ is the radius of H, as otherwise a vertex at distance $rad(H)$ from vertex 0 could buy an edge towards vertex 0 and decrease its cost. Combining these two inequalities with the inequality $s_0 \geq 8k$, which is obtained from (1), we get that $8(rad(H) - 1)k \leq 2 \cdot diam(H)k \leq 4 \cdot rad(H)k$, which is a contradiction.

The second case depicted in Fig. 2(b) is analyzed in the very same way, the only change is that now the heaviest component is $S(u)$. The chain of inequalities is similar to (1):

$$s_u \geq s_0 + s_1 + s_2 + s_3, s_1 \geq s_2 + s_3 + s_4, s_2 \geq s_3 + s_4, s_3 \geq s_4, s_4 \geq k, \quad (2)$$

where the notation is the same as in the first case. We obtain that $s_u \geq 7k$, and subsequently, arguing about the vertex at distance $rad(H)$ from u, the contradiction $7(rad(H) - 1)k \leq 2 \cdot diam(H)k \leq 4rad(H)k$.

Finally, if there is a longer sequence of vertices with degree 2 than the considered sequence of length 5 of edges directed away from u, then we can only

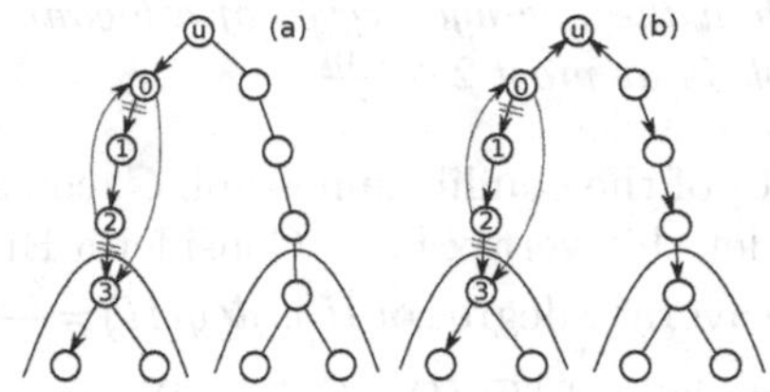

Fig. 3. The 3-Neighborhood $N_3(u)$ of vertex u

consider the last 5 edges (all directed away from u) and apply the very same reasoning. □

We can strengthen the result if we consider stronger version of a Nash equilibrium in which no coalition of two players can change their strategies and improve their overall cost. We call such an equilibrium a 2-*coalitional Nash equilibrium.*

Lemma 3. *The 3-neighborhood $N_3(u)$ of any vertex u of a biconnected component H of a 2-coalitional Nash equilibrium has a vertex of degree at least 3.*

Proof. Assume the converse. Similarly to the proof of Lemma 2, there are two different cases of how the neighborhood of vertex u looks like (see Fig. 3(a) and (b); notation is also the same as in Lemma 2). In both cases consider the coalition of players 0 and 2. Consider the following strategy changes: player 0 deletes edge $(0, 1)$ and instead buys edge $(0, 3)$, whilst player 2 deletes edge $(2, 3)$ and buys edge $(2, 0)$. This strategy change does not change the player coalition's creation cost (in terms of α). Among the vertices $S(0), S(1), S(2)$ and $S(u)$ this strategy change decreases coalition's usage cost by $s_u + s_0 + s_2$ and increases by s_1. Other vertices are partitioned by their shortest distances to vertices 0 and 2, lets assume that for any vertex v which does not belong to $S(0), S(1), S(2)$ or $S(u)$ shortest distance to vertex 0 is x and shortest distance to vertex 2 is y. Obviously $|x - y| \leq 2$. If $|x - y| > 0$ then there is no increase in the usage cost of coalition towards vertex v by this strategy change. The only possibility of increase is when $x = y$, but in that case v is the descendant of vertex 3 in the BFS tree rooted at vertex 1. Similarly to Lemma 2, we denote k to be the number of vertices which are descendants of vertex 3 in the BFS tree rooted at vertex 1. Analogously to the proof of Lemma 2, the following inequalities hold for the case depicted in Fig. 3(a): $s_0 \geq s_1 + s_2, s_1 \geq s_2$ and $s_2 \geq k$, whilst for the case depicted in Fig. 3(b), we have $s_u \geq s_0 + s_1 + s_2, s_1 \geq s_2, s_2 \geq k$. In both cases $s_u + s_0 + s_2 > s_1 + k$, which results in a contradiction. □

The following two lemmas are crucial for proving the main result of the paper. The first lemma has been proven in [9]. The second lemma strengthens a similar lemma from [9]. Its proof uses the result of Theorem 1.

Lemma 4 ([9]). *If the t-neighborhood of every vertex of a biconnected component H of a Nash equilibrium contains a vertex of degree at least* 3, *then the average degree of H is at least $2 + \frac{1}{3t+1}$.*

Lemma 5. *If $\alpha > n$, then the average degree of a biconnected component H of a Nash equilibrium graph is at most $2 + \frac{4n}{\alpha - n}$.*

Proof. Among all vertices of the equilibrium graph G, consider a vertex with the smallest usage cost and let this vertex be v. Consider a BFS tree T rooted in v. Let $T' = T \cap H$. Then the average degree of H is $deg(H) = \frac{2|E(T')| + 2|E(H) \setminus E(T')|}{|V(T')|} \leq 2 + \frac{2|E(H) \setminus E(T')|}{|V(T')|}$. We now bound $|E(H) \setminus E(T')|$. We consider vertices that buy an edge in $E(H) \setminus E(T')$ and call them *shopping vertices*. It is easy to see that no shopping vertex buys more than 1 edge, because if any of them buys two or more edges, it is better for it to delete all of the edges and buy 1 new edge towards v: this decreases its creation cost by at least α, whilst increases its usage cost by at most n. It is thus enough to bound the number of shopping vertices. For this, we prove that the distance in the tree T' between any two shopping vertices is lower bounded by $\frac{\alpha - n}{n}$, which then implies that there can not be too many shopping vertices. Namely, the number of shopping vertices is at most $\frac{2nV(T')}{\alpha - n}$. Assigning every node from H to the closest shopping vertex according to the distance in T' forms a partition of H, where every part contains exactly one shopping vertex. As the distance in T' between shopping vertices is at least $\frac{\alpha - n}{n}$, the size of every part is at least $\frac{\alpha - n}{2n}$.

We assume for contradiction that there is a pair of shopping vertices u_1 and u_2 such that $d_{T'}(u_1, u_2) < \frac{\alpha - n}{n}$. Let $u_1 = x_1, \cdots, x_k = u_2$ be the unique path from u_1 to u_2 in T', and (u_1, v_1) and (u_2, v_2) be the edges bought by u_1 and u_2 in $E(H) \setminus E(T')$. Observe first that vertices v_1 and v_2 are not descendants of any vertex x_i, otherwise paths $v_j - x_i$ and $x_i - u_j$ together with an edge (u_j, v_j) form a cycle of length at most $2(d_{T'}(u_1, u_2) + 1) < \frac{2\alpha}{n} + 2$ which contradicts Theorem 1. Thus, $x_0 := v_1, x_1, \ldots, x_k, x_{k+1} := v_2$ is a path. Since x_1 buys edge (x_0, x_1) and x_k buys edge (x_k, x_{k+1}), there is a vertex x_i such that x_i buys both of its adjacent edges (x_{i-1}, x_i) and (x_i, x_{i+1}). Consider the following strategy change for player x_i: delete the two adjacent edges and buy a new edge to vertex v. In this way x_i decreases its creation cost by α.

We now show that $U_{\text{new}}(x_i)$, the usage cost of x_i in the new graph, is less than $U_G(x_i)$, the usage cost in the original graph, plus α, which gives a contradiction. It is easy to observe that $U_{\text{new}}(x_i) \leq n + U_{\text{new}}(v)$, since x_i can always go through v in the new strategy to any vertex. We now consider $U_{\text{new}}(v)$. Note that only the vertices in the path $u_1 - u_2$ and their descendants can increase their distance to v by the strategy change of x_i. Let y be any such vertex. If the closest ancestor of y on the path is x_i, then $d_{\text{new}}(v, y) \leq d_G(v, y)$, so there is no increase. We assume, without loss of generality, that the closest ancestor (of y) x_j has an index less than i, i.e., $j < i$. Then the following chain of inequalities and equalities hold: $d_{\text{new}}(v, y) \leq d_{\text{new}}(v, x_0) + d_{\text{new}}(x_0, x_j) + d_{\text{new}}(x_j, y) = d_G(v, x_0) + d_G(x_0, x_j) + d_G(x_j, y)$ (the inequality is a triangle inequality, whilst the equality holds because x_0 is not a descendant of any vertex on the path in the new graph). Since $d_G(v, y) = d_G(v, x_j) + d_G(x_j, y)$, the difference between new and initial distances is $d_{new}(v, y) - d_G(v, y) = d_G(v, x_0) + d_G(x_0, x_j) - d_G(v, x_j) \leq 2d_G(x_0, x_j) \leq d_G(u_1, u_2) \leq 2 \cdot d_{T'}(u_1, u_2) \leq \frac{2(\alpha - n)}{n}$ (where the latter inequality is implied by our assumption). We need to bound the number of possible y's. Path $u_1 - u_2$ does not go through vertex v, so the number of possible y's is bounded by the size of the subtree of T of a child of v that contains this path. We prove that the size of any subtree of a child of v in T' is at most $\frac{n}{2}$.

Consider any child t of v in T, and consider the subtree of T rooted in t. Let the b be the number of vertices in the subtree, and let a be the number of other vertices of T. Let c_1 be the usage cost of t in the subtree, and let c_2 be the usage cost of v (!!) in the other part of the tree T. Then the usage cost of t in G is upper bounded by $c_1 + a + c_2$, whilst the usage cost of v is exactly $b + c_1 + c_2$. Since v is the vertex with the minimal usage cost, we have $c_1 + a + c_2 \geq b + c_1 + c_2$. Since $a + b = n$, we get that $b \leq \frac{n}{2}$.

Since y was chosen arbitrarily, the increase of the usage cost for v is less than $\frac{n}{2}\frac{2(\alpha-n)}{n} = \alpha - n$, and therefore $U_{\text{new}}(v) < U_G(v) + \alpha - n$ which is a contradiction. □

Combining Lemmas 2 and 3 with Lemmas 4 and 5 gives the main result.

Theorem 2. *For $\alpha \geq 65n$ every Nash equilibrium graph is a tree.*

Theorem 3. *For $\alpha \geq 41n$ every 2-coalitional Nash equilibrium graph is a tree.*

4 Small Cycles and Experimental Results

In this section we consider equilibrium graphs that have small girth c, and show that they exist only for small values of α. We start with an observation that limits the girth of equilibrium graphs containing exactly one cycle.

Proposition 1. *Let G be a Nash equilibrium graph containing a k-cycle $C = \{v_0, v_1, \ldots, v_{k-1}\}$, and F the graph where the edges of C are removed from G. If F consists of k connected components, then $k < 6$.*

Proof. Assume for contradiction that $k \geq 6$. For $0 \leq i < k$ let $s_i > 0$ denote the number of vertices in the connected component of F which contains v_i. If the edge (v_0, v_{k-1}) is bought by the player v_0, then she could replace (v_0, v_{k-1}) by (v_0, v_{k-2}). By doing this, her creation cost will remain the same, her distances to $s_{k-3} + s_{k-2}$ vertices decrease by 1, but her distances to s_{k-1} vertices increase by 1. If the edge (v_0, v_{k-1}) is bought by the player v_{k-1}, this player could replace (v_{k-1}, v_0) by (v_{k-1}, v_1). By this change of her strategy, her distances to s_0 vertices would increase, but she could decrease her distances to $s_1 + s_2$ vertices.

Since we consider a Nash equilibrium, we deduce that $s_{k-3} + s_{k-2} \leq s_{k-1}$ *or* $s_0 \geq s_1 + s_2$. Applying this reasoning for every edge of C, we get that for every i,

$$s_{i-3} + s_{i-2} \geq s_{i-1} \text{ or } s_i \geq s_{i+1} + s_{i+2}, \tag{3}$$

where $0 \leq i < k$ (recall that indexes are considered modulo c). The two inequalities $s_i \geq s_{i+1} + s_{i+2}$ and $s_{i-1} + s_i \leq s_{i+1}$ cannot hold simultaneously. Yet, 3 forces one of the inequalities $s_{i-1} + s_i \leq s_{i+1}$ and $s_{i+2} \geq s_{i+3} + s_{i+4}$ to be true, so we have that inequality $s_i \geq s_{i+1} + s_{i+2}$ implies $s_{i+2} \geq s_{i+3} + s_{i+4}$ for any $0 \leq i < k$. Without loss of generality we can assume that the edge (v_{k-1}, v_0) was bought by v_0. Then we get the chain of inequalities $s_{2i} \geq s_{2i+1} + s_{2i+2}$ for every i, which is obviously a contradiction. □

We now describe our computer-aided approach for upper-bounding α in case of an existence of small cycles in Nash equilibrium graphs. In our approach, we consider a non-tree Nash equilibrium whose smallest cycle has a fixed length c, and we construct a linear program asking for a maximum α, whilst satisfying inequalities of the type "SAVINGS" $\leq$ "INCREASE", which we create by considering various strategy changes of the players of the cycle. The partition of vertices of a Nash equilibrium graph into vertices V_d, $d \in D$, gives a variable $|V_d|$ for every d. The number of variables is $t = c^c - (c-1)^c$. We enumerate over all possible (meaningful) directions of the edges on the considered cycle, and solve the linear program, which gives us an upper bounds on α for every direction of edges. The largest such value is then obviously an upper bound on α for any direction, and thus for any Nash equilibrium containing a cycle of the fixed size.

The number of all possible directions is equal to 2^c, but this number can be decreased to at most $2^{c-3}+2$ by simple observations that all hold without loss of generality. We can assume that the number of *right* edges is at least the number of *left* edges, where an edge (v_i, v_{i+1}) is called a *right* edge, and (v_{i+1}, v_i) is called a *left* edge. Furthermore, we can also assume that the edge (v_0, v_1) is a right edge. If c is even, every considered cycle can be made (by renaming arguments) to fall into one of the following three classes: (1) the edges along the cycle alternate between right and left, or (2) all edges are right edges, or (3) the first two edges are right edges and the last edge is a left edge. The same holds when c is odd, with the exception of the alternating edges.

Our linear program contains all inequalities implied by the strategy changes described in Theorem 1. We furthermore add inequalities for strategy changes of buying one extra edge, and for swapping an edge of the cycle with a new edge towards an vertex of the cycle. We add the equality $\sum_{d \in D} |V_d| = 1$ (which expresses the fact that the variables should sum up to n). Then, the value of a variable $|V_d|$ expresses the fraction of all vertices (instead of the absolute number of vertices).

We used the GUROBI linear-programming solver to maximize α for every generated linear program. The largest such value thus gives an upper bound on α for which a cycle of size c can exist. Due to the huge number of variables, we could not solve the linear program for $c > 7$, because already for $c = 8$, the number of variables was more than 10^7, while the number of constraints is $\Theta(c^2)$. We have made further tweaks to the code, which allowed us to speed up the computation. We observed that many variables had the same coefficients in every generated constraint, and thus at most one such variable is relevant for obtaining the solution of the linear program. We have considered the variables one by one, and added only those having unique coefficients in the considered constraints. To check for uniqueness, we used hashing, as otherwise just creating the matrix of the linear program was too slow. The obtained compression of the number of variables was huge: for $c = 10$, instead of nearly 10^10 variables we obtained only around 10^5.

The obtained upper bounds on α are quite close to n. For girth $c \leq 7$, we obtain $\alpha \leq 1$, which corresponds to $\alpha \leq n$ if we required that $\sum_{d\in D} |V_d| = n$ (instead of $\sum_{d\in D} |V_d| = 1$). For girth $c = 8$, α is upper bounded by $\frac{191}{185}$, for girth $c = 9$, α is upper bounded by $\frac{13}{12}$, whilst for girth $c = 10$, α is bounded by 1.2.

We have performed further experiments with larger values of c, but did not consider all orientations of edges (as this was out of our computational power). Furthermore, since the number of variables is increasing super-exponentially, instead of considering all variables, for larger values of c we have considered only variables $|V_d|$ that have only 0's and $(c-1)$'s as distances in vector d, that is, we have considered 2^c variables. Additionally, we have taken extra 2^c random variables. We have all values of c up to 15. Upper bounds for α obtained using only these variables are very close to the real bounds for $c \leq 10$ (the difference for $k \leq 10$ is between 0 and 0.01). The largest upper bound of $1.3n$ on α appears for $c = 13$, and then only decreases, which is why we conjecture: the upper-bound of $\alpha \leq 1.3n$ can be proved by the considered strategy changes.

Acknowledgements. This work has been partially supported by the Swiss National Science Foundation (SNF) under the grant number 200021_143323/1.

References

1. Albers, S., Eilts, S., Even-Dar, E., Mansour, Y., Roditty, L.: On Nash equilibria for a network creation game. In: Proc. 17th Annual ACM-SIAM Symposium on Discrete Algorithms (SODA), pp. 89–98. ACM, New York (2006)
2. Alon, N., Demaine, E.D., Hajiaghayi, M.T., Leighton, T.: Basic network creation games. SIAM Journal on Discrete Mathematics 27(2), 656–668 (2013)
3. Bilò, D., Gualà, L., Proietti, G.: Bounded-distance network creation games. In: Goldberg, P.W. (ed.) WINE 2012. LNCS, vol. 7695, pp. 72–85. Springer, Heidelberg (2012)
4. Brautbar, M., Kearns, M.: A clustering coefficient network formation game. In: Persiano, G. (ed.) SAGT 2011. LNCS, vol. 6982, pp. 224–235. Springer, Heidelberg (2011)
5. Demaine, E.D., Zadimoghaddam, M.: Constant price of anarchy in network creation games via public service advertising. In: Kumar, R., Sivakumar, D. (eds.) WAW 2010. LNCS, vol. 6516, pp. 122–131. Springer, Heidelberg (2010)
6. Demaine, E.D., Hajiaghayi, M., Mahini, H., Zadimoghaddam, M.: The price of anarchy in network creation games. ACM Trans. Algorithms 8(2), 1–13 (2012)
7. Ehsani, S., Fazli, M., Mehrabian, A., Sadeghian Sadeghabad, S., Safari, M., Saghafian, M., ShokatFadaee, S.: On a bounded budget network creation game. In: Proc. 23rd ACM Symposium on Parallelism in Algorithms and Architectures (SPAA), pp. 207–214 (2011)
8. Fabrikant, A., Luthra, A., Maneva, E., Papadimitriou, C.H., Shenker, S.: On a network creation game. In: Proc. 22nd Annual Symposium on Principles of Distributed Computing (PODC), pp. 347–351. ACM, New York (2003)
9. Mihalák, M., Schlegel, J.C.: The price of anarchy in network creation games is (mostly) constant. Theory Comput. Syst. 53(1), 53–72 (2013)

Fast Low-Cost Estimation of Network Properties Using Random Walks*

Colin Cooper, Tomasz Radzik, and Yiannis Siantos

Department of Informatics, King's College London, WC2R 2LS, UK

Abstract. We study the use of random walks as an efficient estimator of global properties of large undirected graphs, for example the number of edges, vertices, triangles, and generally, the number of small fixed subgraphs. We consider two methods based on first returns of random walks: the cycle formula of regenerative processes and weighted random walks with edge weights defined by the property under investigation. We review the theoretical foundations for these methods, and indicate how they can be adapted for the general non-intrusive investigation of large online networks.

The expected value and variance of first return time of a random walk decrease with increasing vertex weight, so for a given time budget, returns to high weight vertices should give the best property estimates. We present theoretical and experimental results on the rate of convergence of the estimates as a function of the number of returns of a random walk to a given start vertex. We made experiments to estimate the number of vertices, edges and triangles, for two test graphs.

1 Introduction

Recent developments in technology have allowed the creation of large networks, available globally via personal computers, or more recently mobile phones. The original and most outstanding examples of such networks are the WWW, and the email network. Relatively recently, many On-Line Social Networks (OLSN) such as Twitter and Facebook, or online video repositories such as YouTube have sprung up. The size, structure, and rate of growth of these networks is a question of natural interest. As they are so large and our ability to access to them is often limited by the provider, we need methods to investigate them which are fast relative to the network size, are non-intrusive, and have low storage overheads.

We investigate how effective random walk based sampling methods are for estimating structural properties of a connected undirected graph (a model of a large network), such as the number of vertices, edges and small subgraphs. We collect together existing theoretical facts which are useful in designing a random walk based methods, and evaluate the performance of these methods experimentally. The final application is practical, but we are guided by theory

* Research supported in part by EPSRC grant EP/J006300/1 and Samsung Global Outreach Project "Fast low cost methods to learn structure of large networks."

A. Bonato, M. Mitzenmacher, and P. Prałat (Eds.): WAW 2013, LNCS 8305, pp. 130–143, 2013.
© Springer International Publishing Switzerland 2013

as far as possible. The methods we consider are based on randomized crawling by downloading pages from the network under investigation. Our assumption is that the network cannot be explored systematically by BFS, either because of size or because the number of third party accesses to the network is restricted.

Let $G = (V, E)$ be a connected (undirected) graph with $|V| = n$ vertices and $|E| = m$ edges. The expected first return time T_u^+ of a random walk to a vertex u is given by $\mathbf{E}T_u^+ = 1/\pi_u$, where π_u is the stationary probability of vertex u. For a simple random walk, $\mathbf{E}T_u^+ = 2m/d(u)$, where $d(u)$ is the degree of u. If $d(u)$ is large, then $\mathbf{E}T_u^+$ is small and we can quickly obtain an estimate for m. For example, a graph generated by preferential attachment has $m = cn$ edges, and vertices u with degree as large as $d(u) = \sqrt{n}$. We can use a walk starting from such a vertex to estimate m in $2m/d(u) = O(\sqrt{n})$ expected steps.

An important idea is that for graphs in which there is variation in degree sequence, it is possible to use a simple random walk to quickly and accurately estimate the number of edges based on first returns to high degree vertices. If the graph is near regular, $\mathbf{E}T_u^+ = \Theta(n)$, for any start vertex u This is bad if we want a quick answer. Such graphs may still exhibit variations in local structure which we can exploit. For example the number of triangles at a vertex may vary considerably. If so we could use a random walk with vertex weight proportional to the number of triangles at the vertex. By starting from a high weight vertex we should be able to exploit this structural variation to count properties efficiently.

We discuss the following ideas, both theoretically and experimentally.

1. Global properties of graphs can be estimated using first return times of random walks. The general theory is given in Section 3 with respect to the cycle formula of regenerative processes and weighted random walks. For a given property, these approaches either keep a running total of the number of structures (e.g. triangles) observed by each excursion of the walk (the cycle-formula method), or use first return times of walks with edge weights proportional to the number of structures containing the edge (the weight-random-walk method).
2. The use of the cycle formula of regenerative processes is discussed in Section 3.1. The use of weighted random walks is developed in Section 3.2 with respect to various examples such as the number of triangles, vertices and arbitrary fixed subgraphs.
3. The quality of the methods depends on the distribution of first return times to the start vertex. We review the theory relating to this in Section 3.3. Vertices with high degree (or more generally, with high vertex weight) have smaller expected value and (upper bound on) variance of return time, and should estimate properties more effectively. This is also discussed in Section 3.3.
4. Experimentally, as the walk proceeds, it naturally discovers high weight vertices, and the estimates based on returns to these vertices are efficient after a reasonable number of steps. The performance of random walks on suitable test graphs is assessed in Section 4.
5. The expected value and variance of the first return time T_v^+ of a random walk to a given vertex v are known quantities, given by $\mathbf{E}T_v^+ = 1/\pi_v$ and

Var $T_v^+ = (2Z_{vv} + \pi_v - 1)/\pi_v^2$ respectively, where $Z_{vv} = \sum_{t \geq 0}(P_v(v,t) - \pi_v)$ and $P_v(v,t)$ is the probability that a walk starting from v returns to v at step t. It is difficult to evaluate Z_{vv} directly, but we can bound Z_{vv}, and hence the variance of our estimates, using the eigenvalue gap of the transition matrix. The variance of our estimates can also be estimated directly from the return time data using a result of [5]. See Section 3.3 for details.

The aims of the paper are to collect together available information on random walk based methods for estimating network properties and to compare and develop the techniques. Our original contributions are in the design of weighted random walks to estimate e.g. number of vertices, triangles and fixed motifs, and to detect clustering (see Section 3.2). We also provide theoretical and experimental methods to bound and estimate the variance of the first return time T_v^+ to the start vertex v (3.13); see methods **M1, M2** in Section 3.3.

The complexity measures we use to present our results are somewhat crude, as the processing load per walk step varies both locally and on the remote site for the different walks we use. Our basic measures are the number of steps, and the number of returns to the start vertex. By choosing a high weight start vertex the expected first return time can be made sublinear (see 3, 4 above). The variance of this quantity can also be bounded as outlined above.

2 Network Sampling Methods Based on Random Walks

The simplest way to study a network is to inspect it completely using e.g. breadth first search. Failing this, a simple statistical method of sampling vertices uniformly at random (u.a.r) can be considered. In practice for large networks such as the WWW or OLSN's, neither of these methods is feasible, but the network can still be queried by some limited form of crawling or interaction with the network through its API. Our assumption is that query results are held locally on a single processor. The selection of the next vertex to visit (query) is based on a random walk runs on the query data. The random walk is used as a randomized algorithm to determine the next step in the query process. We measure the computational complexity as the number of steps made by the walk, and our aim is to obtain results in a number of steps sub-linear in the network size, which is assumed unknown.

Methods to estimate network properties based on random walks can be divided into two classes: estimates obtained by using a random walk as a surrogate for uniform sampling (an outline of this is given next), and estimates based on first return times of random walks (this is discussed in detail in Section 3).

Sampling the elements of a set uniformly at random with replacement can be used to estimate the set size in sub-linear time by the method of *sample and collide*. The use of this method to estimate network size is described by Bawa et al. [3]. If we sample uniformly at random with replacement from a population of size n, then, by the Birthday Paradox the expected number of trials required for the first repetition is $\sqrt{2n}$. In general, the expected number of repetitions in s samples is $s(s-1)/2n$ with variance $(1-1/n)\, s(s-1)/2n$. If R is the sample size when the first repetition occurs, then an estimate of the network size is $\overline{n} = R^2/2$.

The method of sample and collide requires u.a.r. samples from the population. To obtain a uniform sample from a network using a random walk, we can do the following. Run the walk for $t \geq T$ steps before sampling, where T is a suitable mixing time. In this case the walk is in near-stationarity and $P_u(X_t = x) \sim \pi_x$, where X_t is the position of the walk at step t, and π_x is the stationary distribution of the walk. For a simple random walk $\pi_x = d(x)/2m$, where $d(x)$ is degree of x, and m is the number of edges. Thus, unless the graph is regular, the sample is not uniform. To use a sample from the stationary distribution, we need to unbias the walk. There are several ways to do this. One method is to use the approach of Massoulie et al. [12], and Ganesh et al. [7], who use a continuous time random walk, random waiting time at a vertex x which is negative exponential with mean $1/d(x)$, and a fixed stopping time T. In this way, the obtained stationary distribution is uniform. The discrete equivalent (re-weighted random walk) is to walk for a fixed number of steps T, sample the vertex, and retain the sample with probability $1/d(x)$. This gives a uniform sampling probability of $1/2m$. Another method is to use a Metropolis-Hastings random walk with target stationary distribution $\pi_x = 1/n$. One way to do this, is to use a transition probability $1/M$ where $M \geq \Delta(G)$, the maximum degree of G. See [13] page 264 for more details.

An alternative approach to uniform sampling is developed by Katzir *et al* [8]. A simple random walk is used in conjunction with the birthday paradox, and the statistical bias arising from the non-uniform stationary distribution is approximately corrected.

3 Estimates Based on First Return Time of a Random Walk

For a random walk starting at vertex v, the first return time to v is defined as

$$T_v^+ = \min\{t > 0 : X_t = v\},$$

where X_t is the position of the walk at step t ($X_0 = v$). If the walk is ergodic, it has a well defined stationary distribution π_v at any vertex v, and the expected value of the first return time $\mathbf{E}T_v^+$ is given by $\mathbf{E}T_v^+ = 1/\pi_v$.

We describe two methods to estimate properties of networks based on first return times of random walks. The methods are in no sense mutually exclusive, and can indeed be used together. The first method, the cycle formula of regenerative processes, has typically been used with simple random walks (e.g. [12]). The second method uses first return times of weighted random walks. Both methods are equally viable to estimate a given property.

An important point for either method, is that high weight vertices perform well as start vertices for random walk property estimators. For a simple random walk, the weight of vertex u is the vertex degree $d(u)$, and this feeds into the first return time $\mathbf{E}T_u^+ = 2m/d(u)$. Thus in regular graphs, all vertices are equivalent start points. By re-weighting the walk we can artificially create high weight vertices suitable for estimating a given property.

To give an example of this, consider a regular graph which contains many triangles (copies of K_3), distributed in non-uniform clusters. In a simple random walk, as vertex weight is proportional to degree, this graph has no high weight vertices. By weighting edges proportional to the number of triangles they are contained in, vertices with many triangles assume a high weight. First returns to these vertices can provide a good estimator for the total number of triangles.

For an (ergodic) weighted random walk, the expected value of the first return time is equal to the reciprocal of the stationary probability, as in the case of the simple un-weighted walk:

$$\mathbf{E}T_v^+ = \frac{1}{\pi_v}, \tag{3.1}$$

but the stationary probability now is $\pi_v = w(v)/w_G$, where $w(v)$ is the weight of vertex v and w_G is the total weight of the graph.

3.1 Estimates Based on the Cycle Formula of Regenerative Processes

The cycle formula of regenerative processes can be summarized as follows. Consider a random walk starting from vertex u and let $f(X_t)$ be a vertex valued function. Then

$$\mathbf{E}_u\left(\sum_{t=0}^{T_u^+-1} f(X_t)\right) = \mathbf{E}T_u^+ \sum_{v\in V} \pi_v f(v). \tag{3.2}$$

This identity is a consequence of the result (see e.g. [1] Chapter 2, Lemma 6) that

$$\mathbf{E}_u(\text{number of visits to } v \text{ before time } T_u^+) = \frac{\pi_v}{\pi_u} = \mathbf{E}T_u^+\,\pi_v.$$

Identity (3.2) was used by Massoulié *et al* [12] to count network size using a simple random walk. Putting $f(v) = 1/d(v)$ removes the degree bias from π_v so that $\sum_{v\in V}\pi_v f(v) = n/2m$, and the RHS of (3.2) equals $n/d(u)$.

Following [12] we maintain the convention $f(v) = \phi(v)/w(v)$ when generalizing to weighted random walks, with $\pi_v = w(v)/w_G$. Denote by R_u the random variable $\sum_{t=0}^{T_u^+-1} f(X_t)$, with expectation $\mathbf{E}R_u$ given by (3.2). Let $\overline{\phi} = \sum_{v\in V}\phi(v)$ be the quantity which we want to estimate. Then, as $\mathbf{E}T_u^+ = 1/\pi_u$,

$$\mathbf{E}R_u = \mathbf{E}T_u^+ \times \sum_{v\in V}\pi_v\frac{\phi_v}{w(v)} = \frac{\overline{\phi}}{w(u)}. \tag{3.3}$$

An important point experimentally, is that $\overline{\phi}$ obtained from (3.3) does not depend on the total weight w_G, but only on $w(u)$ (a known quantity).

3.2 Estimates Based on Return Times of Weighted Random Walks

This technique generalizes the following observation. For a simple random walk, the stationary distribution of vertex u is $\pi_u = d(u)/2m = 1/\mathbf{E}T_u^+$. Thus the first return time T_u^+ can be used to estimate the number of edges m of a graph. Let $Z(k) = \sum_{i=1}^{k} Z_i$ be the time of the k-th return to vertex u. The random variable

$$\widehat{m} = \frac{Z(k)d(u)}{2k} \tag{3.4}$$

estimates the total number of edges m.

The basic idea is to design the stationary distribution to reveal the required network property. To do this we fix the edge weights for the walk transitions at any vertex in such a way that the required answer is contained in the graph weight w_G. The total weight w_G can be obtained from the stationary distribution $\pi_v = w(v)/w_G$ of the start vertex v, which by (3.1) is the reciprocal of the expected return time. The larger $w(v)$, the smaller $\mathbf{E}_v T_v^+$, giving us more rapidly k samples for (3.4).

The remainder of this section is arranged as follows. Firstly, we summarize the properties of weighted random walks. Secondly we give examples of using weighted random walks to estimate the total number of triangles t in the network, (a litmus test for social networks), and to estimate the size n of the network. Thirdly, we explain the general framework for estimating the number of small fixed subgraphs ('motifs' like triangles, cliques, cycles, etc.), and for detecting clustering within a given set of vertices S.

Weighted Random Walks. Given a graph $G = (V, E)$ and a positive weight function $w(u, v)$ on edges $\{u, v\} \in E$, we can define a Markov chain with state space $S = V$ and a transition matrix with elements:

$$p_{uv} = \begin{cases} \frac{w(u,v)}{w(u)}, & \text{if } \{u, v\} \in E, \\ 0, & \text{otherwise,} \end{cases}$$

where $w(u) = \sum_{\{u,v\} \in E} w(u, v)$ is the weight of a vertex u, and $w_G = \sum_{u \in V} w(u) = 2 \sum_{\{u,v\} \in E} w(u, v)$, is the weight of the graph G. See [1] for details.

We refer to this chain as a weighted random walk on G. The stationary distribution is:

$$\pi_u = \frac{w(u)}{w_G}. \tag{3.5}$$

A special, but important case of a weighted random walk is the simple random walk, where $w(u, v) = 1$ for all $\{u, v\} \in E$. For this case:

$$p_{uv} = \begin{cases} \frac{1}{d(u)}, & \{u, v\} \in E \\ 0, & \text{otherwise} \end{cases}$$

$$\pi_u = \frac{d(u)}{2|E|}.$$

Estimating the Number of Triangles. For each edge e we assign the weight $1 + t(e)$, where $t(e)$ is the number of triangles containing e. Let $t(v)$ be the number of triangles containing v and $t(G)$ the total number of triangles in G. Then

$$\pi_u = \frac{w(u)}{w_G} = \frac{d(u) + 2t(u)}{2m + 6t(G)}.$$

Let $Z(k) = \sum_{i=1}^{k} Z_i$ be the time of the k-th return to vertex u. We estimate the number of triangles $t(G)$ by

$$\widehat{t} = \max\left\{0, \frac{Z(k)(d(u) + 2t(u))}{6k} - \frac{m}{3}\right\}, \tag{3.6}$$

where m can be estimated by Equation (3.4).

Estimating the Network Size. We use now inversely degree biassed weighted random walks, setting the edge weight $w(u,v) = \frac{1}{d(u)} + \frac{1}{d(v)}$. It can be shown that $w_G = 2n$, so the stationary distribution is:

$$\pi_u = \frac{w(u)}{w_G} = \frac{1 + \sum_{v \in N(u)} \frac{1}{d(v)}}{2n}. \tag{3.7}$$

Let $Z(k) = \sum_{i=1}^{k} Z_i$ be the time of the k-th return to vertex u, as before, and let $w(u)$ be as shown in Equation (3.7). We use the following estimator for the number of vertices:

$$\widehat{n} = \frac{Z(k)w(u)}{2k}. \tag{3.8}$$

Estimating the Number of Occurrences of an Arbitrary Fixed Subgraph. Using a weighted random walk to estimate the number of edges $m(G)$ or triangles $t(G)$ in a graph G are special cases of the following problem. Let $\mathcal{S}$ be a set of unlabeled graphs. For each $M \in \mathcal{S}$ we want to count the number of distinct labeled copies of M in the graph G. The cases edges and triangles given above correspond to $\mathcal{S} = \{K_2\}$ and $\mathcal{S} = \{K_2, K_3\}$ respectively. For each $e \in E(G)$ we put $w(e) = \sum_{M \in \mathcal{S}} N(M, e)$, where $N(M, e)$ is the number of distinct subgraphs H isomorphic to M which contain e. The simplest case (after $\mathcal{S} = \{K_2\}$) is $\mathcal{S} = \{K_2, M\}$, where M can be any connected subgraph, e.g. K_k, $K_{k,\ell}$, a chordless cycle of length 4, or some specific (small) tree. In this case we have the following:

$$w_G = 2\sum_e w(e) = 2\sum_e (1 + N(M,e)) = 2m + 2\nu\mu(G), \tag{3.9}$$

where $\nu = |E(M)|$ and $\mu(G)$ is the number of distinct copies of M in G. As $\pi_v = w(v)/w_G$, and $w(v)$ and ν are known, we can use the method of first returns to estimate $\mu(G)$.

As an experimental heuristic we can use weighted walks with edge weight

$$w(e) = 1 + cN(M, e). \tag{3.10}$$

We have $c = 1$ in (3.9), but any value of $c > 0$ is valid. The parameter c can be chosen smaller than 1 (e.g. $c = 1/10$) in order to stop large values of $N(M, e)$ distorting the eigenvalue gap, and hence mixing rate of the walk. We adopted this approach with some success for counting triangles in the Google web graph (see Section 4).

Detecting Edges within a Given Set S. This is intended as an experimental measure of evidence for clustering. Let $S \subseteq V$ be given. We use the following edge weights, where $c > 0$ constant. For edge $e = \{u, v\}$, let $w(e) = 1$, if neither vertex is in S, let $w(e) = 1 + c$ if exactly one vertex is in S and let $w(e) = 1 + 2c$ if both vertices are in S. It follows that $w_G = 2m + 2cd(S)$, where $d(S)$ is the degree of S.

3.3 Distributional Properties of First Return Times

As stated earlier, the expected value of the first return time T_v^+ to a vertex v is $\mathbf{E}T_v^+ = 1/\pi_v$; see e.g. [1]. The variance of T_v^+ is given by

$$\mathbf{Var}\ T_v^+ = \frac{2\mathbf{E}_\pi T_v + 1}{\pi_v} - \frac{1}{\pi_v^2}. \qquad (3.11)$$

Here $\mathbf{E}_\pi T_v$ is the expected hitting time of v from stationarity, i.e.:

$$\mathbf{E}_\pi T_v = \sum_{u \in V} \pi_u \mathbf{E}_u T_v,$$

where $\mathbf{E}_u T_v$ is the expected time to hit v starting from u. The quantity $\mathbf{E}_\pi T_v$ can be expressed as $\mathbf{E}_\pi T_v = Z_{vv}/\pi_v$ where

$$Z_{vv} = \sum_{t \geq 0} (P_v(v, t) - \pi_v), \qquad (3.12)$$

and $P_v(v, t)$ is the probability that a walk starting from v returns to v at step t. Thus

$$\mathbf{Var}\ T_v^+ = \frac{2Z_{vv} + \pi_v - 1}{\pi_v^2}. \qquad (3.13)$$

For rapidly mixing random walks (e.g. walks on expander graphs) Z_{vv} is constant C. Indeed it can be bounded by $1 \leq Z_{vv} \leq 1/(1 - \lambda_2)$ where $1 - \lambda_2$ is the eigenvalue gap of the transition matrix (see below for a proof of this). As $\pi_v = w(v)/w_G$, both the expected first return time $1/\pi_v$ and the variance $\sim (2C - 1)/\pi_v^2$ of first return time decrease with increasing vertex weight $w(v)$. This implies that returns to high weight vertices should make the best estimators for w_G, and that they will return sample values more often and more reliably.

The quantity $\mathbf{E}_\pi T_v$ can be bounded in various ways, to give estimates of Z_{vv} and $\mathbf{Var}\ T_v^+$. We give two methods: (M1) an estimate based on eigenvalue gap, and (M2) a direct estimate from the return time data. One standard deviation of the sample mean for estimates of the number of edges was derived by these methods. This is illustrated in Figure 1 (top part): the outer dashed curve is obtained using method (M1) and the inner dashed curve using method (M2). The graph used in those experiments is described in Section 4.

M1. From (3.12) we have

$$Z_{vv} = \sum_{t \geq 0} (P_v(v,t) - \pi_v) \leq \sum_{t \geq 0} |P_v(v,t) - \pi_v|$$

Using the result that $|P_v(v,t) - \pi_v| \leq \lambda_2^t$ (see e.g. [11]), gives

$$Z_{vv} \leq \frac{1}{1 - \lambda_2}. \tag{3.14}$$

M2. We estimate Z_{vv} directly from the first return time data. Let T be a mixing time of a random walk on a graph G. The method described in [5] states (subject to certain technical conditions) that for $t > T$ the probability $\rho(t)$ that a first return to v has not occurred by t is of the form

$$\rho(t) \sim \exp\left(-t/\mathbf{E}_\pi T_v^+\right).$$

Replacing $1 - \rho(t)$ by the proportion $y(t)$ of returns at or before step t, estimates Z_{vv}. For the plot in Figure 1 (top part), an estimate of $\widehat{Z}_{vv} = 1.6$ was obtained.

4 Evaluation of Random Walk Based Methods

Figures 1-3 show the convergence of our experiments as a function of the number of returns k to the start vertex of the walk. The test networks included here are a triangle closing preferential attachment graph, and a sample from the www (the Google Web Sample).

The figures are presented as follows. The horizontal axis is k – the number of returns to the start vertex. The vertical axis is the estimate of the property. The data points plotted are based on 10 independent experiments. The underlying data points appear close to each other because there can be several returns within a short period, followed by a long wait for the next return. In all plots, the thick line "Experiments Average" is our estimate – the sample mean as a function of $10k$ (the k-th return in 10 experiments); the two dotted lines "Experiments Deviation" show one standard deviation of the sample mean; and the horizontal line through the middle of the plot area shows the true value of the property. For the estimates based on first return times of weighted random walks, we plot also the standard deviation of the sample mean estimated using method M2 (the dashed curves "Z_{vv} Deviation"). Finally, for the estimates of the number of edges, we also computed an upper bound on the standard deviation using method M1. This bound is shown in Figure 1 (top part) by the two outer dashed curves "Bound on Deviation," but is outside of the visible area in the plot in Figure 2.

Hybrid Triangle Closing Model. We use a hybrid triangle closing model which generates graphs as follows. At each step we add a new vertex v with r edges to the existing graph. To add a vertex v, we first attach to an existing vertex x chosen preferentially. The remaining $r - 1$ edges from v are added as follows. With probability p we attach to a vertex chosen by preferential attachment, and with probability $1-p$ we add an edge from v to a random neighbour of vertex x. Using this approach we are able to control the number of triangles generated while maintaining the power-law degree distribution to be asymptotic to 3.

We generated a graph using this model with $n = 600,000, m \sim 1.8 * 10^6$ and $p = 0.6$. At each step $r = 3$ edges were added. The total number of triangles was 550,499. The graph has a power law coefficient of 2.9. The second eigenvalue of the transition matrix (simple random walk) is 0.88265, making the eigenvalue gap 0.1733.

Figure 1 shows the convergence of the edge $\widehat{m}$, vertex $\widehat{n}$ and triangle $\widehat{t}$ estimates computed using the weighted random walk method described in Section 3.2. The random walks started at a vertex u of degree $d(u) = 61824$, and belonging to $t(u) = 70045$ triangles. The weights of this vertex are: $w_{SRW}(u) = d(u) = 61824$, for the simple random walk used to estimate the number of edges; $w_{TRW}(u) = d(u) + 2t(u) = 201914$, for the weighted random walk used to estimate the number of triangles; and $w_{VRW}(u) = 15201$, for the weighted random walk used to estimate the number of vertices. The expected first return times to vertex u are 58, 34 and 79, respectively.

All 10 experiments gave reasonable estimates of all three parameters after roughly 100 to 1000 returns to the start vertex, that is after at most $n/10$ samples (visits to a vertex). Figure 1, top part, shows good rate of convergence of the edge estimate $\widehat{m}$, and a good match between the standard deviation of the experimental data (the dotted "Experiments Deviation" lines) and the standard deviation obtained by estimating the parameter Z_{vv} (the "Z_{vv} Deviation" curves). The estimates using the cycle formula as described in Section 3.1 were similar, so we omit the details.

Google Web Sample. We used a sample from the Google web graph which was released for the purposes of the Google programming contest in 2002 [10]. This data set consists of 855,802 vertices, 5,066,842 edges and 31,356,298 triangles. The second eigenvalue of the transition matrix (simple random walk) is 0.99970, making the eigenvalue gap 3×10^{-4}. For this network the estimates converged slower than in the generated test graphs, with much more variation around the expected values. The structure of the graph is very inhomogeneous; presumably this is why the data set is made available.

In our experiments random walks started at a vertex u of degree $d(u) = 6353$, with $t(u) = 53371$ triangles. For the simple random walk, which is used for estimating the number of edges, the expected first return time to the start vertex u is equal to 1595. The computed estimates for the number of edges are given in Figure 2 (top). The convergence is slow and the theoretical standard deviation bound is outside the figure. We have to wait for about 1000 returns to the start vertex to get a reasonable estimate, which means that the number of samples (visits to a vertex) is roughly of the same order as the number of vertices n.

We compare the performance of the cycle-formula method and the weight-random-walk method for estimating the number of vertices and the number of triangles in the Google web graph. (Observe that these two methods are exactly the same when used for estimating the number of edges: $f(v) \equiv 1$ for the

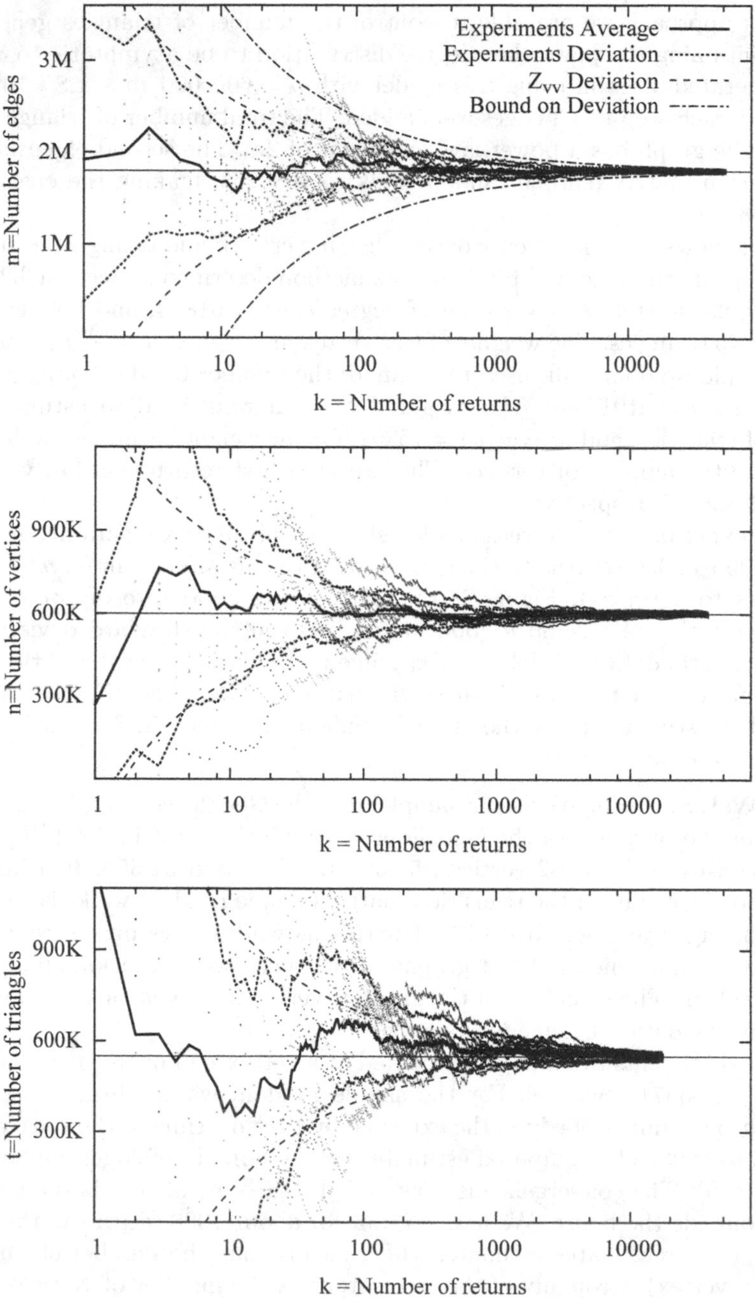

Fig. 1. Triangle-closing preferential-attachment graph. Estimate of number of edges, vertices and triangles. The key from the top plot applies to all plots.

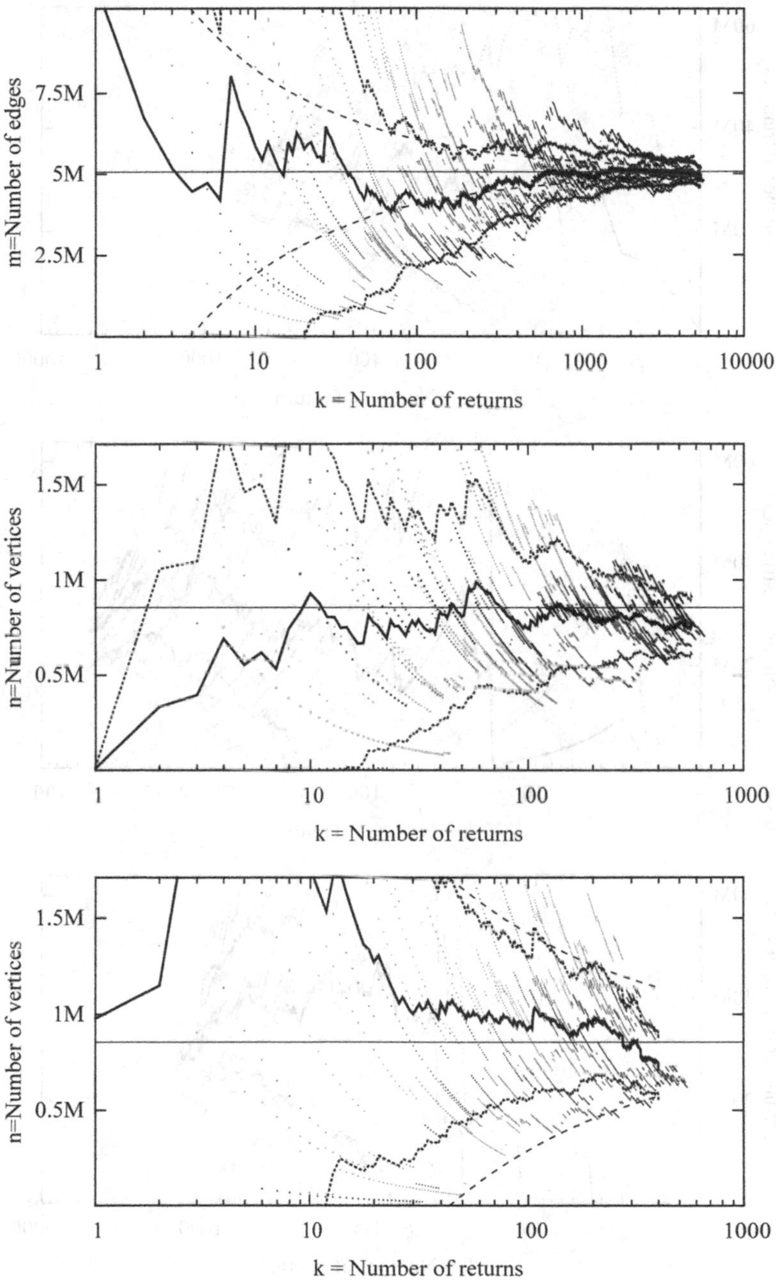

Fig. 2. Google web graph. Estimate of the number of edges (top), the number of vertices by cycle formula (middle), and the number of vertices by return times of weighted random walks (bottom). The key is the same as in Figure 1.

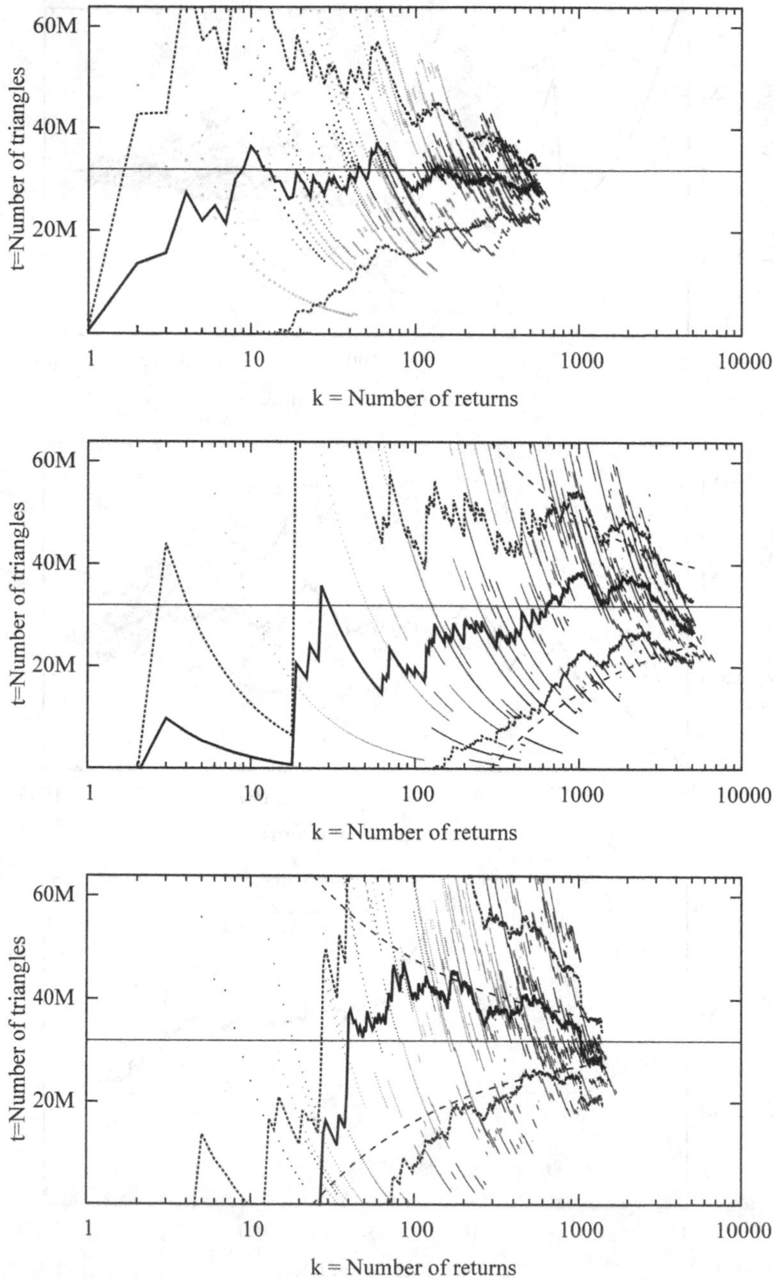

Fig. 3. Google web graph. Estimate of number of triangles: cycle formula (top), return times of weighted random walks with the weight factor $c = 1$ (middle), and $c = 0.1$ (bottom). The key is the same as in Figure 1.

cycle-formula method and $w(v,u) \equiv 1$ for the weighted-random-walks method.) For the weighted-random-walk method, the weights of the start vertex u are $w_{TRW}(u) = d(u) + 2t(u) = 113095$ and $w_{VRW}(u) = 855$. The expected first return times to vertex u are 1753 and 2002, respectively.

Figure 2 (parts 2 and 3) shows the estimates of the number of vertices computed by the cycle-formula method and the weighted-random-walk method. The top and middle parts of Figure 3 show the estimates of the number of triangles. The convergence is slow for both methods, but the estimates given by the cycle formula are more accurate, especially for the number of triangles.

To see if we can improve the performance of the weighted-random-walk method for estimating the number of triangles, we varied the value of the parameter c in the edge-weight fiormula (3.10) to avoid distorting the already small eigenvalue gap even further. The edge weight is $w(e) = 1 + ct(e)$, the vertex weight is $w(u) = d(u) + 2ct(u)$, and the graph weight is $2m + 6ct(G)$. To simplify the experiments, we used the correct value of m. The plots for $c = 1$ and $c = 1/10$ are given in the middle and bottom parts of Figure 3, respectively. The value $c = 1/10$ worked quite well, giving clearly better results than the value $c = 1$, but not matching fully the performance of the cycle-formula method. The value of c can be optimized by further experiments.

References

1. Aldous, D., Fill, J.: Reversible Markov chains and random walks on graphs, `http://stat-www.berkeley.edu/pub/users/aldous/RWG/book.html`
2. Barabasi, A., Albert, R.: Emergence of scaling in random networks. Science 286(5439), 509–512 (1999)
3. Bawa, M., Garcia-Molina, H., Gionis, A., Motwani, R.: Estimating aggregates on a peer-to-peer network. Technical Report, CS Dept, Stanford University (2003)
4. Broder, A., Kumar, R., Maghoul, F., Raghavan, P., Rajagopalan, S., Stata, R., Tomkins, A., Wiener, J.: Graph structure in the web. Computer Networks 33, 309–320 (2000)
5. Cooper, C., Frieze, A.: The cover time of random regular graphs. SIAM Journal of Discrete Mathematics 18(4), 728–740 (2005)
6. Cooper, C., Radzik, T., Siantos, Y.: Estimating Network Parameters Using Random Walks. In: Proc. CASoN 2012 (2012)
7. Ganesh, A., Kermarrec, A.-M., Le Merrer, E., Massoulie, L.: Peer counting and sampling in overlay networks based on random walks. Distrib. Comput. 20, 267–278 (2007)
8. Katzir, L., Liberty, E., Somekh, O.: Estimating sizes of social networks via biased sampling. In: Proc. WWW 2011, pp. 597–606 (2011)
9. Feller, W.: An Introduction to Probability Theory and Its Applications, vol. I. Wiley (1970)
10. Leskovec, J.: Stanford network analysis package (2009), `http://snap.stanford.edu/`
11. Lovasz, L.: Random walks on graphs: A survey. Bolyai Society Mathematical Studies, 353–397 (1996)
12. Massoulie, L., Le Merrer, E., Kermarrec, A.-M., Ganesh, A.: Peer counting and sampling in overlay networks: random walk methods. In: PODC 2006, pp. 123–132 (2006)
13. Mitzenmacher, M., Upfal, E.: Probability and Computing. CUP (2005)

An L_p Norm Relaxation Approach to Positive Influence Maximization in Social Network under the Deterministic Linear Threshold Model

Rupei Xu*

Department of Computer Science
The University of Texas at Dallas
800 W. Campbell Road
Richardson, TX 75080, USA
Rupei.Xu@utdallas.edu

Abstract. In this paper, an *Influence Maximization* problem in *Social Network* under the *Deterministic Linear Threshold* model is considered. The objective is to minimize the number of eventually negatively opinionated nodes in the network in a dynamic setting. The main ingredient of the new approach is the application of the sparse optimization technique. In the presence of inequality constraints and nonlinear relationships, the standard convex relaxation method of the L_1 relaxation does not perform well in this context. Thereforc we propose to apply the L_p relaxation where $0 < p < 1$. The resulting optimization model is therefore non-convex. By means of an interior point method, the model can be solved efficiently and stably, typically yielding robust and sparse solutions in our numerical experiments with the simulated data.

Keywords: Social Network, Influence Maximization, L_p Relaxation, Deterministic Linear Threshold, Interior Point Method.

1 Introduction

The developments of modern information technologies have clearly changed the lifestyle and the social behaviors of people in a profound way that has never been paralleled in human history. There are abundant studies in the literature in this regard; we shall only name here for instance the practice of viral marketing (see [1]). Notably, the popular online social network sites such as Facebook, Myspace and Twitter have opened a new door for the viral marketing. Research shows that people trust the information obtained from their close social circle far more than the information obtained from general advertisement channels such as TV, newspaper and online advertisement [2]. Many people believe that the 'word-of-mouth' marketing is still the most effective marketing strategy [3] after all. In new product adoption or new technology diffusion, people tend to follow the opinions of their friends in the social circle.

* This research was done under the supervision of Professor Shuzhong Zhang.

A. Bonato, M. Mitzenmacher, and P. Prałat (Eds.): WAW 2013, LNCS 8305, pp. 144–155, 2013.
© Springer International Publishing Switzerland 2013

Influence Maximization(IM) problem is to determine the initial seed set to make the spread of the influence maximized in the social network. Given a seed set, how to compute the total influence can be found in [4,5,6]. Domingos [1] and Richardson [7] proposed the problem of node selection, in which they consider the relation of individuals and proposed a probabilistic propagation model. Kempe, Kleinberg, and Tardos [8] first formulate the Influence Maximization as a discrete optimization problem. In [9], they further formulate the problem by optimization with two different models: *independent cascade model* and *linear threshold model*. The *independent cascade model* is first proposed by Goldenberg *et al.* in [10] [11] and the *linear threshold model* is proposed by Granovetter and Schelling in [12] and [13]. Kempe *et al.* [8] proved that the natural greedy algorithm achieves a $(1-1/e)$-approximation simply by showing that the influence spreads under both the two models are submodular. Later, Chen *et al.* [4,5] showed that the problem of exactly computing the influence given a seed set in both the *independent cascade model* and the *linear threshold model* are ♯ P-hard, which indicates that the greedy algorithm is not a polynomial time approximation for the two models. Lu *et al.* [14] study the IM problem in *deterministic linear threshold model* and show that in *deterministic linear threshold model*, there can be no polynomial time $n^{1-\varepsilon}$-approximation unless $P = NP$ even in the simple case that one person needs at most two active neighbors to become active.

2 Problem Formulation

In this section we consider an *influence maximization* problem under the framework of the so-called *deterministic linear threshold model* as proposed in [14]. Essentially, we consider a stable social network, through which the opinions about (or influence of) an object (may be a brand name or even a certain political position) evolve through the interactions modeled by a social network. We further assume that one may invest (in terms of efforts) on the nodes over time, so as to influence the opinions about the object. In the commercial world, this effort maybe in the form of advertisements, or a promotion; in politics, the effort can be the form of campaigning. In particular, we assume that the opinion of a node can be influenced by its neighbors, and can also be influenced by a direct effort.

Mathematically, a *Social Network* is modeled by a directed graph $G(V, E, W)$, which does not contain a self-loop. We consider a dynamic setting where the time periods are denoted by $k = 1, 2, ..., N$; the opinion of $v_i \in V$ is modeled by a numerical value s_i, which can take any real value. Suppose that there are m nodes on the network ($|V| = m$).

Let $s^k \in \mathbb{R}^m$ be such that s_i^k is the opinion of node v_i at time k, with $i = 1, 2, ..., m$ and $k = 1, 2, ..., N$. Let $x^k \in \mathbb{R}_+^m$ be the effort to be spent in time period k, where its i-th component x_i^k is the effort spent on node v_i at time period k. (In our setting x^k is a decision vector, while s^k is a state vector.) The propagation dynamic process is governed by the equation

$$s_j^{k+1} = s_j^k + \sum_{v_i \in \mathcal{N}(v_j)} s_i^k \cdot w_{ij} + x_j^k, \tag{1}$$

where $\mathcal{N}(v_j)$ indicates the set of nodes that are directed to v_j; i.e. $\mathcal{N}(v_j) = \{v_i \mid (v_i, v_j) \in E\}$. The weight w_{ij} measures the level of influence that v_i has over v_j. For a node v_i, if $s_i^N > \theta_i$ then we call the node *positively opinionated* through the process; if $s_i^N < \theta_i$ we call it *negatively opinionated*; otherwise, we call it *neutrally opinionated*. In the above expression, the parameter $\theta_i > 0$ maybe equal cross the nodes or maybe node-specific.

In the literature, mostly combinatorial/graphic methods and simulation methods are applied to solve optimization problems related to social network models. However, there are several benefits to a linear algebraic approach to graph algorithms, such as the syntactic complexity, ease of implementation and performance [15]. There are clearly benefits to apply continuous optimization machineries. In particular, we observe that there is a connection between the social network models and the so-called sparse optimization models such as compressed sensing. For example, in the social network, there are small-world [16] [17] and scale free [18] [19] [20] [21] properties. To some extent, *heavy tail distribution* refers to a sparsity structure for the data of big scale. For instance, about half a billion users are registered on Facebook [22]. Some of the nodes of Twitter corresponding to the celebrities such as Lady Gaga and Justin Bieber have degree over ten million [23]. In such cases, exploring sparsity structure while using continuous optimization relaxation can be advantageous. In many cases, sparsity can also refers to the structure of a solution to some social network problems. In the current paper, we set out to explore one such model, where the problem is to optimally allocate some resources to influence through the social network so as to minimize the *cardinality* of the eventual population who hold a negative opinion about a given stand in question.

Let $A_{ij} = w_{ij} \in \mathbb{R}^{m \times m}$. Suppose that the initial state of the nodes is s^0. Then the dynamic equation (1) can be written as

$$s^{k+1} = As^k + x^k,\ k = 0, 1, 2, ..., N. \tag{2}$$

In matrix notation, $s^N = b + Bx$ where $b = A^N s^0$ and $B = [A^{N-1}, A^{N-2}, ..., A, I]$, the decision vector x is obtained by stacking x^1, x^2,..., x^N on top of each other. Obviously, x being the amount of effort to be spent is naturally nonnegative, while in our model s can take positive or negative values.

Now we assume that the total budget of the efforts spent over time is constrained to a total amount of unit 1. The objective is to find a way to allocate the budget in such a way that the total number of negatively opinionated nodes will be minimized. As a matter of notation, for vector s we denote $s_+ := \max(s, 0)$ to be the *positive* part of s and $s_- := -\min(s, 0)$ be the *negative* part of s ($s = s_+ - s_-$). Our *influence maximization* model is formulated as:

$$
\begin{aligned}
\min\ & f_0(z) := \|s_-\|_0 + \|s_+\|_q^q \\
\text{s.t.}\ & s_+ - s_- = b + Bx - \theta \\
& e^T x = 1 \\
& x \geq 0 \\
& s_+ \geq 0 \\
& s_- \geq 0
\end{aligned} \tag{3}
$$

A few words are in order here to further illustrate the model. The first constraint in (3) is due to the dynamic equation (2) (keeping in mind that $s = s_+ - s_-$); the second constrain is due to the budget; and the last set of constraints are due to the nonnegativity of the decision variables. In the objective, we have two different tasks to achieve. First of all, we wish to minimize the total number of negatively opinionated nodes, which can be modeled by the L_0-norm of the vector representing the negatively opinionated nodes as a whole: $\|s_-\|_0$. The second task is to regulate and evenly distribute the positively opinionated nodes (diffuse the investment to avoid the case that the total budget is wasted on the extreme positive nodes). One way to achieve the regularization is to introduce a convex measure on the vector s_+. Our choice is to incorporate this with the L_q norm. In fact, in our implementation later we shall simply set $q = 2$.

3 L_1 Norm Relaxation

In view of computational complexity, problem (3) is essentially a sparse optimization model (with L_0 norm in the objective), and is NP-hard in general [24]. The recent intensive study on *compressed sensing* made a strong case for sparse optimization. In the case of system of linear equations, the sparsest solution can often be recovered by relaxing L_0-norm to the convex L_1-norm. Specially, in [25] [26], it was shown that if a certain *Restricted Isometry Property*(RIP) condition holds for B, then the solutions of L_p norm minimization for $p = 0$ and $p = 1$ are identical. However, note that the constraint set in (3) involves *inequalities*. Therefore, the validity of the usual L_1-convexification approach in compressed sensing needs to be tested. In other words, we consider the following convexified model as

$$
\begin{aligned}
\min\ & f_1(z) := \|s_-\|_1 + \|s_+\|_q^q \\
\text{s.t.}\ & s_+ - s_- = b + Bx - \theta \\
& e^T x = 1 \\
& x \geq 0 \\
& s_+ \geq 0 \\
& s_- \geq 0
\end{aligned} \tag{4}
$$

The numerical performance of the above L_1-convexification approach will be tested in Numerical Experiments Section.

4 L_p Norm Relaxation

In this part, we use L_p norm relaxation to approximate the L_0 norm in the objective function.

$$\begin{aligned} \min\ & f_p(z) := \|s_-\|_p^p + \|s_+\|_q^q \\ \text{s.t.}\ & s_+ - s_- = b + Bx - \theta \\ & e^T x = 1 \\ & x \geq 0 \\ & s_+ \geq 0 \\ & s_- \geq 0 \end{aligned} \tag{5}$$

The challenge now is that Model (5) is no longer convex. Nevertheless, the feasible set of the model is still convex. In fact, as it turns out, the interior point methods are very effective solution methods for solving such non-convex model. Ge *et al.* [29] proposed an interior-point algorithm to solve the problem of minimizing L_p norm $(0 < p < 1)$ with inequality constraints, for which the potential reduction and the affine scaling methodologies can be applied. The interior-point algorithm is guaranteed to run in polynomial-time and will solve the problem to an ϵ-optimal solution in some well-defined sense. Unfortunately, the method of [29] cannot be directly applied in our context, because our model only a part of the decision variables will need be sparse. The objective of this problem is the combination of both convex and concave functions. Whether the interior-point algorithm still works well? However, some modification will be possible. In the next section, we shall present such an extension.

5 Interior-Point Algorithm

In this section, potential reduction algorithm [30] is reduced to solve this problem. Starting from an interior-point feasible solution such as the analytic center, this algorithm follows an interior feasible path and finally converges to either a global minimizer or a KKT point or local minimizer.

Let $z = (x; s_+; s_-)$ be the overall decision vector for (5), and $\underline{z}$ is a lower bound of the optimal value of (5). The potential function of this problem is defined as:

$$\phi(z) = \rho \log\left(\|s_-\|_p^p + \|s_+\|_2^2 + e^T x - \underline{z}\right) - \sum_{j=1}^{mN} \log x_j - \sum_{j=1}^{m} \log(s_+)_j^2 - \sum_{j=1}^{m} \log(s_-)_j^p ,$$

where the parameter $\rho > mN + 2m$. We may set $\underline{z} = 0$ as an obvious lower bound. Observe that

$$\frac{\sum_{j=1}^{mN} x_j + \sum_{j=1}^{m} (s_+)_j^2 + \sum_{j=1}^{m} (s_-)_j^p}{mN + 2m} \geqslant \left(\prod_{j=1}^{mN} x_j \prod_{j=1}^{m} (s_+)_j^2 \prod_{j=1}^{m} (s_-)_j^p\right)^{1/(mN+2m)} .$$

Therefore,

$$(mN+2m)\log f_p(z) - \sum_{j=1}^{mN}\log x_j - \sum_{j=1}^{m}\log (s_+)_j^2 - \sum_{j=1}^{m}\log (s_-)_j^p$$
$$\geq (mN+2m)\log (mN+2m).$$

Thus, if

$$\phi(z) \leq (\rho - (mN+2m))\log(\epsilon) + (mN+2m)\log(mN+2m),$$

then we have $f_p(z) \leq \epsilon$, which implies that z must be an ϵ-global minimizer.

The remaining task is to investigate how the potential value can be reduced at each iteration. Suppose for simplicity that the linear constraints can be expressed as $Hz = h$ and $z \geq 0$. Consider now how one interior iteration proceeds from $z > 0$ to $z^+ > 0$.

We have

$$\begin{aligned}\phi(z^+) - \phi(z) = {} & \rho\left(\log\left(\|s_-^+\|_p^p + \|s_+^+\|_2^2 + e^T x^+\right) - \log\left(\|s_-\|_p^p + \|s_+\|_2^2 + e^T x\right)\right) \\ & + \left(-\sum_{j=1}^{mN}\log x_j^+ - \sum_{j=1}^{m}\log (s_+^+)_j^2 - \sum_{j=1}^{m}\log (s_-^+)_j^p\right) \\ & + \left(\sum_{j=1}^{mN}\log x_j + \sum_{j=1}^{m}\log (s_+)_j^2 + \sum_{j=1}^{m}\log (s_-)_j^p\right).\end{aligned}$$

Let d_z be a vector such that $Hd_z = 0$, and $z^+ = z + d_z > 0$. By the concavity of $\log f(z)_p$, we have

$$\log f_p(z^+) - \log f_p(z) \leq \frac{1}{f_p(z)}\nabla(f_p(z))^T d_z.$$

By restricting $\|Z^{-1}d_z\| \leq \beta < 1$ where $Z = \mathrm{Diag}(z)$, we have $z + d_z > 0$ and also

$$\begin{aligned}&\left(-\sum_{j=1}^{mN}\log x_j^+ - \sum_{j=1}^{m}\log (s_+^+)_j^2 - \sum_{j=1}^{m}\log (s_-^+)_j^p\right) \\ &+ \left(\sum_{j=1}^{mN}\log x_j + \sum_{j=1}^{m}\log (s_+)_j^2 + \sum_{j=1}^{m}\log (s_-)_j^p\right) \leq -c^T Z^{-1} d_z + \frac{\beta^2}{(1-\beta)}\end{aligned}$$

where $c = (e; 2; p) \in \mathbb{R}^{(mN+2m)\times 1}$, $e \in \mathbb{R}^{mN\times 1}$, $2 \in \mathbb{R}^{m\times 1}$, $p \in \mathbb{R}^{m\times 1}$ (see Section 9.3 in [31] for more detailed), and

$$\phi(z^+) - \phi(z) \leq \left(\frac{\rho}{f_p(z)}\nabla f_p(z)^T Z - c^T\right) Z^{-1} d_z + \frac{\beta^2}{(1-\beta)}.$$

Let $d' = Z^{-1}d_z$. To achieve a potential reduction, one can minimize an affine-scaled linear function subject to a ball constraint.(Readers can see Chapter 1 and 4 in [32] for more details.)

$$\begin{aligned} U(d') := \min\ & (\tfrac{\rho}{f_p(z)} \nabla f_p(z)^T Z - c^T)d' \\ \text{s.t.}\ & HZd' = 0 \\ & \|d'\|^2 \leq \beta^2. \end{aligned} \tag{6}$$

The solution for (6) is explicit:

$$U(d') = -\beta \|g(z)\|$$

with the optimal direction

$$d' = \frac{\beta}{\|g(z)\|} g(z)$$

where

$$g(z) = -\left(I - ZH^T(HZ^2H^T)^{-1}HZ\right)\left(\frac{\rho}{f_p(z)} Z\nabla f_p(z) - c\right).$$

We may also write the above solution as

$$g(z) = c - \frac{\rho}{f_p(z)} Z\left(\nabla f_p(z) - H^T y\right),$$

with

$$y = \left(HZ^2H^T\right)^{-1} HZ\left(\mathrm{Diag}(c)\, Z^p - \frac{f_p(z)}{\rho} c\right)$$

where $Z^p = \mathrm{Diag}\left(\left(x; (s_+)^2; (s_-)^p\right)\right)$. If $\|g(z)\| \geq 1$, then the optimal value of (6) is less than $-\beta$, and so

$$\phi(z^+) - \phi(z) < -\beta + \frac{\beta^2}{(1-\beta)}.$$

Thus, the potential value is reduced by a constant if we set $\beta = 1/10$. This case could occur for at most

$$O\left((\rho - (mN + 2m)) \log(1/\epsilon)\right)$$

iterations before reaching an ϵ-global minimizer (cf. (6)).

On the other hand, if $\|g(z)\| \leq 1$, then, since

$$g(z) = c - \frac{\rho}{f_p(z)} Z\left(\nabla f(z) - H^T y\right),$$

we have

$$\frac{\rho}{f_p(z)} Z\left(\nabla f_p(z) - H^T y\right) \geq 0,$$

and

$$\frac{\rho}{f_p(z)} Z\left(\nabla f_p(z) - H^T y\right) \leq e + c, \forall j.$$

In other words,

$$\left(\nabla f_p(z) - H^T y\right)_j \geq 0, \tag{7}$$

and

$$\frac{z_j}{f_p(z)} \left(\nabla f_p(z) - H^T y\right)_j \leq \frac{2}{\rho}, \forall j = 1, 2, ..., mN; \tag{8}$$

$$\frac{z_j}{f_p(z)} \left(\nabla f_p(z) - H^T y\right)_j \leq \frac{3}{\rho}, \forall j = mN + 1, ..., mN + m; \tag{9}$$

$$\frac{z_j}{f_p(z)} \left(\nabla f_p(z) - H^T y\right)_j \leq \frac{1 + p}{\rho}, \forall j = mN + m + 1, ..., mN + 2m. \tag{10}$$

The first condition (7) shows that the Lagrange multiplier y is feasible. For the inequalities (8),(9),(10), by choosing $\rho \geq \frac{3(mN+2m)}{\epsilon}$ we have

$$\frac{1}{f_p(z)} z^T \left(\nabla f_p(z) - H^T y\right) \leq \epsilon.$$

Therefore,

$$\frac{z^T \left(\nabla f_p(z) - H^T y\right)}{\overline{z} - \underline{z}} \leq \frac{z^T \left(\nabla f_p(z) - H^T y\right)}{f_p(z)} \leq \epsilon,$$

which implies that z is an ϵ-KKT point.

The above analysis leads to the following result:

Theorem 1. *The interior-point algorithm returns an ϵ-KKT or ϵ-global solution in no more than $O\left(\frac{mN+2m}{\epsilon} \log \frac{1}{\epsilon}\right)$ iterations.*

6 Numerical Experiments

The setup of our simulation tests is as follows.

First we use the *ComplexNetworkPackage64bit.v14* [33] to produce a small world network, where we use the parameters $NumberOfNodes = 20$ (Number of nodes), $Alpha = -2.2$ (Alpha of the scale-free graph). Then assigh normal distribution random numbers to each edges. After the two steps, we can get a weighted directed graph as Figure 1. The graph is drawn by Gephi according to the simulation data.

By the data in DBLP of April 6, 2013, the collaborative networks of Professor shuzhong Zhang Figure 2 and Professor Zhiquan Tom Luo Figure 3 can be obtained, which are two real world examples. Although they are all undirected graphs, the directed networks can be created by this way that, each egde is drawn according to the collaborative orders of the main author and his coauthors. For example, in the DBLP dataset, Professor Zhiquan Tom Luo is before Jos F. Sturm in the collaborative order of Professor Shuzhong Zhang, thus when we

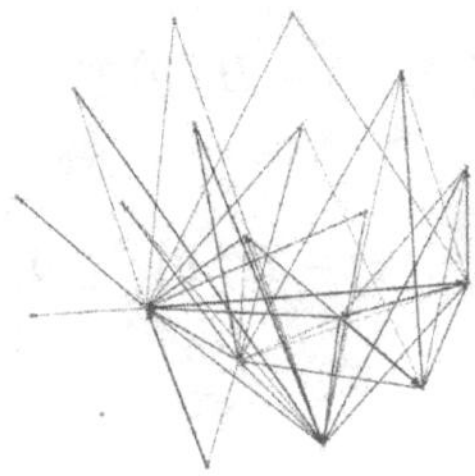

Fig. 1. Small World Network

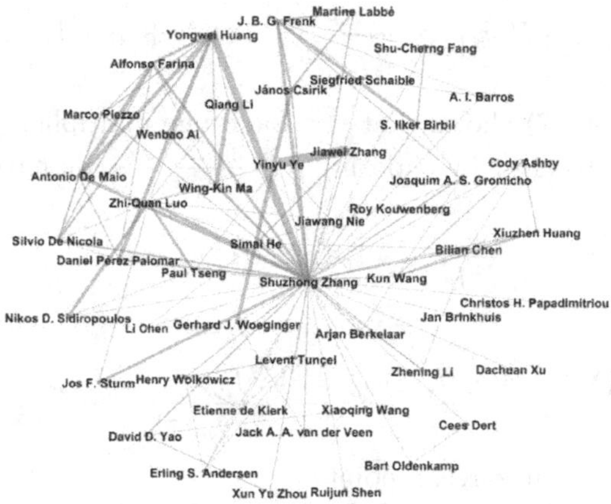

Fig. 2. Collaborative Network of Professor Shuzhong Zhang

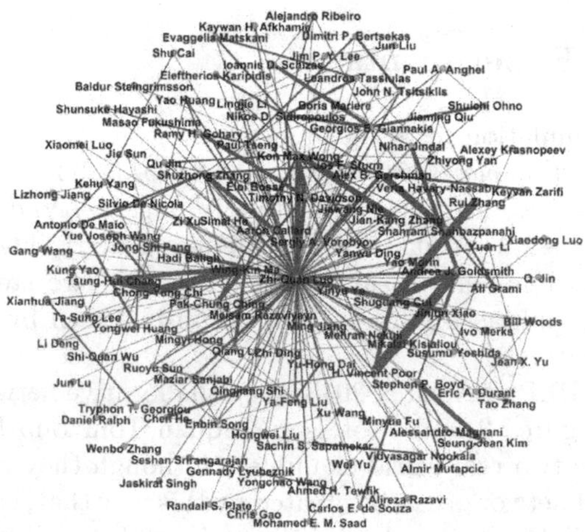

Fig. 3. Collaborative Network of Professor Zhiquan Tom Luo

connect the two nodes of Professor Zhiquan Tom Luo and Jos F. Sturm, the direction is from the first to the latter and the weight of that edge is the total number of papers they have written together.

Then we resort to the software package CVX to solve the L_1-norm (convex) model, and we implement the affine scaling algorithm to solve the L_p-norm (non-convex) model. We use $\alpha = 10^{-10}$ as the tolerance level for zeros: if $(s_-)_j > \alpha$, we count $(s_-)_j$ as non-zero; otherwise, $(s_-)_j$ is counted as zero.

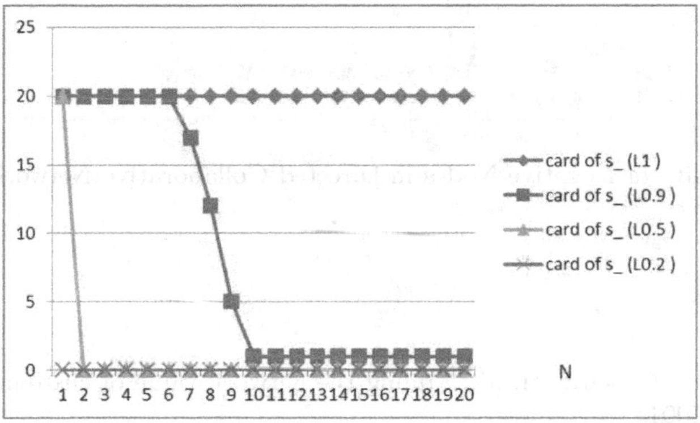

Fig. 4. Cardinality of Negative Nodes in Small World Network

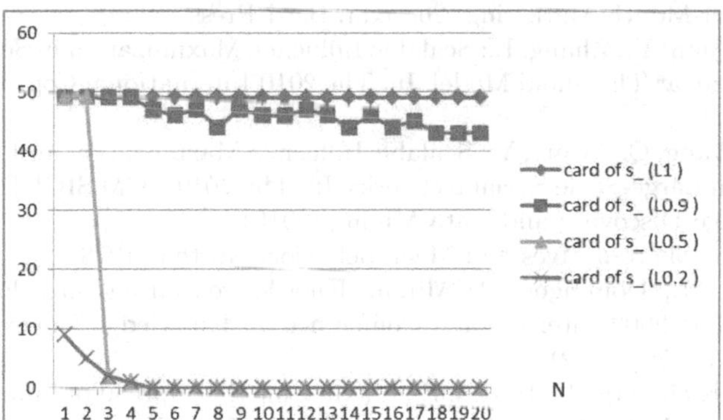

Fig. 5. Cardinality of Negative Nodes in Directed Collaborative Network of Professor Shuzhong Zhang

The simulation results show that L_p-norm $(0 < p < 1)$ non-convex relaxation method performs much better than the L_1-norm convex relaxation approach. The former method has a much sparser solution of negative nodes s_-. This implies that if we implement the resource according to the solution of L_p-norm relaxation model, then in the final stage the nodes with negative opinions are much less than that of the L_1-norm convex relaxation approach. Moreover, the solutions of the L_p-norm model are far more robust and stable.

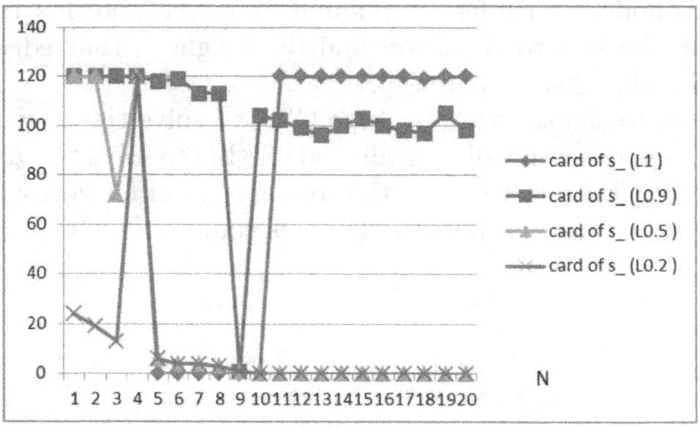

Fig. 6. Cardinality of Negative Nodes in Directed Collaborative Network of Professor Zhiquan Tom Luo

References

1. Domingos, P., Richardson, M.: Mining the network value of customers. In: KDD, pp. 57–66 (2001)
2. Nail, J.: The Consumer Advertising Backlash. Forrester Research and Intelliseek Market Research Report (May 2004)
3. Misner, I.R.: The World's Best Known Marketing Secret: Building Your Business with Word-of-Mouth Marketing, 2nd edn. Bard Press (1999)
4. Chen, W., Yuan, Y., Zhang, L.: Scalable Influence Maximization in Social Networks Under the Linear Threshold Model. In: The 2010 International Conference on Data Mining (2010)
5. Chen, W., Wang, C., Wang, Y.: Scalable Influence Maximization for Prevalent Viral Marketing in Large-Scale Social Networks. In: The 2010 ACM SIGKDD Conference on Knowledge Discovery and Data Mining (2010)
6. Schelling, T.: Micromotives and Macrobehavior. Norton (1978)
7. Richardson, M., Domingos, P.: Mining Knowledge-sharing Sites for Viral Marketing. In: The 2002 International Conference on Knowledge Discovery and Data Mining, pp. 61–70 (2002)
8. Kempe, D., Kleinberg, J., Tardos, É.: Maximizing The Spread of Influence Through a Social Network. In: The 2003 International Conference on Knowledge Discovery and Data Mining, pp. 137–146 (2003)
9. Kempe, D., Kleinberg, J., Tardos, É.: Influential Nodes in a Diffusion Model for Social Networks. In: Caires, L., Italiano, G.F., Monteiro, L., Palamidessi, C., Yung, M. (eds.) ICALP 2005. LNCS, vol. 3580, pp. 1127–1138. Springer, Heidelberg (2005)
10. Goldenberg, J., Libai, B., Muller, E.: Using Complex Systems Analysis to Advance Marketing Theory Development. Academy of Marketing Science Review (2001)
11. Goldenberg, J., Libai, B., Muller, E.: Talk of the Network:A Complex Systems Look at the Underlying Process of Word-of-Mouth. Marketing Letters 12(3), 211–223 (2001)
12. Granovetter, M.: Threshold Models of Collective Behavior. American Journal of Sociology 83(6), 1420–1443 (1978)

13. Zou, F., Willson, J., Wu, W.: Fast Information Propagation in Social Networks. In: MDMAA (2010)
14. Lu, Z., Zhang, W., Wu, W., Fu, B., Du, D.Z.: Approximation and Inapproximation for The Influence Maximization Problem in Social Networks under Deterministic Linear Threshold Model. In: 2011 31st International Conference on Distributed Computing Systems Workshops (2011)
15. Kepner, J., Gilbert, J.: Graph Algorithms in the Language of Linear Algebra, 1st edn. SIAM (2011)
16. Watts, D.J., Strogatz, S.H.: Collective Danamics of 'Small-World' Networks 393, 440–442 (1998)
17. Kleieberg, J.: The Small-World Phenomenon: An Algorithmic Perspective. In: Proceedings of 32rd ACM Symposium on Theory of Computing (2000)
18. Barabási, A.L., Albert, R., Jeong, H.: Mean-Field Theory for Scale-Free Random Networks. Physica A: Statistical Mechanics and its Applications 272(1), 173–187 (1999)
19. Romualdo, P.S., Vespignani, A.: Epidemic spreading in scale-free networks. Physical Review Letters 86(14), 3200–3203 (2001)
20. Bonato, A.: A Course on the Web Graph, Providence, Rhode Island. American Mathematical Society Graduate Studies Series in Mathematics (2008)
21. Kumar, R., Novak, J., Tomkins, A.: Structure and evolution of on-line social networks. In: Proceedings of the 12th ACM SIGKDD International Conference on Knowledge Discovery and Data Mining (2006)
22. Facebook: statistics, `http://www.facebook.com/press/info.php?statistics` (accessed September 1, 2011)
23. Twitaholic, `http://twitaholic.com/` (accessed September 1, 2011)
24. Natarajan, B.K.: Sparse Approximate Solution to Linear Systems. SIAM Journal on Computing 24, 227–234 (1995)
25. Candés, E.J., Tao, T.: Decoding by Linear Programming. IEEE Transaction of Information Theory 51, 4203–4215 (2005)
26. Donoho, D.: For Most Large Underdetermined Systems of Linear Equations the Minimal L_1-norm Solution is Also the Sparsest Solution. Technical Report, Stanford University (2004)
27. Bruckstein, A.M., Donoho, D.L., Elad, M.: From sparse solutions of systems of equations to sparse modeling of signals and images. SIAM Review 51(1), 34–81 (2009)
28. Tropp, J.A., Wright, S.J.: Computational methods for sparse solution of linear inverse problems. Proceedings of the IEEE 98(6), 948–958 (2010)
29. Ge, D., Jiang, X. and Ye, Y., A Note on the Complexity of L_p Minimization, `http://www.stanford.edu/~yyye/lpmin_v14.pdf`
30. Ye, Y.: On the complexity of approximating a KKT point of quadratic programming. Mathematical Programming 80, 195–211 (1998)
31. Bertsimas, D., Tsitsiklis, J.: Introduction to linear Optimization. Athena Scientific, 414 (1997)
32. Ye, Y.: Interior point algorithms: theory and analysis. John Wiley and Sons, Inc., New York (1997)
33. `http://www.levmuchnik.net/Content/Networks/ComplexNetworksPackage.html`

Fast Algorithms for the Maximum Clique Problem on Massive Sparse Graphs

Bharath Pattabiraman[1,*], Md. Mostofa Ali Patwary[1,*], Assefaw H. Gebremedhin[2], Wei-keng Liao[1], and Alok Choudhary[1]

[1] Northwestern University, Evanston, IL.
{bpa342,mpatwary,wkliao,choudhar}@eecs.northwestern.edu
[2] Purdue University, West Lafayette, IN.
agebreme@purdue.edu

Abstract. The maximum clique problem is a well known NP-Hard problem with applications in data mining, network analysis, information retrieval and many other areas related to the World Wide Web. There exist several algorithms for the problem with acceptable runtimes for certain classes of graphs, but many of them are infeasible for massive graphs. We present a new exact algorithm that employs novel pruning techniques and is able to quickly find maximum cliques in large sparse graphs. Extensive experiments on different kinds of synthetic and real-world graphs show that our new algorithm can be orders of magnitude faster than existing algorithms. We also present a heuristic that runs orders of magnitude faster than the exact algorithm while providing optimal or near-optimal solutions.

1 Introduction

A clique in an undirected graph is a subset of vertices in which every two vertices are adjacent to each other. The *maximum* clique problem seeks to find a clique of the largest possible size in a given graph.

The maximum clique problem, and the related *maximal* clique and clique *enumeration* problems, find applications in a wide variety of domains, many intimately related to the World Wide Web. A few examples include: information retrieval [2], community detection in networks [15,29,33], spatial data mining [40], data mining in bioinformatics [37], disease classification based on symptom correlation [7], pattern recognition [31], analysis of financial networks [5], computer vision [19], and coding theory [8]. To get a sense for how clique computation arises in the aforementioned contexts, consider a generic data mining or information retrieval problem. A typical objective here is to retrieve data that are considered similar based on some metric. Constructing a graph in which vertices correspond to data items and edges connect similar items, a clique in the graph would then give a cluster of similar data. More examples of application areas for clique problems can be found in [18,30].

The maximum clique problem is NP-Hard [16]. Most exact algorithms for solving it employ some form of *branch-and-bound* approach. While branching systematically searches for all candidate solutions, bounding (also known as *pruning*) discards fruitless candidates based on a previously computed bound. The algorithm of Carraghan

* Authors contributed equally.

A. Bonato, M. Mitzenmacher, and P. Prałat (Eds.): WAW 2013, LNCS 8305, pp. 156–169, 2013.
© Springer International Publishing Switzerland 2013

and Pardalos [9] is an early example of a simple and effective branch-and-bound algorithm for the maximum clique problem. More recently, Östergård [28] introduced an improved algorithm and demonstrated its relative advantages via computational experiments. Tomita and Seki [35], and later, Konc and Janežič [21] use upper bounds computed using vertex coloring to enhance the branch-and-bound approach. Other examples of branch-and-bound algorithms for the clique problem include [6,34,3]. Prosser [32] in a recent work compares various exact algorithms for the maximum clique problem.

In this paper, we present a new exact branch-and-bound algorithm for the maximum clique problem that employs several new pruning strategies in addition to those used in [9], [28], [35] and [21], making it suitable for massive graphs. We run our algorithms on a large variety of test graphs and compare its performance with the algorithm of Carraghan and Pardalos [9], the algorithm of Östergård [28] and the algorithm of Konc and Janežič [21]. We find our new exact algorithm to be up to orders of magnitude faster on large, sparse graphs and of comparable runtime on denser graphs. We also present a hew heuristic, which runs several orders of magnitude faster than the exact algorithm while providing solutions that are optimal or near-optimal for most cases. We have made our implementations publicly available[1]. Both the exact algorithm and the heuristic are well-suited for parallelization.

2 Related Previous Algorithms

Given a simple undirected graph G, the maximum clique can clearly be obtained by enumerating *all* of the cliques present in it and picking the largest of them. Carraghan and Pardalos [9] introduced a simple-to-implement algorithm that avoids enumerating all cliques and instead works with a significantly reduced partial enumeration. The reduction in enumeration is achieved via a *pruning* strategy which reduces the search space tremendously. The algorithm works by performing at each step i, a *depth first search* from vertex v_i, where the goal is to find the largest clique containing the vertex v_i. At each *depth* of the search, the algorithm compares the number of remaining vertices that could potentially constitute a clique containing vertex v_i against the size of the largest clique encountered thus far. If that number is found to be smaller, the algorithm backtracks (search is pruned).

Östergård [28] devised an algorithm that incorporated an additional pruning strategy to the one by Carraghan and Pardalos. The opportunity for the new pruning strategy is created by *reversing* the order in which the search is done by the Carraghan-Pardalos algorithm. This allows for an additional pruning with the help of some auxiliary bookkeeping. Experimental results in [28] showed that the Östergård algorithm is faster than the Carraghan-Pardalos algorithm on random and DIMACS benchmark graphs [20]. However, the new pruning strategy used in this algorithm is intimately tied to the order in which vertices are processed, introducing an inherent sequentiality into the algorithm.

A number of existing branch-and-bound algorithms for maximum clique also use a vertex-coloring of the graph to obtain an upper bound on the maximum clique. A popular and recent algorithm based on this idea is the algorithm of Tomita and Seiku [35] (known as MCQ). More recently, Konc and Janežič [21] presented an improved

[1] http://cucis.ece.northwestern.edu/projects/MAXCLIQUE/

version of MCQ, known as MaxCliqueDyn (MCQD and MCQD+CS), that involves the use of tighter, computationally more expensive upper bounds applied on a fraction of the search space.

3 The New Algorithms

We describe in this section new algorithms that overcome the shortcomings mentioned earlier; the new algorithms use additional pruning strategies, maintain simplicity, and avoid a sequential computational order. We begin by first introducing the following notations. We identify the n vertices of the input graph $G = (V, E)$ as $\{v_1, v_2, \ldots, v_n\}$. The set of vertices adjacent to a vertex v_i, the set of its neighbors, is denoted by $N(v_i)$. And the degree of the vertex v_i, the cardinality of $N(v_i)$, is denoted by $d(v_i)$.

3.1 The Exact Algorithm

The maximum clique in a graph can be found by computing the largest clique containing each vertex and picking the largest among these. A key element of our exact algorithm is that during the search for the largest clique containing a given vertex, vertices that cannot form cliques larger than the current maximum clique are *pruned*, in a hierarchical fashion. The method is outlined in detail in Algorithm 1. Throughout, the variable max stores the size of the maximum clique found thus far. Initially it is set to be equal to the lower bound lb provided as an input parameter. It gives the maximum clique size when the algorithm terminates.

Algorithm 1. Algorithm for finding the maximum clique of a given graph. *Input*: Graph $G = (V, E)$, lower bound on clique lb (default, 0). *Output*: Size of maximum clique.

```
procedure MAXCLIQUE(G = (V, E), lb)
    max ← lb
    for i : 1 to n do
        if d(v_i) ≥ max then                          ▷ Pruning 1
            U ← ∅
            for each v_j ∈ N(v_i) do
                if j > i then                         ▷ Pruning 2
                    if d(v_j) ≥ max then              ▷ Pruning 3
                        U ← U ∪ {v_j}
            CLIQUE(G, U, 1)
```

– Subroutine

```
procedure CLIQUE(G = (V, E), U, size)
    if U = ∅ then
        if size > max then
            max ← size
        return
    while |U| > 0 do
        if size + |U| ≤ max then                      ▷ Pruning 4
            return
        Select any vertex u from U
        U ← U \ {u}
        N'(u) := {w | w ∈ N(u) ∧ d(w) ≥ max}          ▷ Pruning 5
        CLIQUE(G, U ∩ N'(u), size + 1)
```

To obtain the largest clique containing a vertex v_i, it is sufficient to consider only the neighbors of v_i. The main routine MAXCLIQUE thus generates for each vertex $v_i \in V$ a set $U \subseteq N(v_i)$ (neighbors of v_i that survive pruning) and calls the subroutine CLIQUE on U. The subroutine CLIQUE goes through every relevant clique containing v_i in a recursive fashion and returns the largest. We use $size$ to maintain the size of the clique found at any point through the recursion.

Since we start with a clique of just one vertex, the value of $size$ is set to one initially, when CLIQUE is called (Line 10, MAXCLIQUE).

Our algorithm consists of several pruning steps. Pruning 1 (Line 4, MAXCLIQUE) filters vertices having strictly fewer neighbors than the size of the maximum clique already computed. These vertices can be ignored, since even if a clique were to be found, its size would not be larger than max. While forming the neighbor list U for a vertex v_i, we include only those of v_i's neighbors for which the largest clique containing them has not been found (Pruning 2; Line 7, MAXCLIQUE), to avoid recomputing previously found cliques. Pruning 3 (Line 8, MAXCLIQUE) excludes vertices $v_j \in N(v_i)$ that have degree less than the current value of max, since any such vertex could not form a clique of size larger than max. Pruning 4 (Line 7, CLIQUE) checks for the case where even if all vertices of U were added to get a clique, its size would not exceed that of the largest clique encountered so far in the search, max. Pruning 5 (Line 11, CLIQUE) reduces the number of comparisons needed to generate the intersection set in Line 12. Note that the routine CLIQUE is similar to the Carraghan-Pardalos algorithm [9]; Pruning 5 accounts for the main difference. Also, Pruning 4 is used in most existing algorithms, whereas Prunings 1, 2, 3 and 5 are not.

3.2 The Heuristic

The exact algorithm examines all relevant cliques containing every vertex. Our heuristic, shown in Algorithm 2, considers only the *maximum degree* neighbor at each step instead of recursively considering all neighbors from the set U, and thus is much faster.

Algorithm 2. Heuristic for finding the maximum clique in a graph. *Input*: Graph $G = (V, E)$. *Output*: Approximate size of maximum clique.

```
procedure MAXCLIQUEHEU(G = (V, E))
    for i : 1 to n do
        if d(v_i) ≥ max then
            U ← ∅
            for each v_j ∈ N(v_i) do
                if d(v_j) ≥ max then
                    U ← U ∪ {v_j}
            CLIQUEHEU(G, U, 1)
```

– *Subroutine*

```
procedure CLIQUEHEU(G = (V, E), U, size)
    if U = ∅ then
        if size > max then
            max ← size
        return
    Select a vertex u ∈ U of maximum degree in G
    U ← U \ {u}
    N'(u) := {w | w ∈ N(u) ∧ d(w) ≥ max}
    CLIQUEHEU(G, U ∩ N'(u), size + 1)
```

3.3 Complexity

The exact algorithm, Algorithm 1, examines for every vertex v_i all candidate cliques containing the vertex v_i in its search for the largest clique. Its time complexity is exponential in the worst case. The heuristic, Algorithm 2, loops over the n vertices, each time possibly calling the subroutine CLIQUEHEU, which effectively is a loop that runs until the set U is empty. Clearly, $|U|$ is bounded by the max degree Δ in the graph. The subroutine also includes the computation of a neighbor list, whose runtime is bounded by $O(\Delta)$. Thus, the time complexity of the heuristic is bounded by $O(n \cdot \Delta^2)$.

Table 1. Overview of real-world graphs in the testbed and their origins

Graph	Description
cond-mat-2003 [26]	A collaboration network of scientists posting preprints on the condensed matter archive at www.arxiv.org in the period
email-Enron [23]	A communication network representing email exchanges.
dictionary28 [4]	Pajek network of words.
Fault_639 [14]	A structural problem discretizing a faulted gas reservoir with tetrahedral Finite Elements and triangular Interface Elements.
audikw_1 [11]	An automotive crankshaft model of TETRA elements.
bone010 [39]	A detailed micro-finite element model of bones representing the porous bone micro-architecture.
af_shell [11]	A sheet metal forming simulation network.
as-Skitter [23]	An Internet topology graph from trace routes run daily in 2005.
roadNet-CA [23]	A road network of California. Nodes represent intersections and endpoints and edges represent the roads connecting them.
kkt_power [11]	An Optimal Power Flow (nonlinear optimization) network.

4 Experiments and Results

We present in this section results comparing the performance of our algorithm with the algorithms of Carraghan-Pardalos [9], Östergård algorithm [28], and Konc and Janezik [21]. We implemented the algorithm of [9] ourselves. For the algorithm of [28], we used the publicly available *cliquer* source code [27]. For the algorithm of [21], we used the code *MaxCliqueDyn* (MCQD, available at `http://www.sicmm.org/~konc/maxclique/`). Among the variants available in MCQD, we report results on MCQD+CS (which uses improved coloring and dynamic sorting), since it is the best-performing variant.

The experiments are performed on a Linux workstation running 64-bit Red Hat Enterprise Linux Server release 6.2 with a 2 GHz Intel Xeon E7540 processor. The codes are implemented in C++ and compiled using gcc version 4.4.6 with -O3 optimization.

4.1 Test Graphs

Our testbed is grouped in three categories.

1. Real-world graphs. Under this category, we consider 10 graphs (downloaded from the University of Florida Sparse Matrix Collection [11]) that originate from various real-world applications. Table 1 gives a quick overview of the graphs and their origins.

2. Synthetic Graphs. In this category we consider 15 graphs generated using the R-MAT algorithm [10]. The graphs are subdivided in three categories depending on the structures they represent.

A. Random graphs (5 graphs) – Erdős-Rényi random graphs generated using R-MAT with the parameters (0.25, 0.25, 0.25, 0.25). Denoted with prefix *rmat_er*.

B. Skewed Degree, Type 1 graphs (5 graphs) – graphs generated using R-MAT with the parameters (0.45, 0.15, 0.15, 0.25). Denoted with prefix *rmat_sd1*.

C. Skewed Degree, Type 2 graphs (5 graphs) – graphs generated using R-MAT with the parameters (0.55, 0.15, 0.15, 0.15). Denoted with prefix *rmat_sd2*.

Table 2. Structural properties (the number of vertices, $|V|$; edges, $|E|$; and the maximum degree, Δ) of the graphs, G in the testbed: DIMACS Challenge graphs (upper left); UF Collection (lower and middle left); RMAT graphs (right).

G	$\|V\|$	$\|E\|$	Δ	G	$\|V\|$	$\|E\|$	Δ
cond-mat-2003	31,163	120,029	202	*rmat_sd1_1*	131,072	1,046,384	407
email-Enron	36,692	183,831	1,383	*rmat_sd1_2*	262,144	2,093,552	558
dictionary28	52,652	89,038	38	*rmat_sd1_3*	524,288	4,190,376	618
Fault_639	638,802	13,987,881	317	*rmat_sd1_4*	1,048,576	8,382,821	802
audikw_1	943,695	38,354,076	344	*rmat_sd1_5*	2,097,152	16,767,728	1,069
bone010	986,703	35,339,811	80	*rmat_sd2_1*	131,072	1,032,634	2,980
af_shell10	1,508,065	25,582,130	34	*rmat_sd2_2*	262,144	2,067,860	4,493
as Skitter	1,696,415	11,095,298	35,455	*rmat_sd2_3*	524,288	4,153,043	6,342
roadNet-CA	1,971,281	2,766,607	12	*rmat_sd2_4*	1,048,576	8,318,004	9,453
kkt_power	2,063,494	6,482,320	95	*rmat_sd2_5*	2,097,152	16,645,183	14,066
rmat_er_1	131,072	1,048,515	82	*hamming6-4*	64	704	22
rmat_er_2	262,144	2,097,104	98	*johnson8-4-4*	70	1,855	53
rmat_er_3	524,288	4,194,254	94	*keller4*	171	9,435	124
rmat_er_4	1,048,576	8,388,540	97	*c-fat200-5*	200	8,473	86
rmat_er_5	2,097,152	16,777,139	102	*brock200_2*	200	9,876	114

3. DIMACS graphs. This last category consists of 5 graphs selected from the Second DIMACS Implementation Challenge [20].

The DIMACS graphs are an established benchmark for the maximum clique problem, but they are of rather limited size and variation. In contrast, the real-work networks included in category 1 of the testset and the synthetic (RMAT) graphs in category 2 represent a wide spectrum of large graphs posing varying degrees of difficulty for testing the algorithms. The *rmat_er* graphs have *normal* degree distribution, whereas the *rmat_sd1* and *rmat_sd2* graphs have skewed degree distributions and contain many dense local subgraphs. The *rmat_sd1* and *rmat_sd2* graphs differ primarily in the magnitude of maximum vertex degree they contain; the *rmat_sd2* graphs have much higher maximum degree. Table 2 lists basic structural information (the number of vertices, number of edges and the maximum degree) about all 30 of the test graphs.

4.2 Results

Table 3 shows the size of the maximum clique (ω) and the runtimes of our exact algorithm (Algorithm 1) and the algorithms of Caraghan and Pardalos (CP), Östergård (*cliquer*) and Konc and Janežič (MCQD+CS) for all the graphs in the testbed. The last two columns show the results of our heuristic (Algorithm 2)—the size of the maximum clique returned and its runtime. The columns labeled $P1$, $P2$, $P3$ and $P5$ list the number of vertices/branches pruned in the respective pruning steps of Algorithm 1. Pruning 4 is omitted since it is used by all the algorithms compared in the table. These numbers have been rounded (K stands for 10^3, M for 10^6 and B for 10^9), although the exact numbers can be found in the Appendix (Table 4).

In Table 3, the fastest runtime for each instance is indicated with boldface. An asterisk (*) indicates that an algorithm did not terminate within 25,000 seconds for a

Table 3. Comparison of runtimes (in seconds) of algorithms [9] (*CP*), [28] (*cliquer*), [21] (*MCQD+CS*) and our new exact algorithm (τ_{A1}) for the graphs in the testbed. $P1$, $P2$, $P3$ and $P5$ are the number of vertices/branches pruned in steps Pruning 1, 2, 3 and 5 of our exact algorithm (K stands for 10^3, M for 10^6 and B for 10^9). ω denotes the maximum clique size in the graph, ω_{A2} denotes the clique size returned by our heuristic and τ_{A2} shows its runtime.

Graph	ω	τ_{CP}	$\tau_{cliquer}$	$\tau_{MCQD+CS}$	τ_{A1}	$P1$	$P2$	$P3$	$P5$	ω_{A2}	τ_{A2}
cond-mat-2003	25	4.875	11.17	2.41	**0.011**	29K	48K	6,527	17K	25	<0.01
email-Enron	20	7.005	15.08	3.70	**0.998**	32K	155K	4,060	8M	18	0.261
dictionary28	26	7.700	32.74	7.69	**<0.01**	52K	4,353	2,114	107	26	<0.01
Fault_639	18	14571.20	4437.14	-	**20.03**	36	13M	126	1,116	18	5.80
audikw_1	36	*	9282.49	-	**190.17**	4,101	38M	59K	721K	36	58.38
bone010	24	*	10002.67	-	**393.11**	37K	34M	361K	44M	24	24.39
af_shell10	15	*	21669.96	-	**50.99**	19	25M	75	2,105	15	10.67
as-Skitter	67	24385.73	*	-	**3838.36**	1M	6M	981K	737M	66	27.08
roadNet-CA	4	*	*	-	**0.44**	1M	1M	370K	4,302	4	0.08
kkt_power	11	*	*	-	**2.26**	1M	4M	401K	2M	11	1.83
rmat_er_1	3	256.37	215.18	49.79	**0.38**	780	1M	915	8,722	3	0.12
rmat_er_2	3	1016.70	865.18	-	**0.78**	2,019	2M	2,351	23K	3	0.24
rmat_er_3	3	4117.35	3456.39	-	**1.87**	4,349	4M	4,960	50K	3	0.49
rmat_er_4	3	16419.80	13894.52	-	**4.16**	9,032	8M	10K	106K	3	1.44
rmat_er_5	3	*	*	-	**9.87**	18K	16M	20K	212K	3	2.57
rmat_sd1_1	6	225.93	214.99	50.08	**1.39**	39K	1M	23K	542K	6	0.45
rmat_sd1_2	6	912.44	858.80	-	**3.79**	90K	2M	56K	1M	6	0.98
rmat_sd1_3	6	3676.14	3446.02	-	**8.17**	176K	4M	106K	2M	6	1.78
rmat_sd1_4	6	14650.40	13923.93	-	**25.61**	369K	8M	214K	5M	6	4.05
rmat_sd1_5	6	*	*	-	**46.89**	777K	16M	455K	12M	6	9.39
rmat_sd2_1	26	427.41	213.23	**48.17**	242.20	110K	853K	88K	614M	26	32.83
rmat_sd2_2	35	4663.62	**851.84**	-	3936.55	232K	1M	195K	1B	35	95.89
rmat_sd2_3	39	13626.23	**3411.14**	-	10647.84	470K	3M	405K	1B	37	245.51
rmat_sd2_4	43	*	**13709.52**	-	*	*	*	*	*	42	700.05
rmat_sd2_5	N	*	*	-	*	*	*	*	*	51	1983.21
hamming6-4	4	**<0.01**	**<0.01**	**<0.01**	<0.01	0	704	0	0	4	<0.01
johnson8-4-4	14	0.19	**<0.01**	**<0.01**	0.23	0	1,855	0	0	14	<0.01
keller4	11	22.19	0.15	**0.02**	23.35	0	9,435	0	0	11	<0.01
c-fat200-5	58	0.60	0.33	**0.01**	0.93	0	8,473	0	0	58	0.04
brock200_2	12	0.98	0.02	**<0.01**	1.10	0	9,876	0	0	10	<0.01

particular instance. A hyphen (-) indicates that the publicly available implementation (the *MaxCliqueDyn* code) had to be aborted because the input graph was too large for the implementation to handle. Even for the instances for which the code eventually run successfully, we had to first modify the graph reader to make it able to handle graphs with multiple connected components. For the graph *rmat_sd2_5*, none of the algorithms computed the maximum clique size in a reasonable time; the entry there is marked with N, standing for "Not Known".

We discuss in what follows our observations from this table for the exact algorithm and the heuristic.

Exact Algorithms. As expected, our exact algorithm gave the same size of maximum clique as the other three algorithms for all test cases. In terms of runtime, its relative performance compared to the other three varied in accordance with the advantages afforded by the various pruning steps.

Vertices that are discarded by Pruning 1 are skipped in the main loop of the algorithm, and the largest cliques containing them are not computed. Pruning 2 avoids re-computing previously computed cliques in the neighborhood of a vertex. In the absence of Pruning 1, the number of vertices pruned by Pruning 2 would be bounded by the number of edges in the graph (note that this is more than the total number of vertices in the graph). While Pruning 3 reduces the size of the input set on which the maximum clique is to be computed, Pruning 5 brings down the time taken to generate the intersection set in Line 12 of the subroutine. Pruning 4 corresponds to back tracking. Unlike Pruning steps 1, 2, 3 and 5, Pruning 4 is used by all three of the other algorithms in our comparison. The primary strength of our algorithm is its ability to take advantage of pruning in multiple steps in a hierarchical fashion, allowing for opportunities for one or more of the steps to kick in and impact performance.

As a result of the differences seen in the effects of the pruning steps, as discussed below, the runtime performance of our algorithm (seen in Table 3) compared to the other three algorithms varied in accordance with the difference in the structures represented by the different categories of graphs in the testbed.

Real-World Graphs. For most of the graphs in this category, it can be seen that our algorithm runs several orders of magnitude faster than the other three, mainly due to the large amount of pruning the algorithm enforced. These numbers also illustrate the great benefit of hierarchical pruning. For the graphs *Fault_639*, *audikw_1* and *af_shell10*, there is only minimal impact by Prunings 1, 3 and 5, whereas Pruning 2 makes a big difference resulting in impressive runtimes. The number of vertices pruned in steps Pruning 1 and 3 varied among the graph *within* the category, ranging from 0.001% for *af_shell* to a staggering 97% for *as Skitter* for the step Pruning 1.

Synthetic Graphs. For the synthetic graph types *rmat_er* and *rmat_sd1*, our algorithm clearly outperforms the other three by a few orders of magnitude in all cases. This is also primary due to the high number of vertices discarded by the new pruning steps. In particular, for *rmat_sd1* graphs, between 30 to 37% of the vertices are pruned just in the step Pruning 1. For the *rmat_sd2* graphs, which have relatively larger maximum clique and higher maximum degree than the *rmat_sd1* graphs, our algorithm is observed to be faster than CP but slower than *cliquer*.

DIMACS Graphs. The runtime of our exact algorithm for the DIMACS graphs is in most cases comparable to that of CP and higher than that of *cliquer* and *MCQD+CS*. For these graphs, only Pruning 2 was found to be effective, and thus the performance results agree with one's expectation. We include in the Appendix timing results on a larger collection of DIMACS graphs.

It is to be noted that the DIMACS graphs are intended to serve as challenging test cases for the maximum clique problem, and graphs with such high edge densities and low vertex count are rare in practice. Most of these have between 20 to 1024 vertices with an average edge density of roughly 0.6, whereas, most real world graphs are often very large and sparse. Good examples are Internet topology graphs [13], the web graph [22], social network graphs [12], and the real-world graphs in our testbed.

The Heuristic. It can be seen that our heuristic runs several orders of magnitude faster than our exact algorithm, while delivering either optimal or very close to optimal solution. It gave the optimal solution on 25 out of the 30 test cases. On the remaining 5 cases where it was suboptimal, its accuracy ranges from 83% to 99% (on average

93%). Additionally, we run the heuristic by choosing a vertex randomly in Line 6 of Algorithm 2 instead of the one with the maximum degree. We observe that on average, the solution is optimal only for less than 40% of the test cases compared to 83% when selecting the maximum degree vertex.

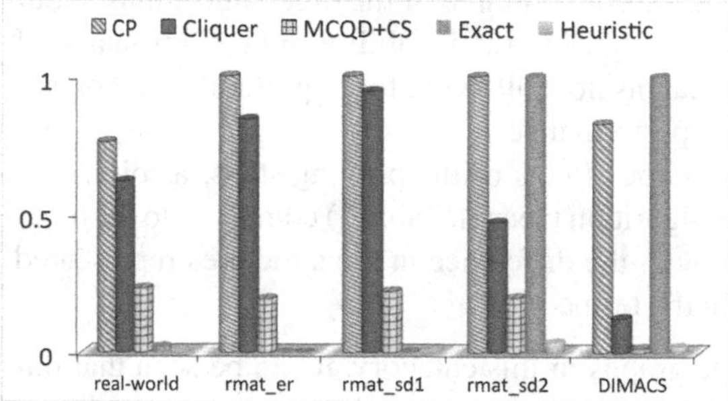

Fig. 1. Runtime (normalized, mean) comparison between various algorithms. For each category of graph, first, all runtimes for each graph were normalized by the runtime of the slowest algorithm for that graph, and then the mean was calculated for each algorithm. Graphs were considered only if the runtimes for at least three algorithms was less than the 25,000 seconds limit set.

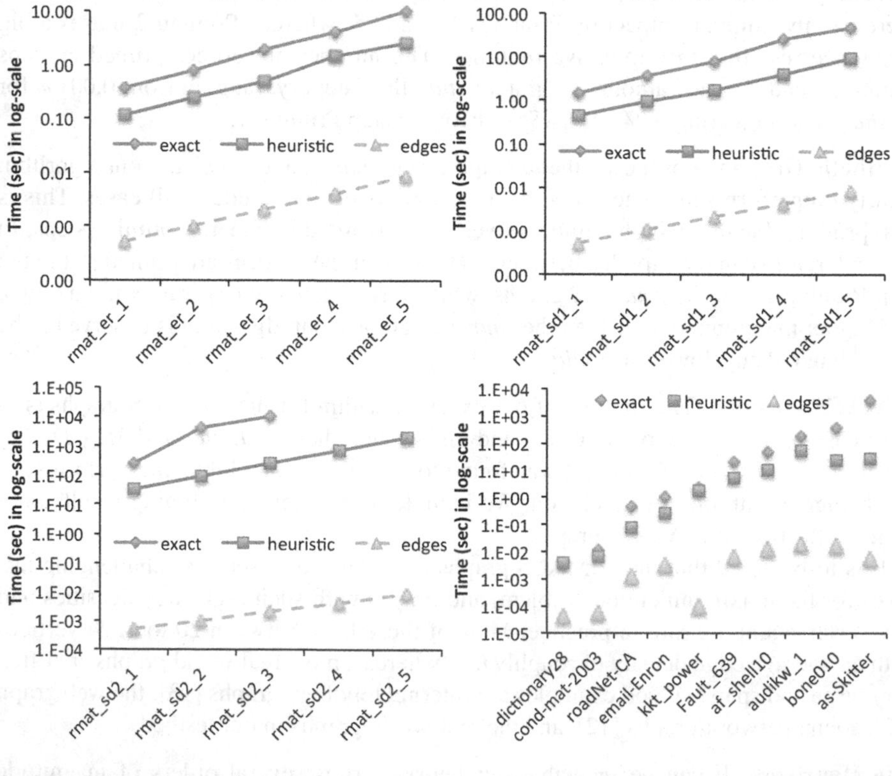

Fig. 2. Run time plots of the new exact and heuristic algorithms. The third curve, labeled *edges*, shows the quantity, number of edges in the graph divided by the clock frequency of the computing platform used in the experiment.

Figure 1 provides an aggregated visual summary of the runtime trends of the various algorithms across the five categories of graphs in the testbed.

To give a sense of runtime growth rates, we provide in Figure 2 plots of the runtime of the new exact algorithm and the heuristic for the synthetic and real-world graphs in the testbed. Besides the curves corresponding to the runtimes of the *exact* algorithm and the *heuristic*, the figures also include a curve corresponding to the number of *edges* in the graph divided by the clock frequency of the computing platform used in the experiment. This curve is added to facilitate comparison between the growth rate of the algorithms with that of a linear-time (in the size of the graph) growth rate. It can be seen that the runtime of the heuristic by and large grows somewhat linearly with the size of a graph. The exact algorithm's runtime, which is orders of magnitude larger than the heuristic, exhibited a similar growth behavior for these test-cases (even though its worst-case complexity suggests exponential growth).

5 Conclusion

We presented a new exact and a new heuristic algorithm for the maximum clique problem. We performed extensive experiments on three broad categories of graphs comparing the performance of our algorithms to the algorithms due to Carraghan and Pardalos (CP) [9], Östergård (*cliquer*) [28] and Konc and Janežič (*MCQD+CS*) [21]. For DIMACS benchmark graphs and certain dense synthetic graphs (*rmat_sd2*), our new exact algorithm performs comparably with the CP algorithm, but slower than *cliquer* and *MCQD+CS*. For large sparse graphs, both synthetic and real-world, our new algorithm runs several orders of magnitude faster than the other three. The heuristic, which runs many orders of magnitude faster than our exact algorithm and the others, gave optimal solution for 83% of the test cases, and when it is sub-optimal, its accuracy ranged between 0.83 and 0.99.

In this work, we did not compare the performance of our algorithm against those for which an implementation is not publicly available such as [36,25]. It would be interesting to implement these and compare in future work. Further, the MCQD implementation uses an adjacency matrix, whereas our algorithm uses an adjacency list to represent the graph. Although it is unlikely for the overall results to be drastically different with a change in the graph representation, it will be interesting to study to what degree the performance will change with the change in graph representation. The heuristic's performance is impressive as presented; still it is worthwhile to compare with other existing heuristics approaches such as [1,17].

Acknowledgements. This work is supported in part by the following grants: NSF awards CCF-0833131, CNS-0830927, IIS-0905205, CCF-0938000, CCF-1029166, and OCI-1144061; DOE awards DE-FG02-08ER25848, DE-SC0001283, DE-SC0005309, DESC0005340, and DESC0007456; AFOSR award FA9550-12-1-0458. The work of Assefaw Gebremedhin is supported by the NSF award CCF-1218916 and by the DOE award DE-SC0010205.

References

1. Andrade, D., Resende, M., Werneck, R.: Fast local search for the maximum independent set problem. Journal of Heuristics 18, 525–547 (2012)
2. Augustson, J.G., Minker, J.: An analysis of some graph theoretical cluster techniques. J. ACM 17, 571–588 (1970)
3. Babel, L., Tinhofer, G.: A branch and bound algorithm for the maximum clique problem. Mathematical Methods of Operations Research 34, 207–217 (1990)
4. Batagelj, V., Mrvar, A.: Pajek datasets (2006), `http://vlado.fmf.uni-lj.si/pub/networks/data/`
5. Boginski, V., Butenko, S., Pardalos, P.M.: Statistical analysis of financial networks. Computational Statistics & Data Analysis 48, 431–443 (2005)
6. Bomze, I.M., Budinich, M., Pardalos, P.M., Pelillo, M.: The Maximum Clique Problem. In: Handbook of Combinatorial Optimization, pp. 1–74. Kluwer Academic Publishers (1999)
7. Bonner, R.E.: On some clustering techniques. IBM J. Res. Dev. 8, 22–32 (1964)
8. Brouwer, A.E., Shearer, J.B., Sloane, N.J.A., Smith, W.D.: A new table of constant weight codes. IEEE Transactions on Information Theory, 1334–1380 (1990)
9. Carraghan, R., Pardalos, P.: An exact algorithm for the maximum clique problem. Oper. Res. Lett. 9, 375–382 (1990)
10. Chakrabarti, D., Faloutsos, C.: Graph mining: Laws, generators, and algorithms, ACM Comput. Surv. 38 (2006)
11. Davis, T.A., Hu, Y.: The university of florida sparse matrix collection. ACM Transactions on Mathematical Software (TOMS) 38, 1:1–1:25 (2011)
12. Domingos, P., Richardson, M.: Mining the network value of customers. In: Proc. of the 7th ACM SIGKDD KDD 2001, KDD 2001, San Francisco, California, pp. 57–66. ACM, New York (2001)
13. Faloutsos, M., Faloutsos, P., Faloutsos, C.: On power-law relationships of the Internet topology. In: Proc. of the Conference on Applications, Technologies, Architectures, and Protocols for Computer Communication, SIGCOMM 1999, Cambridge, Massachusetts, United States, pp. 251–262. ACM (1999)
14. Ferronato, M., Janna, C., Gambolati, G.: Mixed constraint preconditioning in computational contact mechanics. Computer Methods in Applied Mechanics and Engineering 197, 3922–3931 (2008)
15. Fortunato, S.: Community detection in graphs. Physics Reports 486, 75–174 (2010)
16. Garey, M.R., Johnson, D.S.: W. H. Freeman & Co., New York, NY, USA (1979)
17. Grosso, A., Locatelli, M., Pullan, W.: Simple ingredients leading to very efficient heuristics for the maximum clique problem. Journal of Heuristics 14, 587–612 (2008)
18. Gutin, G., Gross, J.L., Yellen, J.: Handbook of graph theory. Discrete Mathematics & Its Applications. CRC Press (2004)
19. Horaud, R., Skordas, T.: Stereo correspondence through feature grouping and maximal cliques. IEEE Trans. Pattern Anal. Mach. Intell. 11, 1168–1180 (1989)
20. Johnson, D., Trick, M.A. (eds.): Cliques, coloring and satisfiability: Second dimacs implementation challenge. DIMACS Series on Discrete Mathematics and Theoretical Computer Science, vol. 26 (1996)
21. Konc, J., Janežič, D.: An improved branch and bound algorithm for the maximum clique problem. MATCH Commun. Math. Comput. Chem. 58, 569–590 (2007)
22. Kumar, R., Raghavan, P., Rajagopalan, S., Tomkins, A.: Extracting Large-Scale Knowledge Bases from the Web. In: VLDB 1999, pp. 639–650 (1999)

23. Leskovec, J., Kleinberg, J., Faloutsos, C.: Graphs over time: densification laws, shrinking diameters and possible explanations. In: Proceedings of the Eleventh ACM SIGKDD International Conference on Knowledge Discovery in Data Mining, KDD 2005, Chicago, Illinois, USA, pp. 177–187. ACM, New York (2005)
24. Leydesdorff, L.: On the normalization and visualization of author co-citation data: Salton's cosine versus the jaccard index. J. Am. Soc. Inf. Sci. Technol. 59, 77–85 (2008)
25. Li, C.-M., Quan, Z.: An efficient branch-and-bound algorithm based on maxsat for the maximum clique problem (2010)
26. Newman, M.E.J.: Coauthorship networks and patterns of scientific collaboration. Proceedings of the National Academy of Sciences of the United States of America 101, 5200–5205 (2004)
27. Niskanen, S., Östergård, P.R.J.: Cliquer user's guide, version 1.0, Tech. Rep. T48, Communications Laboratory, Helsinki University of Technology, Espoo, Finland (2003)
28. Östergård, P.R.J.: A fast algorithm for the maximum clique problem. Discrete Appl. Math. 120, 197–207 (2002)
29. Palla, G., Derenyi, I., Farkas, I., Vicsek, T.: Uncovering the overlapping community structure of complex networks in nature and society. Nature 435, 814–818 (2005)
30. Pardalos, P.M., Xue, J.: The maximum clique problem. Journal of Global Optimization 4, 301–328 (1994)
31. Pavan, M., Pelillo, M.: A new graph-theoretic approach to clustering and segmentation. In: Proc. of the 2003 IEEE Computer Society Conference on Computer Vision and Pattern Recognition, CVPR 2003, pp. 145–152. IEEE Computer Society, Washington, DC (2003)
32. Prosser, P.: Exact algorithms for maximum clique: A computational study, arXiv preprint arXiv:1207.4616v1 (2012)
33. Sadi, S., Öğüdücü, S., Uyar, A.S.: An efficient community detection method using parallel clique-finding ants. In: Proc. of IEEE Congress on Evol. Comp., pp. 1–7 (July 2010)
34. San Segundo, P., Rodríguez-Losada, D., Jiménez, A.: An exact bit-parallel algorithm for the maximum clique problem. Comput. Oper. Res. 38, 571–581 (2011)
35. Tomita, E., Seki, T.: An efficient branch-and-bound algorithm for finding a maximum clique. In: Calude, C.S., Dinneen, M.J., Vajnovszki, V. (eds.) DMTCS 2003. LNCS, vol. 2731, pp. 278–289. Springer, Heidelberg (2003)
36. Tomita, E., Sutani, Y., Higashi, T., Takahashi, S., Wakatsuki, M.: A simple and faster branch-and-bound algorithm for finding a maximum clique. In: Rahman, M. S., Fujita, S. (eds.) WALCOM 2010. LNCS, vol. 5942, pp. 191–203. Springer, Heidelberg (2010)
37. Matsunaga, T., Yonemori, C., Tomita, E., Muramatsu, M.: Clique-based data mining for related genes in a biomedical database. BMC Bioinformatics 10, 205 (2009)
38. Turner, J.: Almost all k-colorable graphs are easy to color. Journal of Algorithms 9, 63–82 (1988)
39. van Rietbergen, B., Weinans, H., Huiskes, R., Odgaard, A.: A new method to determine trabecular bone elastic properties and loading using micromechanical finite-element models. Journal of Biomechanics 28, 69–81 (1995)
40. Wang, L., Zhou, L., Lu, J., Yip, J.: An order-clique-based approach for mining maximal co-locations. Information Sciences 179, 3370–3382 (2009)

Appendix

Table 4. $P1$, $P2$, $P3$, $P4$ and $P5$ are the number of vertices pruned in steps Pruning 1, 2, 3, 4, and 5 of Algorithm 1. An asterisk (*) indicates that the algorithm did not terminate within 25,000 seconds for that instance. ω denotes the maximum clique size.

G	ω	$P1$	$P2$	$P3$	$P4$	$P5$
cond-mat-2003	25	29,407	48,096	6,527	2,600	17,576
email-Enron	20	32,462	155,344	4,060	110,168	8,835,739
dictionary28	26	52,139	4,353	2,114	542	107
Fault_639	18	36	13,987,719	126	10,767,992	1,116
audikw_1	36	4,101	38,287,830	59,985	32,987,342	721,938
bone010	24	37,887	34,934,616	361,170	96,622,580	43,991,787
af_shell10	15	19	25,582,015	75	40,629,688	2,105
as-Skitter	67	1,656,570	6,880,534	981,810	26,809,527	737,899,486
roadNet-CA	4	1,487,640	1,079,025	370,206	320,118	4,302
kkt_power	11	1,166,311	4,510,661	401,129	1,067,824	1,978,595
rmat_er_1	3	780	1,047,599	915	118,461	8,722
rmat_er_2	3	2,019	2,094,751	2,351	235,037	23,908
rmat_er_3	3	4,349	4,189,290	4,960	468,086	50,741
rmat_er_4	3	9,032	8,378,261	10,271	933,750	106,200
rmat_er_5	3	18,155	16,756,493	20,622	1,865,415	212,838
rmat_sd1_1	6	39,281	1,004,660	23,898	151,838	542,245
rmat_sd1_2	6	90,010	2,004,059	56,665	284,577	1,399,314
rmat_sd1_3	6	176,583	4,013,151	106,543	483,436	2,677,437
rmat_sd1_4	6	369,818	8,023,358	214,981	889,165	5,566,602
rmat_sd1_5	6	777,052	16,025,729	455,473	1,679,109	12,168,698
rmat_sd2_1	26	110,951	853,116	88,424	1,067,824	614,813,037
rmat_sd2_2	35	232,352	1,645,086	195,427	81,886,879	1,044,068,886
rmat_sd2_3	39	470,302	3,257,233	405,856	45,841,352	1,343,563,239
rmat_sd2_4	43	*	*	*	*	*
rmat_sd2_5	N	*	*	*	*	*
hamming6-4	4	0	704	0	583	0
johnson8-4-4	14	0	1855	0	136,007	0
keller4	11	0	9435	0	8,834,190	0
c-fat200-5	58	0	8473	0	70449	0
brock200_2	12	0	9876	0	349,427	0

Table 5. Comparison of runtimes of algorithms [9] (*CP*), [28] (*cliquer*) and [21] (*MCQD+CS*) with that of our new exact algorithm (τ_{A1}) for DIMACS graphs. An asterisk (*) indicates that the algorithm did not terminate within 10,000 seconds for that instance. ω denotes the maximum clique size, ω_{A2} the maximum clique size found by our heuristic and τ_{A2}, its runtime.

G	$\lvert V\rvert$	$\lvert E\rvert$	ω	τ_{CP}	$\tau_{cliquer}$	$\tau_{MCQD+CS}$	τ_{A1}	ω_{A2}	τ_{A2}
brock200_1	200	14,834	21	*	10.37	0.75	*	18	0.02
brock200_2	200	9,876	12	0.98	0.02	0.01	1.1	10	<0.01
brock200_3	200	12,048	15	14.09	0.16	0.03	14.86	12	<0.01
brock200_4	200	13,089	17	60.25	0.7	0.12	65.78	14	<0.01
c-fat200-1	200	1,534	12	<0.01	<0.01	<0.01	<0.01	12	<0.01
c-fat200-2	200	3,235	24	<0.01	<0.01	<0.01	<0.01	24	<0.01
c-fat200-5	200	8,473	58	0.6	0.33	0.01	0.93	58	0.04
c-fat500-1	500	4,459	14	<0.01	<0.01	<0.01	<0.01	14	<0.01
c-fat500-2	500	9,139	26	0.02	<0.01	0.01	0.01	26	0.01
c-fat500-5	500	23,191	64	3.07	<0.01	<0.01	*	64	0.11
hamming6-2	64	1,824	32	0.68	<0.01	<0.01	0.33	32	<0.01
hamming6-4	64	704	4	<0.01	<0.01	<0.01	<0.01	4	<0.01
hamming8-2	256	31,616	128	*	0.01	0.01	*	128	0.67
hamming8-4	256	20,864	16	*	<0.01	0.1	*	16	0.03
hamming10-2	1,024	518,656	512	*	0.31	-	*	512	95.24
johnson8-2-4	28	210	4	<0.01	<0.01	<0.01	<0.01	4	<0.01
johnson8-4-4	70	1,855	14	0.19	<0.01	<0.01	0.23	14	<0.01
johnson16-2-4	120	5,460	8	20.95	0.04	0.42	22.07	8	<0.01
keller4	171	9,435	11	22.19	0.15	0.02	23.35	11	<0.01
MANN_a9	45	918	16	1.73	<0.01	<0.01	2.5	16	<0.01
MANN_a27	378	70,551	126	*	*	3.3	*	125	1.74
p_hat300-1	300	10,933	8	0.14	0.01	<0.01	0.14	8	<0.01
p_hat300-2	300	21,928	25	831.52	0.32	0.03	854.59	24	0.03
p_hat500-1	500	31,569	9	2.38	0.07	0.04	2.44	9	0.02
p_hat500-2	500	62,946	36	*	159.96	1.2	*	34	0.14
p_hat700-1	700	60,999	11	12.7	0.12	0.13	12.73	9	0.04
p_hat1000-1	1,000	122,253	10	97.39	1.33	0.41	98.48	10	0.11
san200_0.7_1	200	13,930	30	*	0.99	<0.01	*	16	0.01

A Faster Algorithm to Update Betweenness Centrality after Node Alteration

Keshav Goel[2], Rishi Ranjan Singh[1], Sudarshan Iyengar[1], and Sukrit[3]

[1] Department of Computer Science and Engineering, Indian Institute of Technology, Ropar, Punjab, India
{rishirs,sudarshan}@iitrpr.ac.in

[2] Department of Computer Engineering, National Institute of Technology, Kurukshetra, India
keshavgoel1993@gmail.com

[3] Department of Computer Science and Engineering, PEC University of Technology, Chandigarh, India
sukritkumar.becse11@pec.edu.in

Abstract. Betweenness centrality is a centrality measure that is widely used, with applications across several disciplines. It is a measure which quantifies the importance of a vertex based on its occurrence in shortest paths between all possible pairs of vertices in a graph. This is a global measure, and in order to find the betweenness centrality of a node, one is supposed to have complete information about the graph. Most of the algorithms that are used to find betwenness centrality assume the constancy of the graph and are not efficient for *dynamic networks*. We propose a technique to update betweenness centrality of a graph when nodes are added or deleted. Our algorithm experimentally speeds up the calculation of betweenness centrality (after updation) from 7 to 412 times, for real graphs, in comparison to the currently best known technique to find betweenness centrality.

Keywords: Betweenness Centrality, Minimum Cycle Basis, Bi-connected Components.

1 Introduction

Network Centrality measures are used to quantify the intuitive notion of nodes' importance in a network. There are several application centric definitions of network centrality measures, the popular ones being *degree centrality, closeness centrality, eigenvector centrality* and *betweenness centrality*. For background and description of centrality measures please refer to the excellent books by Newman [20] and Jackson [13].

There are a number of centrality indices based on the *shortest path lengths* (closeness centrality [22], graph centrality [10]) and the *number of shortest paths* (stress centrality [24], betweenness centrality [7, 1]) in a graph. Each centrality measure signifies a particular characteristic that is exhibited by a node. *Closeness centrality* of a vertex indicates the distance of a vertex from other vertices.

A. Bonato, M. Mitzenmacher, and P. Prałat (Eds.): WAW 2013, LNCS 8305, pp. 170–184, 2013.
© Springer International Publishing Switzerland 2013

Graph centrality denotes the difference between closeness centrality of the vertex under consideration and the vertex with the highest closeness centrality. *Stress centrality* simply denotes the total number of shortest paths passing through a vertex.

The idea of betweenness centrality was proposed by Freeman [7] and Anthonisse [1]. Betweenness centrality of a node v is defined as $BC(v) = \sum_{s \neq t \neq v \epsilon V} \frac{\sigma_{st}(v)}{\sigma_{st}}$, where σ_{st} is the total number of shortest paths from vertex s to vertex t and $\sigma_{st}(v)$ is the total number of shortest paths from vertex s to vertex t passing through vertex v.

Betweenness centrality insinuates a more *global* characteristic unlike the degree centrality which considers the number of links from a node - which is clearly a local characteristic. Betweenness centrality has found many important applications in diverse fields. It has been used in the identification of sensitive nodes in biological networks[18]. Similarly, it can be used in electronic communication system networks, public transit system networks, gas pipeline networks, waste-water disposal system networks, etc. In protein-protein interaction (PPI) networks, essential proteins can be identified by their high betweenness centrality [14]. This characteristic of proteins can be used in selecting suitable drug targets [28] for various ailments including cancer[27], tuberculosis [23], zoonotic cutaneous leishmaniasis [6], etc. Betweenness centrality score of a person, on popular social networking sites, like 'Facebook' or 'Twitter', is being used by advertisers to choose him/her as an ambassador for their organization. Betweenness centrality is also used to identify nodes which are crucial for information flow in a brain network [12] where different regions of the brain represent nodes in the network and white matter represents the links.

Brandes [3] suggested an algorithm to calculate betweenness centrality that reduced the time complexity from $O(|V|^3)$ to $O(|V||E|)$ for unweighted graphs. Since, real world networks tend to be large and transient, such algorithms are found to be impractical if one requires to compute the betweenness centrality of nodes in a dynamic network. Work has been done by Lee et al. [15] and Green et al. [9] to find out betweenness centrality for edge updation in a graph. Most of the literature are found only for edge updation case. They assume that deletion of a node from a graph is equivalent to deleting all edges incident on that node. It is easy to analyze that algorithm proposed here is $deg(v)$-times faster than the algorithms with above mentioned concept for updation after node deletion where $deg(v)$ is the degree of the deleted vertex. The ranking of vertices on the basis of betweenness centrality is of use in various applications which are mentioned in the subsequent sections. To ascertain that the order of vertices in terms of their betweenness centralities before updation of graph is not the same as after updation, we performed experiments and got positive results. We present situations which demand a better way of updating betweenness centrality in changing networks.

1.1 Motivation

It is sometimes necessary to calculate betweenness centrality for a network at every stage of transition. With a large network and the current algorithms in use, recalculation becomes difficult. Some examples of such networks are given below.

- Complex communication networks are continuously growing and evolving. Each node in a communication network has a maximum capacity for carrying load[1], after which the node shuts down and its load is distributed among the remaining nodes. Due to increased load, other nodes may shut down and the network may become disconnected. This phenomena is commonly known as *cascading failure*. It has been found through experiments conducted by S. Narayanan [18] that breakdown of nodes with higher betweenness centrality causes greater harm. In such networks we can compute a sequence of nodes as following: We start with the given network. At each step, we delete the node with the highest betweenness centrality, add that node to the sequence and then repeat this process until the network becomes disconnected. This sequence can be used to decide the order in which security should be provided to the nodes in the network and that can ensure that if a node in the present network fails, the node with the highest betweenness centrality in the resulting network have enough security and resources. This requires repetitive calculation of betweenness centrality which when done with the conventional Brandes[3] algorithm will be highly inefficient. Similarly, points which have excess load in power grid systems and computer networks can be provided with more resources; stations with excess traffic in public transit systems can be provided with more measures to redistribute traffic and sewer lines with higher betweenness centrality can be provided with more frequent maintenance to prevent blockades. This exercise can also be done after the failure of some random node in a graph and appropriate actions on the nodes in the network may be taken thereafter.
- In social networking websites like 'Twitter' and 'Facebook', betweenness centrality of a node denotes the number of heterogeneous groups of nodes, the node under consideration links [25]. Since these nodes are involved in passing of information between heterogeneous groups of nodes, they're more important than a node with just a higher degree.[2] Also, we may want to determine the next important actor in case the current social network is altered. Such networks are highly dynamic due to the continuous addition and removal of actors.
- In a network composed of nations, the betweenness centrality of a nation describes its potential to act as information broker and provides information

[1] The amount of information flowing through a node in a communication network, is called its load.

[2] In the study conducted by A. Hanna [11] on the uprising in Egypt, where the social networking site 'Twitter' played an important role in the formation of public opinion against Mr. Hosni Mubarak (the then President of Egypt), it was found that these nodes played an important role in shaping public opinion.

about its overall activity level in the network. Thus nations can analyze the variation in their eminence with changes in network. They can analyze how other nations affect them and what actions will benefit or harm them. For example: Suppose in a network, countries are represented by nodes and a link signifies that there is a trade relation between the two countries. Countries may want to know what the effect of formation/removal of links or nodes of other countries with them or of other countries with other countries will be on their trade relations.
- A similar application can be in case a new actor wants to join a network. He will want to form links such that his prominence is maximum. He will have to form links to nodes accordingly. For example: A professional wants to join a network of other professionals in his area, he will try to connect with other actors considering what effect that will have on him. This requires repetitive calculation of betweenness centrality for a wide number of cases.

We tested our algorithm for both real and synthetic graphs and got positive results. For synthetic graphs, we achieved speedups ranging from 1.78 to 14 times and for real graphs we got speedups ranging from 7 times to 412 times, in comparison with Brandes algorithm [3]. In section 2, we present some basic definitions and concepts used in the paper. Section 3 contains the algorithm with explanation. Implementation and results are presented in section 4. We've further discussed the previous work conducted on betweenness centrality in section 5. We conclude in section 6.

2 Preliminary

In this section we define some terms which have been used throughout the paper. We also explain the basic concepts which provide basis for developing algorithm in section 3.

2.1 Terminology

We use following terms interchangeably throughout the paper; node or vertex and graph or network. A (simple) *path* in a graph is a sequence of edges connecting a sequence of vertices without any repetition of vertices. Thus a path between two vertices v_i and v_j (called terminal vertices) can be denoted as a sequence of vertices, $\{v_i, ..., v_j\}$ such that $v_i \neq v_j$ and no vertices in the sequence are repeated. The *length* of a path is the sum of the weights of edges in the path (edge weight is taken as one for unweighted graphs). A *shortest path* between two vertices is the smallest length path between them. An *end vertex* is a vertex with degree one. A graph is said to be connected if there exists a path between each pair of vertices. An *articulation vertex* is a vertex whose deletion will leave the graph disconnected. A *biconnected graph* is a connected graph having no articulation vertex. A *cycle* in a graph is a path having the same terminal vertices. A *cycle basis* of a graph is defined as a maximal set of linearly independent cycles. Weight of a cycle basis is the sum of the lengths

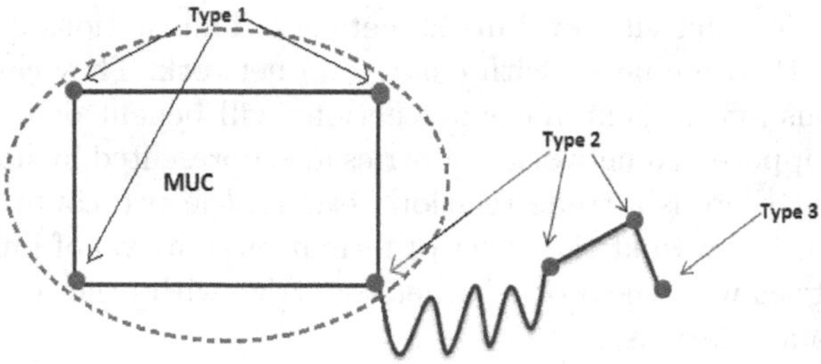

Fig. 1. Type 1: Vertex belongs to a *MUC* but is not an articulation vertex. Type 2: Vertex is an articulation vertex. Type 3: Vertex does not belong to a *MUC* and is an end vertex.

of all cycles in the cycle basis. A cycle basis of minimum total weight is called *minimum cycle basis (MCB)*.

The set achieved by repetitive merging (taking union) of all the elements of the MCB that have at least one vertex in common is called as *MUCset*. Each element of MUCset is termed as *Minimum Union Cycle (MUC)*. Thus, two MUCs can not have any vertex in common. A *connection vertex c* in a *MUC* (say MUC_i) is an articulation vertex such that it is adjacent to a vertex which does not belong to MUC_i. On removal of the connection vertex c, the graph will become disconnected and the components that are disconnected from MUC_i are together termed as *disconnected subgraph* G_c.

2.2 Basic Concept

Throughout the paper, we have considered only the case of vertex deletion in undirected unweighted connected graphs. For the case when a vertex is added, all lemmas, observations, and results hold with a slight modification. On the basis of the method used for updation of betweenness centrality after deletion of a vertex, we can categorize the vertices of the graph into three groups as mentioned in Figure 1.

In this paper, we explain the updation process after alteration of vertices of Type 1. Deletion of a vertex of Type 2 will leave the graph disconnected and concept of betweenness centrality will no longer be valid for the graph. So, we can not consider this case for updation. After deletion of vertices of Type 3, we can use a procedure similar to Algorithm 2 to update the centality scores. Now, we define few more terminologies, give lemmas and establish a theorem which provides basis to develop our algorithm.

Pair dependency of a pair of vertices (s, t) on a vertex v is defined as: $\delta_{st}(v) = \frac{\sigma_{st}(v)}{\sigma_{st}}$ where σ_{st} is the number of shortest paths from vertex s to vertex t and $\sigma_{st}(v)$ is the number of shortest paths from vertex s to vertex t passing through vertex v. Betweenness centrality of a vertex v can be defined in terms of pair dependency as: $BC(v) = \sum_{s \neq v \neq t \in V} \delta_{st}(v)$. Let BFT_r denotes the breadth-first traversal (BFT) of the graph rooted on vertex r. [3] *Dependency* of a vertex s on a vertex v is defined as: $\delta_{s\bullet}(v) = \sum_{t \in V \setminus \{s,v\}} \delta_{st}(v)$. Let us define a set $P^s(w)$ $= \{v : v \in V,\ w \textit{ is a successor of } v \textit{ in } BFT_s\}$. Brandes [3] proved that:

$$\delta_{s\bullet}(v) = \sum_{w : v \in P^s(w)} \frac{\sigma_{sv}}{\sigma_{sw}}(1 + \delta_{s\bullet}(w)). \tag{1}$$

Let $SP(v_i, v_j)$ be the set of all shortest paths from vertex v_i to v_j. Let MUC_U be the MUC where alteration has been made. Let G_i be the subgraph made of the components that will be disconnected from MUC_U after removal of connection vertex $c_i \in MUC_U$. Let $V(G_i)$ denote the set of vertices in subgraph G_i. Then we can establish following lemmas and theorem.

Lemma 1. *If v lies on all shortest paths between s and t, where $s \neq t \neq v \in V$, then:*

$$\sigma_{st} = \sigma_{sv}.\sigma_{vt}.$$

Lemma 2. *For $u \in V(G_k)$ and $v \in MUC_U$, every element of $SP(u, v)$ must contain c_k.*

Proof. Since c_k (connection vertex) is the only vertex that links G_k with MUC_U. So, every path between vertices in G_k and MUC_U must pass through c_k. □

Lemma 3. *Betweenness centrality of a vertex v can be changed only due to the shortest paths that had the altered vertex as one of their terminal vertices, where $v \in V \setminus MUC_U$.*

Proof. Assume $s, t \in V$ and $s \neq t \neq v$. Betweenness centrality of vertex v, $BC(v)$ will be influenced by following types of shortest paths:

1. Shortest paths that do not pass through the MUC_U. They start and end outside MUC_U, without passing through it. Naturally when an alteration is made in a graph, these paths remain unchanged. An example is shown in Fig. 2.
2. Shortest paths that have one terminal vertex in one disconnected subgraph G_i and the other in a disconnected subgraph G_j: These shortest paths may change. For one such shortest path, suppose the terminal vertices s and t are in different disconnected subgraphs, which pass through connection vertices c_1 and c_2, respectively (v lies in the disconnected subgraph where s lies). According to Lemma 1 and Lemma 2, $\delta_{st}(v) = \frac{\sigma_{st}(v)}{\sigma_{st}} = \frac{\sigma_{sc_1}(v).\sigma_{c_1c_2}.\sigma_{c_2t}}{\sigma_{sc_1}.\sigma_{c_1c_2}.\sigma_{c_2t}} = \frac{\sigma_{sc_1}(v)}{\sigma_{sc_1}}$. We can observe that the shortest paths from s to c_1 remain same after node deletion and so this factor doesn't change.
3. Shortest paths that have one terminal vertex in one disconnected subgraph G_i and the other in MUC_U: Out of these shortest paths, the paths where deleted vertex is not a terminal vertex, we can get a relation similar to the one obtained above. When the deleted vertex is a terminal vertex (i.e. either s or t is deletion vertex), a factor of $\delta_{st}(v) = \frac{\sigma_{st}(v)}{\sigma_{st}}$ should be deleted from

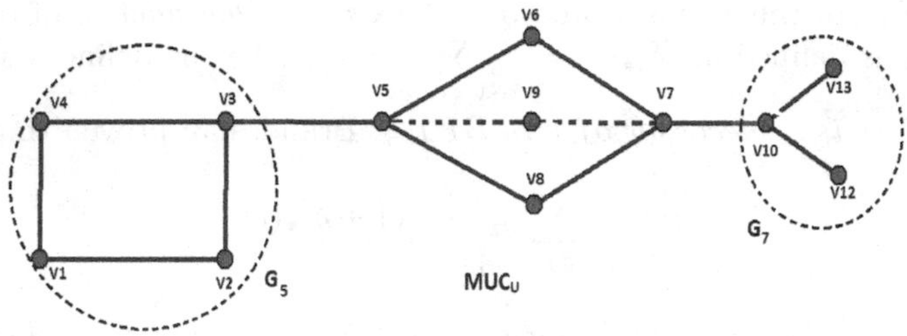

Fig. 2. Vertex $V9$ deleted. No effect on the shortest paths starting and ending in G_5 or G_7 itself. Other shortest paths may be altered.

betweenness centrality score of vertex v. This is because existing shortest paths from s to t are nonexistent now.

Thus, only one type of shortest paths (with altered vertex as one of terminal vertex) can change the betweenness centrality of the vertex v. □

Theorem 1. *Let v_d be the vertex to be deleted. Let $BC(v)$ be the betweenness centrality of the vertex v and dependency of the vertex v_d on the vertex $v \in V \setminus MUC_U$ is $\delta_{v_d \bullet}(v)$. Then the updated betweenness centrality of the vertex v after deletion of the vertex v_d can be calculated as:*

$$BC'(v) = BC(v) - 2\delta_{v_d \bullet}(v)$$

Proof. By the definition of dependency, $\delta_{v_d \bullet}(v)$ gives the effect of all shortest paths starting at vertex v_d in the betweenness centrality of node v. According to Lemma 3, shortest paths with v_d as terminal vertex (start vertex or end vertex on the path) are only affecting the change in centrality of vertices outside the MUC_U. After deletion of v_d, all such shortest paths will be deleted. Since the graph is undirected, $\sigma_{v_d t} = \sigma_{t v_d}$, so, we will subtract the dependency $\delta_{v_d \bullet}(v)$ twice. □

3 Algorithm

After deletion of a vertex which belonged to a MUC and was not an articulation vertex, we will update the betweenness centrality in different ways for the two types of vertices: vertices outside MUC_U and vertices in MUC_U. We use Theorem 1 to update betweenness centrality for vertices outside MUC_U and the algorithm is explained in detail in section 3.2. When vertices in MUC_U are considered, we observe that several shortest paths that were passing through altered vertex changed after deletion. So we recompute betweenness centrality using the idea given by Lee et al. [15] which is explained in brief in section 3.3. We assume that the betweenness centrality score of all vertices is available before proceeding with the preprocessing step of our algorithm.

Algorithm 1. Preprocesssing Step: Calculating MUCs in the Graph

Use Tarjan's Algorithm to calculate a set of biconnected components, C.
for each $C_i \in C$ **do**
  if $|C_i| = 2$ **then**
    Remove C_i from C.
  end if
end for
while $\exists$ C_i, $C_j \in C$ where C_i and C_j have at least one common vertex **do**
  Remove C_i *and* C_j from C.
  Insert $C_i \cup C_j$ in C.
end while
$MUCset \leftarrow C$
for each $MUC_j \in MUCset$ **do**
  Find all the connection vertices and corresponding disconnected subgraphs.
end for

3.1 Preprocessing Step

Every time a change is made in the graph, updating the MUCset becomes necessary. We can do it in two ways, either by updation of MUCset (approach used in [15]) or by recalculation of MUCset. Approach for updation of MUCset takes longer time than recalculation. So, instead of updating MUCset, we recalculate it using the output of Tarjan's biconnected components algorithm[26] (commonly known as Tarjan's algorithm) as explained in algorithm 1. The time complexity for recalculation is $O(|V| + |E|)$. We use the following procedure for calculating MUCset.

Every graph can be decomposed into a set of biconnected components, C, where the elements of C are denoted by C_i using Tarjan's algorithm. Let $|C_i|$ denote the number of vertices in the biconnected component C_i. Each biconnected component contains at least one edge (two vertices) and may share vertices (articulation vertex) with other biconnected components. We remove components which contain only one edge because single edge can not form a MUC. Since, the elements of a MUCset are disjoint, we take repetitive union of the components that have atleast one vertex in common. We form the MUCset in this fashion. Then for each MUC, we calculate connection vertices and disconnected subgraph(s) associated with each connection vertex.

3.2 Calculating Changes in Betweenness Centrality for Vertices outside MUC_U

Effect of the altered vertex on betweenness centrality of vertices outside MUC_U can be found by forming breadth-first traversal (BFT) for the vertex that was deleted. The BFT can be calculated with a time complexity of $O(|E|)$. We then calculate the dependency of each vertex with respect to deletion vertex, starting from the vertices in the bottom level and recursively calculate the dependency for vertices in subsequent higher levels using equation 1. Then we use Theorem 1 to update the centrality values. The complete procedure is shown in Algorithm 2. In case of vertex addition, we will add the dependency to the betweenness centrality scores of each vertex outside MUC_U.

3.3 Calculating Betweenness Centrality for Vertices in MUC_U

This section briefly describes the idea suggested by Lee et al. [15] for recomputation of betweenness centrality for vertices in MUC_U. In the disconnected subgraph G_j, let $V(G_j)$ denote the vertex set and let $|V(G_j)|$ denote the number of vertices. Let $|SP(u,v)|$ denote the number of shortest paths between vertex u and vertex v. Here, we will explain the basic steps of the algorithm, in brief. For detailed concept and used algorithm, please refer to the QUBE algorithm [15]. Let betweenness centrality of vertex v, $BC(v)$ for all $v \in MUC_U$ be initialized with 0. Let c_j be a connection vertex of MUC_U and G_j be the corresponding disconnected subgraph. Now we calculate the betweenness centrality of vertex v by calculating and adding the effect of following three types of shortest paths on vertex v:

1. The shortest paths with both source and destination in MUC_U: For counting the effect of these paths, we use the algorithm suggested by Brandes [3] for only the vertices in MUC_U and compute local betweenness centrality $BC_0^{MUC_U}(v)$, for all vertices $v \in MUC_U$.
2. The shortest paths with either source or destination (but not both) in MUC_U: Let $< s,..,t >$ be a shortest path from $s \in V(G_j)$ to $t \in MUC_U$. In this case, $\frac{\sigma_{st}(v)}{\sigma_{st}} = \frac{\sigma_{c_jt}(v)}{\sigma_{c_jt}}$. So, to calculate the total effect of such paths, for each shortest path $< c_j, ..., t >$, we add the following factor to $BC(v)$:

$$BC_1^{<c_j,\ldots,t>}(v) = \begin{cases} \frac{|V(G_j)|}{|SP(c_j,t)|}, & \text{if } v \in < c_j, ..., v_t > \backslash \{v_t\} \\ 0, & \text{otherwise} \end{cases}$$

3. The shortest paths with neither source nor destination in MUC_U: Let $< s,..,t >$ be a shortest path from $s \in V(G_j)$ to $t \in V(G_k)$ where $j \neq k$. In this case, $\frac{\sigma_{st}(v)}{\sigma_{st}} = \frac{\sigma_{c_jc_k}(v)}{\sigma_{c_jc_k}}$. So, to calculate the total effect of such paths, for each shortest path $< c_j, ..., c_k >$, we add the following factor to $BC(v)$:

$$BC_2^{<c_j,\ldots,c_k>}(v) = \begin{cases} \frac{|V(G_j)||V(G_k)|}{|SP(c_j,c_k)|}, & \text{if } v \in < c_j, ..., c_k > \\ 0, & \text{otherwise} \end{cases}$$

When either of the subgraphs is disconnected, an additional factor:

$$BC_3^i(c_i) = \begin{cases} |V(G_i)|^2 - \sum_{l=1}^{x}(|V(G_i^l)|^2), & \text{if } G_i \text{ is disconnected} \\ 0, & \text{otherwise} \end{cases}$$

is added to the betweenness centrality calculations (c_i is a connection vertex). Where ${G_j}^l$ is the lth component of G_i and x is the number of connected components in G_i.

So, we have the following formula to calculate the betweenness centrality score of a vertex in MUC_U:

$$BC(v) = BC_0^{MUC_U}(v) + 2\sum_{G_j,t}\sum_{x\in SP(c_j,t)} BC_1^x(v) +$$

$$\sum_{G_j,G_k(j\neq k)}\sum_{y\in SP(c_j,c_k)} BC_2^y(v) \qquad (+\ BC_3^i(c_i) \quad if\ v = c_i).$$

4 Implementation and Results

We have implemented the algorithm for deletion of vertices, however it can be easily modified to implement node addition. The algorithm can work faster for updating betweenness centrality, since it forms a subset of vertices of the graph, MUC_U, for which recalculation of betweenness centrality is to be done. For the rest of vertices, algorithm 2 updates the betweenness centrality in negligible time as compared to recalculation step. The recalculation procedure includes local Brandes algorithm so it is directly proportional to number of vertices inside MUC_U. We have compared our results with Brandes algorithm [3] since that is the best known algorithm, according to our knowledge, for calculation of betweenness centrality after vertex updation. The experiments were performed on an Intel i5-2450M C.P.U. with 2.5 GHz clock speed and 4 G.B. main memory.

We use a similar measure used by authors in [15] termed as *proportion* to compare the algorithms. Proportion can be calculated as:

$$\left(\frac{Number\ of\ vertices\ in\ MUC_U}{Total\ number\ of\ vertices\ in\ the\ graph}\right).100$$

Proportion is a direct function of the number of vertices in MUC_U and thus speedup achieved by our algorithm is directly affected by the proportion. So, a smaller proportion would mean betweenness centrality for lesser number of nodes will have to be recomputed and so this should achieve greater speed-up which we achieved in our experimental result. We considered the following strategy to compute the *average proportion* of a graph. We randomly start deleting vertices from the graph till either the graph becomes disconnected or k vertices are deleted. Then we take the average of proportion values for each deletion. The *average speed-up* is calculated in a similar way. We consider $k = 500$ for real networks. In general, average speed-up on a graph or average proportion of a graph depends on the number of MUCs formed by biconnected components and the fraction of total number of vertices belongs to these MUCs. If a graph consists of most of the MUCs with small number of vertices with respect to the total number of vertices in the graph, the average proportion will be small and thus average speed-up on that graph will be large.

4.1 Results for Synthetic Graphs

We derived three groups of synthetic graphs with 1000, 2000 and 3000 nodes. In each group we generated graphs with proportion x, for each $x \in A$, where $A = \{10, 20, 30, 40, 50, 60, 70, 80\}$. These graphs were random graphs based on the Erdös Rényi graph model. To implement the algorithm, we initially calculated the betweenness centrality of each vertex using Brandes [3] algorithm. Then we ran the preprocessing step (Algorithm 1) to calculate the MUCs, connection

Algorithm 2. Calculating BFT and updating the vertices outside MUC_U accordingly

v_d: Vertex to be deleted.
Input: $BC[v]$ of each vertex of original graph ($v \in V$).
$S \leftarrow$ Empty Stack
$P[w] \leftarrow$Empty List, $w \in V$
$\sigma[t] \leftarrow 0$, $t \in V$, $\sigma[v_d] = 1$
$d[t] \leftarrow -1$, $t \in V$, $d[v_d] = 0$
$Q \leftarrow$ Empty Queue
Enqueue $v_d \rightarrow Q$
while Q not empty **do**
  Dequeue $v \leftarrow Q$
  push $v \rightarrow S$
  for each neighbour of v **do**
    if $d[w] < 0$ **then**
      enqueue $w \rightarrow Q$
      $d[w] \leftarrow d[v] + 1$
    end if
    if $d[w] = d[v] + 1$ **then**
      $\sigma[w] \leftarrow \sigma[w] + \sigma[v]$
      append $v \rightarrow P[w]$
    end if
  end for
end while
$\delta[v] \leftarrow 0$, $v \in V$
while S not empty **do**
  pop $w \leftarrow S$
  for $v \in P[w]$ **do**
    $\delta[v] \leftarrow \delta[v] + \frac{\sigma[v]}{\sigma[w]}(1 + \delta[w])$
  end for
end while
for $v \in V \setminus MUC_U$ **do**
  $BC[v] \leftarrow BC[v] - 2.\delta[v]$
end for

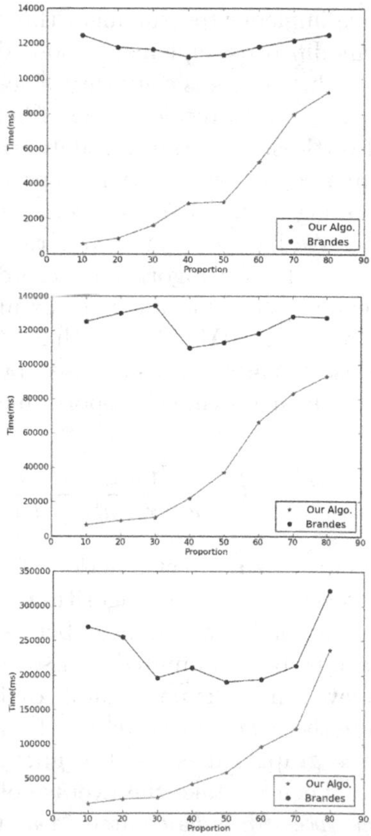

Fig. 3. Comparison of ours and Brandes' result on synthetic graphs with 1000, 2000, and 3000 vertices respectively

vertices and disconnected subgraphs for our graphs. 50 vertices were removed from each x proportion graph of each group and average updation times were calculated for different proportions.

The results are plotted with average updation time (in ms) at the y-axis and the proportion of the graphs at the x-axis for each group and are shown in Fig. 3. We get different speed-ups for different proportion graphs in each group. For synthetic graphs with 1000 nodes, we achieve a speedup of 13.26, 3.90 and 2.25 for graph with proportion 20, 40 and 60 respectively. For synthetic graphs with 2000 nodes, we achieve a speedup of 14.27, 5.01 and 1.78 for graph with proportion 20, 40 and 60 respectively. Similarly, for synthetic graphs with 3000 nodes, we achieve a speedup of 12.54, 5.02 and 2.02 for graph with proportion 20, 40 and 60 respectively as shown in Fig. 3.

4.2 Results for Real Graphs

We also tested our algorithm on real world networks. We chose different types of real datasets to depict the flexibility of our algorithm. We converted directed graphs into undirected graphs. Collaboration networks generally depict the collaborations an author has made while writing research papers. Interaction networks are networks where nodes are connected due to their features. An ownership network suggests transfer of resources between two nodes. Trust network is a network of individuals with kindred interests and connections. The yeast protein-protein interaction network was also taken as an input.[3]

For such various real networks like Geom[3], Erdos02[4], Erdos972[5], etc., information about networks, average proportion and average speed-ups over Brandes algorithm are summarized in Table 1.

Table 1. Figures for Real Data sets

Name of Dataset	Type	$\|V\|$	$\|E\|$	Avg. Proportion	Avg. Speed-up
YeastL	Interaction	2361	6646	31.16	64.88
YeastS	Interaction	2361	6646	28.05	72.76
Geom	Collaboration	7343	11898	17.08	7.63
Erdos02	Collaboration	6927	11850	7.13	323.75
Edros972	Collaboration	5488	8972	9.49	411.835
ODLIS[21]	Dictionary data	2909	16377	67.68	18.52
Wiki-vote[16]	Trust	8297	100762	38.452	28.472

5 Related Work

The idea of betweenness centrality was first introduced by Freeman [7] and Anthonisse [1]. Newman [19] defined another measure that considered random

[3] Dataset available at `http://vlado.fmf.uni-lj.si/pub/networks/data`
[4] Available at `http://www.cise.u.edu/research/sparse/matrices/Pajek/`
[5] `http://www.cise.ufl.edu/research/sparse/matrices/`

walks on any arbitrary length rather than just shortest paths between two vertices. Brandes [5] considered other types of betweenness centrality like edge betweenness and group betweenness and algorithms to compute them efficiently. Betweenness centrality was earlier calculated by finding the number and length of shortest paths between two vertices and then adding up pair dependencies for all pairs. Brandes [3] suggested an algorithm which introduces a recursive way to sum the dependencies in graphs. Although the algorithm proposed by Brandes[3] was faster than the one used previously, it was still too costly for large graphs. So, several approximation algorithms were proposed by Bader et al. [2], Brandes et al. [4], Geisberger et al. [8] and Makarychev [17]. Real world networks tend to be large and transient. Work has been done by Lee et al. [15] and Green et al. [9] to find out betweenness centrality after updation in a graph. The algorithm suggested by [15] selects a subset of vertices whose betweenness centrality is updated. However, it works only in the case of edge removal and addition. The algorithm suggested by [9] takes into consideration different instances that may arise due to edge addition in a graph and speeds up the algorithm in these cases. This algorithm works only for streaming graphs i.e. only in case of edge additions.

6 Conclusion

In this paper, we formulated an algorithm that efficiently calculates betweenness centrality when vertices in a graph are updated. We did not consider the traditional way for updating betweenness centrality after node alteration which considers a node alteration event as a series of edge alteration event. We achieved the speedups by calculating two sets of vertices; one for which we need to update betweenness scores and the other for which we need to recompute the betweenness score. We achieve an average speedup of 4.5 for a proportion of 40 for synthetic graphs as compared to Brandes algorithm. For real graphs, we get an average speedup of around 133 for a proportion of 29. The speedup will increase futher when proportion decreases.

Acknowledgement. The authors would like to thank the anonymous reviewers for comments that helped with the clarification of some concepts. They would also like to thank M.J. Lee and Yayati Gupta for their invaluable help.

References

[1] Anthonisse, J.M.: The rush in a directed graph. Technical Report BN 9/71, Stichting Mathematisch Centrum, Amsterdam (1971)

[2] Bader, D.A., Madduri, K.: Parallel algorithms for evaluating centrality indices in real-world networks. In: Proceedings of the 2006 International Conference on Parallel Processing, ICPP 2006, pp. 539–550 (2006)

[3] Brandes, U.: A Faster Algorithm for Betweenness Centrality. Journal of Mathematical Sociology 25(2), 163–177 (2001)

[4] Brandes, U., Pich, C.: Centrality estimation in large networks. International Journal of Bifurcation and Chaos 17(7), 2303 (2007)
[5] Brandes, U.: On variants of shortest-path betweenness centrality and their generic computation. Social Networks 30(2), 136–145 (2008)
[6] Florez, A.F., Park, D., Bhak, J., Kim, B.C., Kuchinsky, A., Morris, J.H., Espinosa, J., Muskus, C.: Protein network prediction and topological analysis in Leishmania major as a tool for drug target selection. BMC Bioinformatics 11, 484 (2010)
[7] Freeman, L.A.: set of measures of centrality based on betweenness. Sociometry 40, 35–41 (1977)
[8] Geisberger, R., Sanders, P., Schultes, D.: Better approximation of betweenness centrality. In: Proceedings of the Tenth Workshop on Algorithm Engineering and Experiments (ALENEX), pp. 90–100. SIAM (2008)
[9] Green, O., McColl, R., Bader, D.A.: A fast algorithm for streaming betweenness centrality. In: 2012 ASE/IEEE International Conference on Social Computing and 2012 ASE/IEEE International Conference on Privacy, Security, Risk and Trust (2012)
[10] Hage, P., Harary, F.: Eccentricity and centrality in networks. Social Networks 17, 57–63 (1995)
[11] Hanna, A.: Revolutionary Making and Self-Understanding: The Case of #Jan25 and Social Media Activism. Presented at Meeting of the International Studies Association, San Diego, CA (2012)
[12] Iturria-Medina, Y., Sotero, R.C., Canales-Rodríguez, E.J., Alemán-Gómez, Y., Melie-García, L.: Studying the human brain anatomical network via diffusion-weighted MRI and Graph Theory. NeuroImage 40(3), 1064–1076 (2008)
[13] Jackson, M.O.: Social and Economic Networks. Princeton University Press (2010)
[14] Joy, M.P., Brock, A., Ingber, D.E., Huang, S.: High-Betweenness Proteins in the Yeast Protein Interaction Network. J. Biomed. Biotechnol. 2, 96–103 (2005)
[15] Lee, M.J., Lee, J., Park, J.Y., Choi, R.H., Chung, C.W.: QUBE: a Quick algorithm for Updating BEtweenness centrality. In: Proceedings of the 21st International Conference on World Wide Web, pp. 351–360 (2012)
[16] Leskovec, J., Kleinberg, J., Faloutsos, C.: Graph evolution: Densification and shrinking diameters. ACM Trans. Knowl. Discov. Data 1(1), 2 (2007)
[17] Makarychev, Y.: Simple linear time approximation algorithm for betweenness. Operations Research Letters 40(6), 450–452 (2012)
[18] Narayanan, S.: The Betweenness Centrality of Biological Networks. M. Sc. Thesis submitted to the Faculty of the Virginia Polytechnic Institute and State University (2005)
[19] Newman, M.E.J.: A measure of betweenness centrality based on random walks. Social Networks 27(1), 39–54 (2005)
[20] Newman, M.: Networks: An Introduction. Oxford University Press, Oxford (2010)
[21] Reitz, J.M.: ODLIS: Online Dictionary of Library and Information Science (2002)
[22] Sabidussi, G.: The centrality index of a graph. Psychometrika 31, 581–603 (1966)
[23] Shanmugham, B., Pan, A.: Identification and Characterization of Potential Therapeutic Candidates in Emerging Human Pathogen Mycobacterium abscessus: A Novel Hierarchical In Silico Approach. PLoS ONE 8(3), e59126 (2013), doi:10.1371/journal.pone.0059126
[24] Shimbel, A.: Structural parameters of communication networks. Bulletin of Mathematical Biophysics 15, 501–507 (1953)

[25] Spiliotopoulos, T., Oakley, I.: Applications of Social Network Analysis for User Modeling. In: International Workshop on User Modeling from Social Media / IUI 2012 (2012)
[26] Tarjan, R.: Depth-First Search and Linear Graph Algorithms. SIAM J. Comput. 1(2), 146–160 (1972)
[27] Website of National Cancer Institute, `http://www.cancer.gov`
[28] Yu, H., Kim, P.M., Sprecher, E., Trifonov, V., Gerstein, M.: The Importance of Bottlenecks in Protein Networks: Correlation with Gene Essentiality and Expression Dynamics. PLoS Comput. Biol. 3(4), e59 (2007), doi:10.1371/journal.pcbi.0030059

Generalized Preferential Attachment: Tunable Power-Law Degree Distribution and Clustering Coefficient*

Liudmila Ostroumova[1,2], Alexander Ryabchenko[1,3], and Egor Samosvat[1,3]

[1] Yandex, Moscow, Russia
[2] Moscow State University, Moscow, Russia
[3] Moscow Institute of Physics and Technology, Moscow, Russia

Abstract. We propose a common framework for analysis of a wide class of preferential attachment models, which includes LCD, Buckley–Osthus, Holme–Kim and many others. The class is defined in terms of constraints that are sufficient for the study of the degree distribution and the clustering coefficient. We also consider a particular parameterized model from the class and illustrate the power of our approach as follows. Applying our general results to this model, we show that both the parameter of the power-law degree distribution and the clustering coefficient can be controlled via variation of the model parameters. In particular, the model turns out to be able to reflect realistically these two quantitative characteristics of a real network, thus performing better than previous preferential attachment models. All our theoretical results are illustrated empirically.

Keywords: networks, random graph models, preferential attachment, power-law degree distribution, clustering coefficient.

1 Introduction

Numerous random graph models have been proposed to reflect and predict important quantitative and topological aspects of growing real-world networks, from Internet and society [1,5,8] to biological networks [2]. Such models are of use in experimental physics, bioinformatics, information retrieval, and data mining. An extensive review can be found elsewhere (e.g., see [1,5,6]). Though largely successful in capturing key qualitative properties of real-world networks, such models may lack some of their important characteristics.

The simplest characteristic of a vertex in a network is the degree, the number of adjacent edges. Probably the most extensively studied property of networks is their vertex degree distribution. For the majority of studied real-world networks, the portion of vertices with degree d was observed to decrease as $d^{-\gamma}$, usually with $2 < \gamma < 3$, see [3,5,9,16]. Such networks are often called scale-free.

Another important characteristic of networks is their clustering coefficient, a measure capturing the tendency of a network to form clusters, densely interconnected sets of vertices. Various definitions of the clustering coefficient can be

* The authors are given in alphabetical order.

A. Bonato, M. Mitzenmacher, and P. Prałat (Eds.): WAW 2013, LNCS 8305, pp. 185–202, 2013.
© Springer International Publishing Switzerland 2013

found in the literature, see [6] for a discussion on their relationship. We consider the most popular two: the global clustering coefficient and the average local clustering coefficient (see Section 3.3 for definitions). For the majority of studied real-world networks, the average local clustering coefficient varies in the range from 0.01 to 0.8 and does not change much as the network grows [5]. Modeling real-world networks with accurately capturing not only their power-law degree distribution, but also clustering coefficient, has been a challenge.

In order to combine tunable degree distribution and clustering in one model, some authors [2,20,21] proposed to start with a concrete prior distribution of vertex degrees and clustering and then generate a random graph under such constraints. However, adjusting a model to a particular graph seems to be not generic enough and can be suspected in "overfitting". A more natural approach is to consider a graph as the result of a random process defined by certain reasonable realistic rules guaranteeing the desired properties observed in real networks. Perhaps the most widely studied realization of this approach is preferential attachment. In Section 2, we give a background on previous studies in this field.

In this paper, we propose a new class of preferential attachment random graph models thus generalizing some previous approaches. We study this class theoretically: we prove the power law for the degree distribution and approximate the clustering coefficient. We demonstrate that in preferential attachment graphs two definitions of the clustering coefficient give quite different values. We also propose a concrete parameterized model from our class where both the power-law exponent and the clustering coefficient can be tuned. All our theoretical results are illustrated experimentally.

The remainder of the paper is organized as follows. In Section 2, we give a background on previous studies of preferential attachment models. In Section 3, we propose a definition of a new class of models, and obtain some general results for all models in this class. Then, in Section 4, we describe one particular model from the proposed class. We demonstrate results obtained for graphs generated in this model in Section 5. Section 6 concludes the paper.

2 Preferential Attachment Random Graph Models

In 1999, Barabási and Albert observed [3] that the degree distribution of the World Wide Web follows the power law with the parameter ~ 2.1. As a possible explanation for this phenomenon, they proposed a graph construction stochastic process, which is a Markov chain of graphs, governed by the *preferential attachment*. At each time step in the process, a new vertex is added to the graph and is joined to m different vertices already existing in the graph chosen with probabilities proportional to their degrees.

Denote by d_v^n the degree of a vertex v in the growing graph at time n. At each step m edges are added, so we have $\sum_v d_v^n = 2mn$. This observation and the preferential attachment rule imply that

$$\mathbf{P}(d_v^{n+1} = d + 1 \mid d_v^n = d) = \frac{d}{2n}, \quad (1)$$

where **P** denotes the probability of an event. Note that the condition (1) on the attachment probability does not specify the distribution of m vertices to be joined to, in particular their dependence. Therefore, it would be more accurate to say that Barabási and Albert proposed not a single model, but a class of models. As it was shown later, there is a whole range of models that fit the Barabási–Albert description, but possess very different behavior.

Theorem 1 (Bollobás, Riordan [6]). *Let $f(n)$, $n \geq 2$, be any integer valued function with $f(2) = 0$ and $f(n) \leq f(n+1) \leq f(n)+1$ for every $n \geq 2$, such that $f(n) \to \infty$ as $n \to \infty$. Then there is a random graph process $T(n)$ satisfying (1) such that, with probability 1, $T(n)$ has exactly $f(n)$ triangles for all sufficiently large n.*

In [7], Bollobás and Riordan proposed a concrete precisely defined model of the Barabási–Albert type, known as the LCD-model, and proved that for $d < n^{\frac{1}{15}}$, the portion of vertices with degree d asymptotically almost surely obeys the power law with the parameter 3. Recently Grechnikov substantially improved this result [17] and removed the restriction on d. It was shown also that the expectation of the global clustering coefficient in the model is asymptotically proportional to $\frac{(\log n)^2}{n}$ and therefore tends to zero as the graph grows [6].

One obtains a natural generalization of the LCD-model, requiring the probability of attachment of a new vertex $n+1$ to a vertex v to be proportional to $d_v^n + m\beta$, where β is a constant representing the *initial attractiveness* of a vertex. Buckley and Osthus [10] proposed a precisely defined model with a nonnegative integer β. Móri [19] generalized this model to real $\beta > -1$. For both models, the degree distribution was shown to follow the power law with the parameter $3 + \beta$ in the range of small degrees. The recent result of Eggemann and Noble [15] implies that the expectation of the global clustering coefficient in the Móri model with $\beta > 0$ is asymptotically proportional to $\frac{\log n}{n}$. For $\beta = 0$, the Móri model is almost identical to the LCD-model. Therefore the authors of [15] emphasize the confusing difference between clustering coefficients ($\frac{(\log n)^2}{n}$ versus $\frac{\log n}{n}$).

The main drawback of the described preferential attachment models is unrealistic behavior of the clustering coefficient. In fact, for all discussed models the clustering coefficient tends to zero as a graph grows, while in the real-world networks the clustering coefficient is approximately a constant [5].

A model with asymptotically constant (average local) clustering coefficient was proposed by Holme and Kim [18]. The idea is to mix preferential attachment steps with the steps of triangle formation. This model allows to tune the clustering coefficient by varying the probability of the triangle formation step. However, experiments and empirical analysis show that the degree distribution in this model obeys the power law with the fixed parameter close to 3, which does not suit most real networks. RAN (random Apollonian network) proposed in [22] is another interesting example of a Barabási-Albert type model with asymptotically constant (average local) clustering.

There is a variety of other models, not mentioned here, that are also based on the idea of preferential attachment. Analyses of properties of all these models are often very similar. In the next section, we consider theorems aimed at simplifying these analyses and providing a general framework for them. In order to do this, we define a new class of preferential attachment models that generalizes models mentioned above, as well as many others. We also propose a new parameterized model which belongs to this class that allows to tune both the power-law exponent and the clustering coefficient by adjusting the parameters.

3 Theoretical Results

In this section, we define a general class of preferential attachment models. For all models in this class we are able to prove the power-law degree distribution. If an additional property is fulfilled, we are able to analyze the behavior of the clustering coefficient as the network grows.

3.1 Definition of the PA-class

Let G_m^n $(n \geq n_0)$ be a graph with n vertices $\{1, \ldots, n\}$ and mn edges obtained as a result of the following random graph process. We start at the time n_0 from an arbitrary graph $G_m^{n_0}$ with n_0 vertices and mn_0 edges. On the $(n+1)$-th step $(n \geq n_0)$, we make the graph G_m^{n+1} from G_m^n by adding a new vertex $n+1$ and m edges connecting this vertex to some m vertices from the set $\{1, \ldots, n, n+1\}$. Denote by d_v^n the degree of a vertex v in G_m^n. If for some constants A and B the following conditions are satisfied

$$\mathbf{P}\left(d_v^{n+1} = d_v^n \mid G_m^n\right) = 1 - A\frac{d_v^n}{n} - B\frac{1}{n} + O\left(\frac{\left(d_v^n\right)^2}{n^2}\right), \ 1 \leq v \leq n\,, \quad (2)$$

$$\mathbf{P}\left(d_v^{n+1} = d_v^n + 1 \mid G_m^n\right) = A\frac{d_v^n}{n} + B\frac{1}{n} + O\left(\frac{\left(d_v^n\right)^2}{n^2}\right), \ 1 \leq v \leq n\,, \quad (3)$$

$$\mathbf{P}\left(d_v^{n+1} = d_v^n + j \mid G_m^n\right) = O\left(\frac{\left(d_v^n\right)^2}{n^2}\right), \ 2 \leq j \leq m, \ 1 \leq v \leq n\,, \quad (4)$$

$$\mathbf{P}(d_{n+1}^{n+1} = m + j) = O\left(\frac{1}{n}\right), \ 1 \leq j \leq m\,, \quad (5)$$

then we say that the random graph process G_m^n is a model from the PA-class. Condition (5) means that the probability to have a self-loop in the added vertex is small. As we will show later, certain minor details of the models from this class, such as whether loops and multiple edges are allowed, are irrelevant.

Since we add m edges at each step, summing up the equalities (3)-(5) (with corresponding coefficients) over all vertices and neglecting error terms we get $2mA + B = m$. It is possible to prove that the sum of error terms in this case is

0, but for simplicity we just set $2mA+B=m$. Furthermore, we have $0 \leq A \leq 1$ (for (3) we need $mA+B \geq 0$ and we set $2mA+B=m$, therefore $A \leq 1$).

Here we want to emphasize that we indeed defined not a single model but a class of models. Even fixing values of parameters A and m does not specify a concrete procedure for constructing a network. What this definition lacks is the precise description of the distribution of vertices a new incoming vertex is being connected to, and therefore there is a range of models possessing very different properties and satisfying the conditions (2–5). For example, the LCD, the Holme–Kim and the RAN models belong to the PA-class with $A=1/2$ and $B=0$. The Buckley–Osthus (Móri) model also belongs to the PA-class with $A=\frac{1}{2+\beta}$ and $B=\frac{m\beta}{2+\beta}$. Another example is considered in detail in Sections 4 and 5. This situation is somewhat similar to that with the definition of the Barabási–Albert models, though our class is wider in a sense that the exponent of the power-law degree distribution is tunable.

In mathematical analysis of network models, there is a tendency to consider only fully and precisely defined models. In contrast, we provide results about general properties for the whole PA-class in the next two subsections.

3.2 Power Law Degree Distribution

Even though the precise description of the distribution of vertices a new incoming vertex is going to be connected to is not specified, we are still able to describe the degree distribution of the network.

First, we estimate $N_n(d)$, the number of vertices with given degree d in G_m^n. We prove the following result on the expectation $\mathsf{E}N_n(d)$ of $N_n(d)$.

Theorem 2. *For every $d \geq m$ we have $\mathsf{E}N_n(d) = c(m,d)\left(n + O\left(d^{2+\frac{1}{A}}\right)\right)$, where*

$$c(m,d) = \frac{\Gamma\left(d+\frac{B}{A}\right)\Gamma\left(m+\frac{B+1}{A}\right)}{A\Gamma\left(d+\frac{B+A+1}{A}\right)\Gamma\left(m+\frac{B}{A}\right)} \overset{d\to\infty}{\sim} \frac{\Gamma\left(m+\frac{B+1}{A}\right)d^{-1-\frac{1}{A}}}{A\Gamma\left(m+\frac{B}{A}\right)},$$

and $\Gamma(x)$ is the gamma function.

Second, we show that the number of vertices with given degree d is highly concentrated around its expectation.

Theorem 3. *For every model from the PA-class and for every $d = d(n)$ we have*

$$\mathsf{P}\left(|N_n(d) - \mathsf{E}N_n(d)| \geq d\sqrt{n}\log n\right) = O\left(n^{-\log n}\right).$$

Therefore, for any $\delta > 0$ there exists a function $\varphi(n) \in o(1)$ such that

$$\lim_{n\to\infty} \mathsf{P}\left(\exists\, d \leq n^{\frac{A-\delta}{4A+2}} : |N_n(d) - \mathsf{E}N_n(d)| \geq \varphi(n)\,\mathsf{E}N_n(d)\right) = 0\,.$$

These two theorems mean that the degree distribution follows (asymptotically) the power law with the parameter $1+\frac{1}{A}$.

Theorem 2 is proved by induction on d and n. It is easy to see that given a graph G_m^n, we can express the conditional expectation of the number of vertices with degree d in G_m^{n+1} (i.e., $\mathsf{E}(N_{n+1}(d) \mid G_m^n)$) in terms of $N_n(d), N_n(d-1), \ldots, N_n(d-m)$. Here we use the fact that the probability of having an edge between the vertex $n+1$ and a vertex v depends on the degree of v (see (2)). Using the law of total expectation we obtain the recurrent relation for $\mathsf{E}N_{n+1}(d)$ and prove the statement of Theorem 2 by induction.

We use the Azuma–Hoeffding inequality to prove the concentration result of Theorem 3. In order to do this, we consider the martingale $X_i(d) = \mathsf{E}(N_n(d) \mid G_m^i)$, $i = 0, \ldots, n$. The complete proofs of these theorems are technical and are placed in Appendix due to space constraints.

3.3 Clustering Coefficient

Here we consider the clustering coefficient in models of the PA-class. There are two popular definitions of the clustering coefficient. The *global clustering coefficient* $C_1(n)$ is the ratio of three times the number of triangles to the number of pairs of adjacent edges in G. The *average local clustering coefficient* is defined as follows: $C_2(n) = \frac{1}{n}\sum_{i=1}^{n} C(i)$, where $C(i)$ is the local clustering coefficient for a vertex i: $C(i) = \frac{T^i}{P_2^i}$, where T^i is the number of edges between neighbors of the vertex i and P_2^i is the number of pairs of neighbors. Results for some classical preferential attachment models (LCD and Móri) are mentioned in Section 2.

Here we generalize these results. First, we study the random variable $P_2(n)$ equal to the number of P_2's in a random graph G_m^n from an arbitrary model that belongs to the PA-class. In the theorems below, we use the following notation. By **whp** ("with high probability") we mean that for some sequence A_n of events, $P(A_n) \to 1$ as $n \to \infty$. We say $a_n \sim b_n$ if $a_n = (1+o(1))b_n$, and we say $a_n \propto b_n$ if $C_0 b_n \le a_n \le C_1 b_n$ for some constants $C_0, C_1 > 0$.

Theorem 4. *For every model from the PA-class, we have*

(1) if $2A < 1$, *then* ***whp*** $P_2(n) \sim \left(2m(A+B) + \frac{m(m-1)}{2}\right)\frac{n}{1-2A}$,
(2) if $2A = 1$, *then* ***whp*** $P_2(n) \sim \left(2m(A+B) + \frac{m(m-1)}{2}\right) n\log(n)$,
(3) if $2A > 1$, *then for any* $\varepsilon > 0$ ***whp*** $n^{2A-\varepsilon} \le P_2(n) \le n^{2A+\varepsilon}$.

The ideas of the proof of Theorem 4 are given in Appendix. Here it is worth noting that the value $P_2(n)$ in scale-free graphs is usually determined by the power-law exponent γ. Indeed, we have $P_2(n) = \sum_{d=1}^{d_{\max}} N_n(d)\frac{d(d-1)}{2} \propto \sum_{d=1}^{d_{\max}} nd^{2-\gamma}$, where $d_{\max}$ is the maximum possible degree of a vertex in G_m^n. Therefore if $\gamma > 3$, then $P_2(n)$ is linear in n. However, if $\gamma \le 3$, then $P_2(n)$ is superlinear.

Next, we study the random variable $T(n)$ equal to the number of triangles in G_m^n. Note that in any model from the PA-class we have $T(n) = O(n)$ since at each step we add at most $\frac{m(m-1)}{2}$ triangles. If we combine this fact with the previous observation, we see that if $\gamma \le 3$, then in any preferential attachment model (in which the out-degree of each vertex equals m) the global clustering coefficient tends to zero as n grows.

Our aim is to find models with constant clustering coefficient. Let us consider a subclass of the PA-class with the following property:

$$\mathbf{P}\left(d_i^{n+1} = d_i^n + 1, d_j^{n+1} = d_j^n + 1 \mid G_m^n\right) = e_{ij}\frac{D}{mn} + O\left(\frac{d_i^n d_j^n}{n^2}\right). \tag{6}$$

Here e_{ij} is the number of edges between vertices i and j in G_m^n and D is a positive constant. Note that this property still does not define the correlation between edges completely.

Theorem 5. *Let G_m^n satisfy the condition (6). Then* **whp** $T(n) \sim D n$.

The proof of this theorem is straightforward. The expectation of the number of triangles we add at each step is $D+o(1)$. The fact that the sum of $O\left(\frac{d_i^n d_j^n}{n^2}\right)$ over all *adjacent* vertices is $o(1)$ can be shown by induction using the conditions (2–5). It is also possible to first prove that the maximum degree grows as n^A and then use this fact to estimate the sum of error terms. Therefore $\mathsf{E}T(n) = Dn + o(n)$. The Azuma–Hoeffding inequality can be used to prove concentration.

As a consequence of Theorems 4 and 5, we get the following result on the global clustering coefficient $C_1(n)$ of the graph G_m^n.

Theorem 6. *Let G_m^n belong to the PA-class and satisfy the condition (6). Then*

(1) If $2A < 1$ then **whp** $C_1(n) \sim \frac{3(1-2A)D}{\left(2m(A+B)+\frac{m(m-1)}{2}\right)}$,
(2) If $2A = 1$ then **whp** $C_1(n) \sim \frac{3D}{\left(2m(A+B)+\frac{m(m-1)}{2}\right)\log n}$,
(3) If $2A > 1$ then for any $\varepsilon > 0$ **whp** $n^{1-2A-\varepsilon} \le C_1(n) \le n^{1-2A+\varepsilon}$.

Theorem 6 shows that in some cases ($2A \ge 1$) the global clustering coefficient $C_1(n)$ tends to zero as the number of vertices grows. We empirically show in Section 5 that the average local clustering coefficient $C_2(n)$ behaves differently.

The theoretical analysis in this case is much harder, but we can easily show why $C_2(n)$ does not tend to zero if the condition (6) holds. From Theorems 2 and 3 it follows that **whp** the number of vertices with degree m in G_m^n is greater than cn for some positive constant c. The expectation of the number of triangles we add at each step is $D + o(1)$. Therefore **whp** $C_2(n) \ge \frac{1}{n}\sum_{i:\deg(i)=m} C(i) \ge \frac{2cD}{m(m+1)}$.

In the next section we introduce a concrete nontrivial model from the PA-class.

4 Polynomial Model

In this section, we consider *polynomial random graph models* that belong to the general PA-class defined above. Applying our theoretical results to polynomial models, we find the model to be very flexible: one can tune the parameter of the degree distribution and the clustering coefficient.

Definition of Polynomial Model. Let us define the *polynomial model.* As in the random graph process from Subsection 3.1, we construct a graph G_m^n step by step. On the $(n+1)$-th step the graph G_m^{n+1} is made from the graph G_m^n by adding a new vertex $n+1$ and sequentially drawing m edges (multiple edges and self-loops are allowed).

We say that an edge ij is directed from i to j if $i \geq j$, so the out-degree of each vertex equals m. We also say that i and j are respectively *source* and *target* ends of ij. We consider different approaches to add new edges from the vertex $n+1$. We first choose an edge from the existing graph G_m^n uniformly and independently at random and then have three options:

- Preferential attachment (PA): draw one edge from $n+1$ to the *target* end of the chosen edge
- Uniform (U): draw one edge from $n+1$ to the *source* end of the chosen edge
- Triangle formation (TF): draw two edges from $n+1$ to *target* and *source* ends of the chosen edge

Let us now specify how to draw m edges from the vertex $n+1$. Consider a collection of positive parameters $\{\alpha_{k,l}\}$ for $0 \leq k \leq m/2$ and $0 \leq l \leq m-2k$ such that $\sum_{k,l} \alpha_{k,l} = 1$, these parameters fully define our model. At the beginning of the $n+1$ step with probabilities $\{\alpha_{k,l}\}$ we choose some $k = k_0$ and $l = l_0$, then we draw l_0 edges using PA, $2k_0$ edges using TF and $(m - l_0 - 2k_0)$ edges using U. This random graph process defines the polynomial model and from the definition it follows that graphs in this model can be generated in linear time. This model belongs to the PA-class. Indeed, one can formally show by simple calculations that the conditions (2–5) hold for this model.

At this point the model is defined but let us explain why we call it polynomial. Denote by $\widehat{d_i^n} = d_i^n - m$ the in-degree of a vertex i in G_m^n. Let us recall that by e_{ij} we denote the number of edges between vertices i and j. For every k, l such that $0 \leq k \leq m/2$ and $0 \leq l \leq m-2k$, let $M_{k,l}^{n,m}(i_1, \ldots, i_m) = \frac{1}{n^{m-l-2k}} \prod_{x=1}^{k} \frac{e_{i_{2x} i_{2x-1}}}{2mn} \prod_{y=2k+1}^{2k+l} \frac{\widehat{d_{i_y}^n}}{mn}$. This is a monomial depending on $\widehat{d_{i_y}^n}$ and $e_{i_{2x} i_{2x-1}}$. We define the polynomial $\sum_{k,l} \alpha_{k,l} M_{k,l}^{n,m}(i_1, \ldots, i_m)$. It is easy to check that

$$\mathbf{P}\left(\text{edges } e_1, \ldots, e_m \text{ go to vertices } i_1, \ldots, i_m, \text{ respectively}\right) =$$

$$= \sum_{k=0}^{m/2} \sum_{l=0}^{m-2k} \alpha_{k,l} M_{k,l}^{n,m}(i_1, \ldots, i_m) \,. \quad (7)$$

Many models are special cases of the polynomial model. If we consider the polynomial $\prod_{y=1}^{m} \frac{\widehat{d_{i_y}^n}+m}{2mn}$, then we obtain a model that is practically identical to the LCD-model. The Buckley–Osthus model can be also interpreted in terms of the polynomial model.

Properties. It is easy to check that the parameters $\alpha_{k,l}$ from (7) and A from (2) are related in the following way:

$$A = \sum \alpha_{k,l} \frac{l+k}{m} . \tag{8}$$

This means that we can use an arbitrary value of $A \in [0,1]$ and any power-law exponent $\gamma \in (2,\infty)$ in the graph generation. Also note that $D = \sum_{k,l} k\alpha_{k,l}$.

In the next section we analyze experimentally some properties of graphs in the polynomial model. We generate polynomial graphs and compare their properties with theoretical results we obtained.

5 Experiments

In this section, we choose a three-parameter model from the family of polynomial graph models defined in Section 4 and analyze the properties of the generated graphs depending on the parameters.

5.1 Description of Empirically Studied Polynomial Model

We study empirically graphs in the polynomial model with $m = 2p$ and the probability to draw edges to vertices $i_1, \ldots, i_{2m}$ equals

$$\prod_{k=1}^{p} \left(\alpha \frac{\widehat{d^n_{i_{2k}}\, d^n_{i_{2k-1}}}}{(mn)^2} + \beta \frac{e_{i_{2k} i_{2k-1}}}{2mn} + \frac{\delta}{(n)^2} \right) .$$

Here we need $\alpha, \beta, \delta \geq 0$ and $\alpha+\beta+\delta = 1$, therefore, we have three independent model parameters: m, α, and β. Note that here we write the polynomial in a symmetric form as we ignore the order of edges.

Based on our theoretical results, we have certain expectations about the properties of generated graphs. From (8) we obtain that in this model $A = \alpha + \frac{\beta}{2}$, $B = m(\delta - \alpha)$, $D = p\beta = \frac{m\beta}{2}$, therefore, due to Theorem 2 and Theorem 6, we get that

$$C_1(n) \sim \frac{3(1-2\alpha-\beta)\beta}{5m-1-2(2m-1)(2\alpha+\beta)},\ \gamma = 1 + \frac{2}{2\alpha+\beta} . \tag{9}$$

5.2 Empirical Results

Degree Distribution and Clustering Coefficient. First, we study two polynomial graphs with $n = 10^7$, $m = 2$, and $A = 0.4$, assigning $\alpha = 0.4, \beta = 0$ for the first graph and $\alpha = 0, \beta = 0.8$ for the second one. The observed degree distributions are almost identical and follow the power law with the expected parameter $\gamma = 3.5$, see Fig. 1a.

For both cases, we also study the behavior of the global and the average local clustering coefficients of generated graphs, 40 samples for each $n = \left[10^{1+0.06i}\right]$, $i = 0, \ldots, 100$, see Fig. 1bc. In the first case we observe $C_1(n) \to 0$, $C_2(n) \to 0$

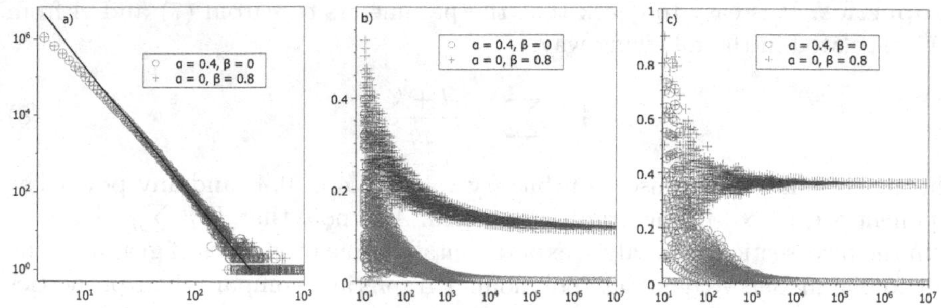

Fig. 1. a) The degree distribution of polynomial graphs with $n = 10^7$ and $m = 2$. b) The global clustering coefficient of polynomial graphs with $m = 2$ depending on n. c) The average local clustering coefficient of polynomial graphs with $m = 2$ depending on n.

(as $\beta = 0$) and in the second case $C_1(n) \to \frac{2}{15}$ (as was expected due to (9)) and $C_2(n) \to \text{const} > 0$.

We also generate graphs with $n = 10^6$, $m = 2$, and varying A (we took $\beta = 0.5$ and $\alpha \in (0, 0.5)$). In other words, we fix the probability of a triangle formation and vary the parameter of the power-law degree distribution. The obtained results are shown in Fig. 2a. The behavior of the clustering coefficients is quite different. If A grows, then $P_2(n)$ grows (therefore $C_1(n) \to 0$), the number of vertices with small degrees and hence high local clustering also grows (therefore $C_2(n)$ increases).

To demonstrate the difference between the global clustering and the average local clustering we generated graphs with $m = 2$, $\alpha = 0.5$, $\beta = 0.2$ and varying n (Fig. 2b). In this case we have $A = \alpha + \frac{\beta}{2} > 0.5$ and $C_1(n) \to 0$, as expected. However, for the local clustering we obtain $C_2(n) \to \text{const} > 0$.

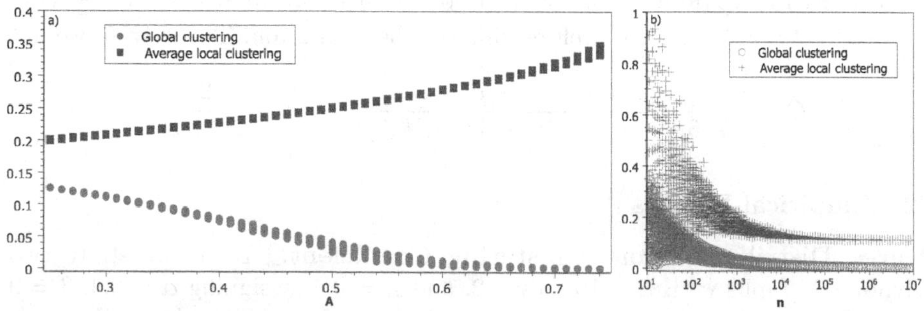

Fig. 2. a) Average local and global clustering in polynomial graphs with $n = 10^6$, $m = 2$, $\beta = 0.5$ depending on A. b) The global and the average local clustering coefficients of polynomial graphs with $m = 2$, $\alpha = 0.5$, $\beta = 0.2$ depending on n.

Comparison with Other Models. The following table summarizes our results for the polynomial model in comparison with other mentioned preferential attachment models:

	A	D	γ	Global clustering	Average local clustering
LCD	1/2	0	3	tends to zero	tends to zero
BO/Móri	$1/(2+\beta)$	0	$(2,\infty)$	tends to zero	tends to zero
HK	1/2	m_t	3	tends to zero	constant
RAN	1/2	3	3	tends to zero	constant
Polynomial	$\sum \alpha_{k,l} \frac{l+k}{m}$	$\sum k\alpha_{k,l}$	$(2,\infty)$	constant for $A < \frac{1}{2}$	constant

The polynomial model seems to be the only model where one can control the exponent in the power law of the degree distribution, and at the same time guarantee a positive clustering coefficient.

6 Conclusions

In this paper, we introduced the PA-class of random graph models that generalizes previous preferential attachment approaches. We proved that any model from the PA-class possesses the power-law degree distribution with tunable parameter. We also estimated its clustering coefficient. Next, we described one particular model from the proposed class (with tunable both the degree distribution parameter and the clustering coefficient). Experiments with generated graphs illustrated our theoretical results. We also demonstrated different behavior of two versions of the clustering coefficient in preferential attachment models.

As the degree distribution of a preferential attachment model allows adjustment to reality, the clustering coefficient still gives rise to a problem in some cases. For most real-world networks the parameter γ of their degree distribution belongs to [2, 3]. As we showed in Section 3, once $\gamma \leq 3$ in a preferential attachment model, the global clustering coefficients decreases as the graph grows, which does not correspond to the majority of real-world networks. The reason is that the number of edges added with a new vertex at each step is a constant and consequently the number of triangles grows too slowly.

Fortunately, there are many ways to overcome this obstacle. Cooper proposed a model in which the number of added edges is a random variable [11]. In collaboration with Prałat he also considered a modification of the Barabási–Albert model, where a new vertex added at time t generates t^c edges [13]. Preferential attachment models with random initial degrees were considered in [14]. Also there are models with adding edges between already existing nodes (e.g. [12]). Using one of these ideas for the PA-class is a topic for future research.

Acknowledgements. Special thanks to Evgeniy Grechnikov, Gleb Gusev, Andrei Raigorodskii and anonymous reviewers for the careful reading and useful comments.

References

1. Albert, R., Barabási, A.-L.: Statistical mechanics of complex networks. Reviews of Modern Physics 74, 47–97 (2002)
2. Bansal, S., Khandelwal, S., Meyers, L.A.: Exploring biological network structure with clustered random networks. BMC Bioinformatics 10, 405 (2009)
3. Barabási, A.-L., Albert, R.: Science 286, 509 (1999); Barabási, A.-L., Albert, R., Jeong, H.: Physica A 272, 173 (1999); Albert, R., Jeong, H., Barabási, A.-L.: Nature 401, 130 (1999)
4. Batagelj, V., Brandes, U.: Efficient generation of large random networks. Phys. Rev. E 71, 036113 (2005)
5. Boccaletti, S., Latora, V., Moreno, Y., Chavez, M., Hwang, D.-U.: Complex networks: Structure and dynamics. Physics Reports 424(45), 175–308 (2006)
6. Bollobás, B., Riordan, O.M.: Mathematical results on scale-free random graphs. In: Handbook of Graphs and Networks: From the Genome to the Internet, pp. 1–3 (2003)
7. Bollobás, B., Riordan, O.M., Spencer, J., Tusnády, G.: The degree sequence of a scale-free random graph process. Random Structures and Algorithms 18(3), 279–290 (2001)
8. Borgs, C., Brautbar, M., Chayes, J., Khanna, S., Lucier, B.: The power of local information in social networks (2012) (preprint)
9. Broder, A., Kumar, R., Maghoul, F., Raghavan, P., Rajagopalan, S., Stata, R., Tomkins, A., Wiener, J.: Graph structure in the web. Computer Networks 33(16), 309–320 (2000)
10. Buckley, P.G., Osthus, D.: Popularity based random graph models leading to a scale-free degree sequence. Discrete Mathematics 282, 53–63 (2004)
11. Cooper, C.: Distribution of Vertex Degree in Web-Graphs. Combinatorics, Probability and Computing 15, 637–661 (2006)
12. Cooper, C., Frieze, A.: A General Model of Web Graphs. Random Structures and Algorithms 22(3), 311–335 (2003)
13. Cooper, C., Prał, P.: at, Scale-free graphs of increasing degree. Random Structures and Algorithms 38(4), 396–421 (2011)
14. Deijfen, M., van den Esker, H., van der Hofstad, R., Hooghiemstra, G.: A preferential attachment model with random initial degrees. Ark. Mat. 47, 41–72 (2009)
15. Eggemann, N., Noble, S.D.: The clustering coefficient of a scale-free random graph. Discrete Applied Mathematics 159(10), 953–965 (2011)
16. Faloutsos, M., Faloutsos, P., Faloutsos, C.: On power-law relationships of the Internet topology. In: Proc. SIGCOMM 1999 (1999)
17. Grechnikov, E.A.: An estimate for the number of edges between vertices of given degrees in random graphs in the Bollobás–Riordan model. Moscow Journal of Combinatorics and Number Theory 1(2), 40–73 (2011)
18. Holme, P., Kim, B.J.: Growing scale-free networks with tunable clustering. Phys. Rev. E 65(2), 026107 (2002)
19. Móri, T.F.: The maximum degree of the Barabási-Albert random tree. Combinatorics, Probability and Computing 14, 339–348 (2005)
20. Serrano, M.Á., Boguñá, M.: Tuning clustering in random networks with arbitrary degree distributions. Phys. Rev. E 72(3), 036133 (2005)
21. Volz, E.: Random Networks with Tunable Degree Distribution and Clustering. Phys. Rev. E 70(5), 056115 (2004)
22. Zhou, T., Yan, G., Wang, B.-H.: Maximal planar networks with large clustering coefficient and power-law degree distribution journal. Phys. Rev. E 71(4), 46141 (2005)

Appendix: Proofs

Proof of Theorem 2

In this proof we use the notation $\theta(\cdot)$ for error terms. By $\theta(X)$ we denote a function such that $|\theta(X)| < X$. We also need the following notation:

$$\mathbf{P}\left(d_v^{n+1} = d \mid d_v^n = d\right) = 1 - A\frac{d}{n} - B\frac{1}{n} + O\left(\frac{d^2}{n^2}\right), \tag{10}$$

$$p_n^1(d) := \mathbf{P}\left(d_v^{n+1} = d+1 \mid d_v^n = d\right) = A\frac{d}{n} + B\frac{1}{n} + O\left(\frac{d^2}{n^2}\right), \tag{11}$$

$$p_n^j(d) := \mathbf{P}\left(d_v^{n+1} = d+j \mid d_v^n = d\right) = O\left(\frac{d^2}{n^2}\right),\ 2 \le j \le m\,. \tag{12}$$

$$p_n := \sum_{k=1}^{m} \mathbf{P}(d_{n+1}^{n+1} = m+k) = O\left(\frac{1}{n}\right). \tag{13}$$

Note that the remainder term of $p_n^j(d)$ can depend on v. We omit v in notation $p_n^j(d)$ for simplicity of proofs.

Put $p_v(d) = \sum_{j=1}^m p_v^j(d)$. Note that $\frac{Ad+B+1}{Ad-A+B}p_v^1(d-1) - p_v(d) = \frac{1}{v} + O\left(\frac{d^2}{v^2}\right)$. We use this equality several times in this proof.

We want to prove that $\mathsf{E}N_n(d) = c(m,d)\left(n + \theta\left(Cd^{2+\frac{1}{A}}\right)\right)$ with some constant C and some function θ. The proof is by induction on d and then on i. First, we prove the theorem for $d = m$ and all i. Then, if we proved the theorem for some $d = d_0$ and all i, we are able to prove it for $d = d_0 + 1$ and for all i.

We use the following equalities

$$\mathsf{E}(N_{i+1}(m) \mid N_i(m)) = N_i(m)\,(1 - p_i(m)) + 1 - p_i\,, \tag{14}$$

$$\begin{aligned}\mathsf{E}(N_{i+1}(d) \mid N_i(d), N_i(d-1), \ldots, N_i(d-m)) = N_i(d)\,(1 - p_i(d)) +\\ + N_i(d-1)p_i^1(d-1) + \sum_{j=2}^{m} N_i(d-j)p_i^j(d-j) + O(p_i)\,.\end{aligned} \tag{15}$$

Consider the case $d = m$. For constant number of small i we obviously have $\mathsf{E}N_i(m) = \frac{i}{Am+B+1} + \theta(C_1)$ with some C_1. Assume that $\mathsf{E}N_i(m) = \frac{i}{Am+B+1} + \theta(C_1)$. From (14) we obtain

$$\begin{aligned}\mathsf{E}N_{i+1}(m) &= \mathsf{E}N_i(m)\,(1 - p_i(m)) + 1 - p_i = \\ &= \left(\frac{i}{Am+B+1} + \theta(C_1)\right)(1 - p_i(m)) + 1 + \theta(C_2/i) = \\ &= \frac{i+1}{Am+B+1} + \theta(C_1)\,(1 - p_i(m)) + \theta\left(\frac{C_3}{i}\right)\frac{1}{Am+B+1} + \theta(C_2/i)\,.\end{aligned}$$

It remains to show that

$$C_1 p_i(m) \geq \frac{C_3}{i(Am+B+1)} + \theta(C_2/i)\,.$$

We have $p_i(m) \geq \frac{mA+B}{i} - \frac{C_0}{i^2}$. It gives us

$$C_1(Am+B) \geq \frac{C_1 C_0}{i} + \frac{C_3}{Am+B+1} + C_2\,.$$

This equality holds for large i and C_1. This completes the proof for $d = m$.

Remind that the proof is by induction on d and i. Consider $d > m$ and assume that we can prove the theorem for all smaller degrees. Now we use induction on i.

We have $N_i(d) \leq \frac{2mi}{d}$, therefore $N_i(d) = O\left(ic(m,d)d^{1/A}\right)$. In particular, for $i < 2\,C_7\,d^2$, where the constant C_7 depends only on the parameters of the model and will be defined later, we have $\mathsf{E}N_i(d) = c(m,d)\left(i + \theta\left(Cd^{2+1/A}\right)\right)$ with some C. Assume that

$$\mathsf{E}N_i(d) = c(m,d)\left(i + \theta\left(Cd^{2+1/A}\right)\right)\,.$$

From (15) we obtain

$$\begin{aligned}
\mathsf{E}N_{i+1}(d) &= \mathsf{E}N_i(d)\,(1 - p_i(d)) + \mathsf{E}N_i(d-1)p_i^1(d-1) + \\
&\quad + \sum_{j=2}^{m} \mathsf{E}N_i(d-j)p_i^j(d-j) + O(p_i) = \\
&= c(m,d)\left(i + \theta\left(Cd^{2+1/A}\right)\right)(1 - p_i(d)) + \\
&+ c(m,d-1)\left(i + \theta\left(C(d-1)^{2+1/A}\right)\right)p_i^1(d-1) + \theta\left(\frac{C_4 c(m,d)d^2 i d^{1/A}}{i^2}\right) = \\
&= c(m,d)(i+1) + c(m,d-1)ip_i^1(d-1) - \\
&- c(m,d)ip_i(d) - c(m,d) + c(m,d)\theta\left(Cd^{2+1/A}\right)(1 - p_i(d)) + \\
&+ \frac{c(m,d)(Ad+B+1)}{Ad-A+B}\theta\left(C(d-1)^{2+1/A}\right)p_i^1(d-1) + \theta\left(\frac{C_4 c(m,d)d^2 d^{1/A}}{i}\right) = \\
&= c(m,d)(i+1) + c(m,d)\theta\left(Cd^{2+1/A}\right)(1 - p_i(d)) + \\
&+ \frac{c(m,d)(Ad+B+1)}{Ad-A+B}\theta\left(C(d-1)^{2+1/A}\right)p_i^1(d-1) + \theta\left(\frac{C_5 c(m,d)d^2 d^{1/A}}{i}\right)\,.
\end{aligned}$$

We need to prove that there exists a constant C that

$$Cd^{2+1/A}p_i(d) \geq \frac{C(Ad+B+1)}{Ad-A+B}(d-1)^{2+1/A}p_i^1(d-1) + \frac{C_5 d^{2+1/A}}{i}\,,$$

$$Cd^{2+1/A}p_i(d) \geq \frac{C(Ad+B+1)}{Ad-A+B}\left(d^{2+1/A} - (2+1/A)d^{1+1/A} + C_6 d^{1/A}\right) \cdot$$
$$\cdot\, p_i^1(d-1) + \frac{C_5 d^{2+1/A}}{i}\,,$$

$$\frac{Cd^{2+1/A}}{i}\left(2A+\frac{(B-A)(2A+1)}{Ad}+O\left(\frac{d}{i^2}\right)\right)\geq Cd^{2+1/A}O\left(\frac{d^2}{i^2}\right)+$$
$$+\frac{C(Ad+B+1)}{Ad-A+B}C_6d^{1/A}\left(A\frac{d-1}{i}+B\frac{1}{i}+O\left(\frac{d^2}{i^2}\right)\right)+\frac{C_5d^{2+1/A}}{i}\,,$$

$$\frac{Cd^{2+1/A}}{i}\geq\frac{C_7Cd^{4+1/A}}{i^2}+\frac{C_8Cd^{1+1/A}}{i}+\frac{C_9d^{2+1/A}}{i}\,.$$

This inequality holds for large $C \geq C_{10}$ and $d \geq d_1$. For constant number of small $d < d_1$ there exists a function $f(d) > 0$ such that

$$f(d)d^{2+1/A}p_i(d)\geq f(d-1)\frac{Ad+B+1}{Ad-A+B}(d-1)^{2+1/A}p_i^1(d-1)+\frac{C_5d^{2+1/A}}{i}\,.$$

Thus the final C is $\max\{C_{10}, \max_{d<d_1}\{f(d)\}\}$. This concludes the proof.

Proof of Theorem 3

To prove Theorem 3 we need the Azuma–Hoeffding inequality:

Theorem 7 (Azuma, Hoeffding). *Let $(X_i)_{i=0}^n$ be a martingale such that $|X_i - X_{i-1}| \leq c_i$ for any $1 \leq i \leq n$. Then*

$$\mathsf{P}\left(|X_n - X_0| \geq x\right) \leq 2e^{-\frac{x^2}{2\sum_{i=1}^n c_i^2}}$$

for any $x > 0$.

Suppose we are given some $\delta > 0$. Fix n and d: $1 \leq d \leq n^{\frac{A-\delta}{4A+2}}$. Consider the random variables $X_i(d) = \mathsf{E}(N_n(d) \mid G_m^i)$, $i = 0, \ldots, n$.

Let us explain the meaning of the random variable $\mathsf{E}(N_n(d) \mid G_m^i)$. For any $t \leq n$ let $\mathsf{E}(N_n(d) \mid G_m^t)$ be the expectation of the number of vertices with degree d we may have at the step n of the process G_m^t if we fix first t steps of the evolution and allow the rest $n-t$ steps to be arbitrary. Note that $X_0(d) = \mathsf{E}N_n(d)$ and $X_n(d) = N_n(d)$. It is easy to see that $X_n(d)$ is a martingale.

We will prove below that for any $i = 0, \ldots, n-1$

$$|X_{i+1}(d) - X_i(d)| \leq Md,$$

where $M > 0$ is some constant. Theorem follows from this statement immediately. Put $c_i = Md$ for all i. Then from Azuma–Hoeffding inequality it follows that

$$\mathsf{P}\left(|N_n(d) - \mathsf{E}N_n(d)| \geq d\sqrt{n}\,\log n\right) \leq 2\exp\left\{-\frac{n\,d^2\,\log^2 n}{2\,n\,M^2d^2}\right\} = O\left(n^{-\log n}\right).$$

If $d \leq n^{\frac{A-\delta}{4A+2}}$, then the value of $\frac{n}{d^{1+1/A}}$ is considerably greater than $d\log n\sqrt{n}$. This is exactly what we need.

It remains to estimate the quantity $|X_{i+1}(d)-X_i(d)|$. The proof is by a direct calculation.

Fix $0 \le i \le n-1$ and some graph G_m^i. Note that

$$\left|\mathsf{E}\left(N_n(d) \mid G_m^{i+1}\right) - \mathsf{E}\left(N_n(d) \mid G_m^i\right)\right| \le$$
$$\le \max_{\tilde{G}_m^{i+1}\supset G_m^i} \left\{\mathsf{E}\left(N_n(d) \mid \tilde{G}_m^{i+1}\right)\right\} - \min_{\tilde{G}_m^{i+1}\supset G_m^i} \left\{\mathsf{E}\left(N_n(d) \mid \tilde{G}_m^{i+1}\right)\right\}.$$

Put $\hat{G}_m^{i+1} = \arg\max \mathsf{E}(N_n(d) \mid \tilde{G}_m^{i+1})$, $\bar{G}_m^{i+1} = \arg\min \mathsf{E}(N_n(d) \mid \tilde{G}_m^{i+1})$. We need to estimate the difference $\mathsf{E}(N_n(d) \mid \hat{G}_m^{i+1}) - \mathsf{E}(N_n(d) \mid \bar{G}_m^{i+1})$.

For $i+1 \le t \le n$ put

$$\delta_t^i(d) = \mathsf{E}(N_t(d) \mid \hat{G}_m^{i+1}) - \mathsf{E}(N_t(d) \mid \bar{G}_m^{i+1}).$$

First let us note that for $t \le C_{11}d^2$, then we have $\delta_t^i(d) \le \frac{2mt}{d} \le Md$ for some constant M.

Now we want to prove that $\delta_n^i(d) \le Md$ by induction. Suppose that $n = i+1$. Fix G_m^i. Graphs $\hat{G}_m^{i+1}$ and $\bar{G}_m^{i+1}$ are obtained from the graph G_m^i by adding the vertex $i+1$ and m edges. Therefore $\delta_{i+1}^i(d) \le 2m$.

Now consider t: $i \le t \le n-1$, $t > C_{11}d^2$. Note that

$$\mathsf{E}\left(N_{t+1}(m) \mid G_m^i\right) = \mathsf{E}\left(N_t(m) \mid G_m^i\right)(1-p_t(m)) + 1 + O(1/t)\,,$$

$$\mathsf{E}\left(N_{t+1}(d) \mid G_m^i\right) = \mathsf{E}\left(N_t(d) \mid G_m^i\right)(1-p_t(d)) +$$

$$+\mathsf{E}\left(N_t(d-1)\,|\,G_m^i\right)p_t^1(d-1)+\sum_{j=2}^{m}\mathsf{E}\left(N_t(d-j) \mid G_m^i\right)p_t^j(d-j)+O(1/t),\;\; d \ge m+1\,.$$

We obtained the same equalities in the proof of Theorem 2, see (14)-(15). Replace G_m^i by $\hat{G}_m^i$ or $\bar{G}_m^i$ in these equalities. Substracting the equalities with $\bar{G}_m^i$ from the equalities with $\hat{G}_m^i$ we get (for $d > m$)

$$\delta_{t+1}^i(d) = \delta_t^i(d)\,(1-p_t(d)) + \delta_t^i(d-1)p_t^1(d-1) + O\left(\frac{\mathsf{E}N_t(d)d^2}{t^2}\right) + O\left(\frac{1}{t}\right) =$$
$$= \delta_t^i(d)\,(1-p_t(d)) + \delta_t^i(d-1)p_t^1(d-1) + \theta\left(\frac{C_{12}d}{t}\right)\,. \quad (16)$$

Here we used that $\mathsf{E}N_t(d) = O\left(td^{-1-1/A}+d\right) = O(t/d)$. From this recurrent relation it is easy to obtain by induction that $\delta_n^i(d) \le Md$ for some M.

$$\delta_{t+1}^i(d) \le Md\,(1-p_t(d)) + M(d-1)p_t^1(d-1) + \frac{C_{12}d}{t} \le$$
$$\le Md - \frac{MA(2d-1)}{t} - \frac{MB}{d} + \frac{C_{13}Md^3}{t^2} + \frac{C_{12}d}{t} \le Md$$

for sufficiently large M. This concludes the proof of Theorem 3.

Proof of Theorem 4

Let us give the sketch of the proof of Theorem 4. We can prove this theorem by induction. Note that

$$P_2(n) = \sum_{d=m}^{\infty} N_n(d)\frac{d(d-1)}{2}.$$

Therefore

$$\mathsf{E}P_2(i+1) = \sum_{d=m}^{\infty} \mathsf{E}N_{i+1}(d)\frac{d(d-1)}{2} = \mathsf{E}P_2(i) + \frac{m(m-1)}{2} + \sum_{d=m}^{\infty} \mathsf{E}N_i(d)p_i(d)d \sim$$

$$\sim \mathsf{E}P_2(i) + \frac{m(m-1)}{2} + \sum_{d=m}^{\infty} \frac{(Ad+B)d\mathsf{E}N_i(d)}{i} = \mathsf{E}P_2(i)\left(1+\frac{2A}{i}\right) + \frac{m(m-1)}{2} +$$

$$+ \sum_{d=m}^{\infty} \frac{(A+B)d\mathsf{E}N_i(d)}{i} = \mathsf{E}P_2(i)\left(1+\frac{2A}{i}\right) + 2m(A+B) + \frac{m(m-1)}{2}.$$

So we obtain

$$\mathsf{E}P_2(n) \sim \left(2m(A+B) + \frac{m(m-1)}{2}\right)\sum_{t=1}^{n}\prod_{i=t+1}^{n}\left(1+\frac{2A}{i}\right) \sim$$

$$\sim \left(2m(A+B) + \frac{m(m-1)}{2}\right)\sum_{t=1}^{n}\frac{n^{2A}}{t^{2A}}.$$

If $2A < 1$ then

$$\mathsf{E}P_2(n) \sim \left(2m(A+B) + \frac{m(m-1)}{2}\right)\frac{n}{1-2A}.$$

If $2A = 1$ then

$$\mathsf{E}P_2(n) \sim \left(2m(A+B) + \frac{m(m-1)}{2}\right) n\log(n).$$

If $2A > 1$ then

$$\mathsf{E}P_2(n) = O\left(n^{2A}\right).$$

Note that if $2A \le 1$, then the structure of an arbitrary graph $G_m^{n_0}$ does not affect the asymptotic of $\mathsf{E}P_2(n)$. If $2A > 1$, then $G_m^{n_0}$ affects only the constant in $O\left(n^{2A}\right)$.

We computed the expectation of P_2. One can prove concentration using standard martingale methods, although the proof is not trivial in this case. Here we need the fact that the maximum degree Δ_n grows as n^A, which can be shown using an induction. Let us consider the case $1-2A > 0$. The intuition behind this proof is the following. If we draw an edge to some vertex then this edge increase the expected final degree of this vertex by $(n/i)^A$. Finally, the expected number of P_2 increases by at most $n^A(n/i)^A = n^{2A}/i^A$ (we multiply the number of extra edges by the maximum possible degree of a vertex). Now, the sum of the squares of these values (see $\sum_{i=1}^{n} c_i^2$ in Theorem 7) is of order n^{1+2A}. So, in Azuma's inequality we can take x growing faster than $n^{1/2+A}$. Note that in this case x can be taken smaller than $\mathsf{E}P_2(n)$ which gives concentration. In the case $1 - 2A < 0$ we are not able to get concentration, but it is possible to get asymptotic from Theorem 4.

Solving Linear Systems with Boundary Conditions Using Heat Kernel Pagerank

Fan Chung and Olivia Simpson

Department of Computer Science and Engineering,
University of California, San Diego
La Jolla, CA 92093
{fan,osimpson}@ucsd.edu

Abstract. We present an efficient algorithm for solving linear systems with a boundary condition by computing the Green's function of a connected induced subgraph S of a graph. Different from previous linear solvers, we introduce the method of using the Dirichlet heat kernel pagerank of the induced graph to approximate the solution to diagonally dominant linear systems satisfying given boundary conditions. Our algorithm runs in time $\tilde{O}(1)$, with the assumption that a unit time allows a step in a random walk or a sampling of a specified distribution, where the big-O term depends on the error term and the boundary condition.

Keywords: graph Laplacian, heat kernel, pagerank, symmetric diagonally dominant linear systems, boundary conditions.

1 Introduction

A number of linear systems have been developed which model flow over nodes of a graph with given boundary conditions. A classical example is the case of an electrical network. Flow can be captured by measuring the passage of electrical current between points in the network, and the amount that is injected and removed from the system. Here, the points at which current is measured can be represented by nodes in a graph, and edges are associated to the ease with which current passes between two points. The injection and ejection points can be viewed as the boundary of the system. The total effective resistance of the network can then be evaluated by solving a system of linear equations over the measurement points.

Another example is a decision-making process among a network of agents. Each agent decides on a value, but may be influenced by the decision of other agents in the network. Over time, the goal is to reach consensus among all the agents, in which each agrees on an common value. Agents are represented by nodes, and each node has an associated value. The amount of influence an agent has on a fellow agent is modeled by a weighted edge between the two representative nodes, and the communication dynamics can be modeled by a linear system. In this case, agents which are free of influence can be viewed as the boundary.

A. Bonato, M. Mitzenmacher, and P. Prałat (Eds.): WAW 2013, LNCS 8305, pp. 203–219, 2013.
© Springer International Publishing Switzerland 2013

In both these cases, the linear systems are equations formulated in the graph Laplacian. Spectral properties of the Laplacian are closely related to reachability and the rate of diffusion across nodes in a graph [9]. Indeed, Laplacian systems have been used to concisely characterize qualities such as edge resistance and the influence of communication on edges [33]. In this paper, we will show how systems of linear equations formulated in the graph Laplacian can be effectively solved and approximated by using a diffusion process over the graph. Namely, we simulate a series of random walks and approximate the solution with sufficiently many samples of the resulting distribution.

In practice, two classes of algorithms exist for computing solutions to a linear system. First are iterative methods, such as the conjugate gradient method or the Chebyshev method [20]. Second are preconditioned iterative methods in which an approximation of the matrix is used [5,19,35]. A good preconditioner is one that requires fewer iterations than the original but still achieves a close approximation to the true solution. Both methods approximate the solution by iteratively testing and improving the solution. That is, for a system of the form $Ax = b$, each iteration yields a new vector $\hat{x}$ for which $A\hat{x}$ is closer to b than the previous. The fastest iterative method known is given in [27] which runs in time $O(m \log^{3/2} n \sqrt{\log\log n} \log(\log n/\epsilon))$ in the unit-cost RAM model. However, the current fastest known algorithm for solving symmetric, diagonally dominant (SDD) linear systems is by a combinatorial method of [24] which runs in $O(m \log^2 n \log\log n \log(1/\epsilon))$-time in the unit-cost RAM model.

In this paper, we present a new approach for solving Laplacian linear systems with a boundary condition by computing the Green's function of a connected induced subgraph on a subset S of vertices. (If the induced subgraph on S is not connected, we can then deal with each connected component separately so this assumption on connectivity is not essential.) The Green's function is basically the inverse of the restricted Laplacian on S. Namely, for a given a system $Lx = b$, where $L = D - A$, we consider the subset S consisting of all vertices v with $b(v) = 0$. The goal is to find a solution x so that $Lx(v) = 0$ for $v \in S$ and x satisfies the boundary condition b in the sense that $x(u) = b(u)$ for $u \notin S$. When the induced subgraph S is connected and b is non-trivial, the solution x can be determined by $\Gamma b'$ where Γ is the inverse of L_S, (which is the restriction of L to rows and columns indexed by vertices in S), and $b' \in \mathbb{R}^S$ is determined by b, as detailed in Section 2.1. To compute b' from b, it takes time no more than the size of the edge boundary of S. With b', our algorithm approximates the solution in time $O\left(\frac{(\log s)^2 (\log(1/\epsilon))^2}{\epsilon^5 \log\log(1/\epsilon)}\right) = \tilde{O}(1)$, where ϵ is the error bound and s denotes the size of S. Note that in our computation, we do not intend to compute or approximate the pseudoinverse of L.

The speedup exhibited by our algorithm is due to our fast algorithm for approximating heat kernel pagerank, which dominates the running time. The PageRank was first introduced by Brin and Page [7] in web search algorithms. In [3,6], sublinear time algorithms are given for approximating PageRank by simulating random walks. Our heat kernel pagerank algorithm is related to the PageRank algorithm in the sense that, while PageRank can be interpreted as a geometric sum of random walks, heat kernel pagerank is an exponential sum [10]. Based on random walks, we may expect more rapid convergence for heat kernel pagerank.

1.1 Previous Work

An early result in scientific computing is the approximation algorithm of the preconditioned Chebyshev method. In [20], Golub and Overton show that for a positive, semi-definite matrix A and a preconditioner B, the preconditioned Chebyshev method finds ϵ-accurate solutions to the system $Ax = b$ in time $O(m\sqrt{\kappa_f(A,B)}S(B)\log(\kappa_f(A)/\epsilon))$. Here, $S(B)$ is the time it takes to solve linear systems in B and

$$\kappa_f(A,B) = \left(\max_{x:Ax\neq 0} \frac{x^TAx}{x^TBx}\right)\left(\max_{x:Ax\neq 0} \frac{x^TBx}{x^TAx}\right).$$

This can be pretty good with a clever choice of preconditioner matrix B.

A large work on finding fast linear solvers for systems of equations in the graph Laplacian was presented by Spielman and Teng in 2004 [35]. Spielman and Teng improve the preconditioned (or inexact) Chebyshev method by exploiting the insight of Vaidya that matrices of subgraphs serve as good preconditioners. In the manuscript [37], Vaidya proves that a maximum spanning tree of a matrix A nm-approximates A and that by adding t^2 edges, one can obtain a sparse graph that $O(nm/t^2)$-approximates A. The result of this is an algorithm for solving SDD linear systems with non-positive off-diagonal entries of degree d in time $O((dn)^{1.75}\log(\kappa_f(A)/\epsilon))$, where $\kappa_f(A)$ is the ratio of largest to smallest eigenvalue of A. This is a huge improvement from the previous worst-case $O(nm)$-time bound for the Chebyshev iterative method.

With these tools, Spielman and Teng's major result is an $m\log^{O(1)} n$-time linear solver for SDD matrices where n is the dimension of the matrix and m the number of non-zero entries.

Prior to the contributions of Spielman and Teng, Joshi [23] showed how to recursively apply the results of Vaidya to achieve a $O(n\log(n/\epsilon))$ linear solver. This was improved for planar linear systems by Reif [31] to $O(n^{1+\beta}\log^{O(1)}(\kappa_f(A)/\epsilon))$ for any $\beta > 0$. The above techniques use spanning trees as the preconditioner. Spielman and Teng later improve upon the results of Boman and Hendrickson [4,5], which apply spanning trees to construct ultra-sparsifiers in time $m^{1.31+o(1)}\log(\kappa_f(A)/\epsilon)$ [34].

Koutis et al. [25] give a nearly optimal linear solver for SDD linear systems; an improvement on the Spielman-Teng linear solver. Their algorithm uses an incremental graph sparsification algorithm as a main tool and outputs an approximate vector $\hat{x}$ satisfying $||\hat{x} - A^+b||_A < \epsilon||A^+b||_A$ in time $O(m\log^2 n\log(1/\epsilon))$. Here A^+ denotes the pseudoinverse of A. In [26], the authors improve this bound to $\tilde{O}(m\log n\log(1/\epsilon))^2$. The faster run time results from an improvement in the incremental sparsifier. As mentioned above, the fastest existing iterative method is $O(m\log^{3/2} n\sqrt{\log\log n}\log(\log n/\epsilon))$-time in the unit-RAM model, due to Lee and Sidford [27]. Their improvement is due to an accelerated randomized coordinate descent method which limits the cost per iteration and also achieves faster convergence.

An early result is the Monte Carlo method for solving linear systems [18], in which the inverse of a matrix is computed by translating the system into a random walk model. Our methods use a similar stochastic process and random sampling procedure. In [32], Sachdeva and Vishnoi show that the inverse of a positive semi-definite matrix can be approximated by taking a weighted sum of matrix exponentials. This method is also closely related to ours. However, rather than taking matrix exponentials explicitly, we are adding small contributions of the matrix exponential multiplied with a specified vector. In this way we avoid explicit computation of the matrix inverse and spare a final matrix vector multiplication in computing the solution.

1.2 Our Contributions

In this paper, we consider a direct method for solving systems of Laplacian linear equations with boundary condition by way of the discrete Green's function. Specifically, our contributions are:

1. We introduce a new approach for solving a specific class of SDD linear systems subject to boundary conditions which avoids solution iterations.
2. We present an algorithm for approximating the product of the discrete Green's function of a graph with a specified vector.
3. We make use of an algorithm introduced in [13] for approximating the Dirichlet heat kernel pagerank of a graph in sublinear time.
4. Using contributions (2) and (3), we present an algorithm which takes as input a graph and a boundary conditions, and outputs a vector which is a close approximation of the solution to the SDD linear system with boundary condition.
5. We improve the running time of existing linear solvers with boundary condition. Our solver for approximating the solution for a linear system of the form $Lx = b$, with x satisfying the boundary condition, runs in sublinear time $\tilde{O}(1)$ where the constant depending on the error bound and the boundary vector b. Here we assume a 'unit' time for each basic 'operation' which includes (i) taking a random walk from a specified vertex for a finite, specified number of steps, (ii) sampling a random vertex according to a specified distribution.
6. We present a number of applications for our fast linear solver such as solving for effective resistance, maximum flow, coupled oscillators, and consensus of multi-agent networks.

Our approach is a revisitation to direct methods for solving linear equations with boundary condition, but rather than using algorithms based on Gaussian elimination, we take advantage of diffusion and random walks for computing a heat kernel pagerank vector. The solution to the system is then a sum of random samples of heat kernel pagerank values.

The algorithm presented herein for approximating Dirichlet heat kernel pagerank has a similar flavor to that for approximating PageRank using random walks in [6]. This heat kernel PageRank approximation is an improvement upon several previous results [2,3,14,11]. The Dirichlet heat kernel pagerank algorithm is related to the rate of convergence of exponential random walks and is of independent interest on its own.

1.3 Organization

The remainder of this paper is organized as follows. Preliminaries definitions and basic facts are given in Section 2. We introduce Dirichlet heat kernel pagerank and present some crucial relationships between Dirichlet heat kernel pagerank and the Green's function. Our main result, an efficient linear solver algorithm, is given in Section 3 and the analysis is given in Section 4. The algorithm for approximating heat kernel pagerank is given in Section 5. Finally we discuss some applications for the linear solver in Section 6.

2 Basic Definitions and Facts

In this section we introduce important results of spectral graph theory which are crucial to the analysis of our algorithm.

2.1 The Laplacian and Green's Function

Let G be an undirected graph given by vertex set $V = V(G)$ and edge set $E = E(G)$. (Although we will mainly focus on simple, unweighted undirected graphs, the definitions and results can easily generalized to general weighted graphs.) We say a graph is size n when $|V| = n$. Let A be the indicator adjacency matrix $A \in \{0,1\}^{V \times V}$ for which $A_{uv} = 1$ if and only if $\{u, v\} \in E$. The *degree* of a vertex v is the number of vertices adjacent to it, $d_v = |\{u \in V | A_{uv} = 1\}|$. Let D be the diagonal degree matrix with entries $D_{vv} = d(v)$ on the diagonal and zero entries elsewhere.

The *Laplacian* is defined $L = D - A$. The problem of interest is a fast way to solve systems of linear equations of the form

$$Lx = b \tag{1}$$

while the solution x is required to satisfy the boundary condition b in the sense that $x(v) = b(v)$ for v in the support of b. Let S denote the set

$$S = \{v \in V : b(v) = 0\}.$$

Let $\delta(S)$ denote the *vertex boundary* of S, which is the set of vertices not in S but which are adjacent to a vertex in S.

$$\delta(S) = \{u \in V : \{u, v\} \in E \text{ for some } v \in S\}.$$

The support of b contains $\delta(S)$.

The problem of solving the linear system in (1) with a given boundary condition is to find $x \in \mathbb{R}^V$ such that

$$x(v) = \begin{cases} \frac{1}{d_v} \sum_{u \sim v} x(u) & \text{if } v \in S \\ b(v) & \text{if } v \notin S \end{cases} \tag{2}$$

Here $u \sim v$ denotes $\{u, v\} \in E$. In other words, the solution x is harmonic in S while satisfying the boundary function b. We remark that in this set up, we *do not* place any condition on b. The entries of b can be either positive or negative.

In many applications involving solving systems of linear equations $Lx = b$, the boundary vector b has a relatively small support as in the example of an electrical network. In the graph representation of an electrical network, nodes are points in the network at which current may be injected or removed and at which potentials are measured. For all nodes except the injection and removal points, the in-current matches the out-current. Then, if we inject one unit of current at the point which we call s, and remove one unit of current from the point which we call t, we have boundary conditions at these points. If c is the vector of currents over nodes in the system, we can encompass the boundary conditions in c:

$$c(i) = \begin{cases} 1 & \text{if } v_i = s \\ -1 & \text{if } v_i = t \\ 0 & \text{otherwise} \end{cases}$$

The solution x of the linear system $Lx = c$, with x satisfying the boundary condition c, can then be used to determine the effective resistance of the system. This belongs to one of the classical problems, called the Dirichlet problem, for finding a harmonic function on a set of points satisfying given boundary conditions. There is a large literature on this topic for various domains [17]. The discrete version of the Dirichlet problem can be reformulated as follows:

Let $A_{\delta(S)}$ be the $s \times |\delta(S)|$ matrix by restricting the columns of A to $\delta(S)$ and rows to S. Also, let A_S be the matrix A restricted to rows and columns indexed by S. For a boundary vector $b \in \mathbb{R}^{\delta(S)}$, we wish to find a vector $x \in \mathbb{R}^S$ so that

$$D_S x = A_S x + A_{\delta S} b$$

$$\text{or, equivalently} \qquad x_S = (D_S - A_S)^{-1}(A_{\delta S} b)$$

provided the inverse L_S^{-1} exists and x_S denotes the restriction of x to vertices in S. Indeed, if the induced subgraph on S is connected and the vertex boundary $\delta(S)$ is not empty, then it is known [9] that the Green function $\Gamma = (D_S - A_S)^{-1} = L_S^{-1}$ exists. We can then write

$$x = \Gamma b'$$

where $\Gamma = L_S^{-1}$ and $b' = A_{\delta S} b$.

Therefore we have established the following:

Theorem 1. *In a graph G, suppose b is a nontrivial vector in $\mathbb{R}^V$ and the induced subgraph on $S = V \setminus supp\ b$ is connected. Suppose x is the solution for a linear system $Lx = b$, satisfying the boundary condition b as in (2). Then the restriction of x to S, denoted by x_S, satisfies*

$$x_S = L_S^{-1}(A_{\delta S}b).$$

For given $b \in \mathbb{R}^{\delta(S)}$, the computation of $b' = A_{\delta S}b$ takes time proportional to the *edge boundary* of S, denoted by $\partial(S)$.

$$\partial(S) = \{\{u, v\} \in E : u \in S, v \notin S\}.$$

Hence the computational complexity of solving the linear system with a boundary condition on $\delta(S)$ is the combination of the running time $|\partial(S)|$ for computing b' plus the running time for computing $\Gamma b'$.

2.2 Dirichlet Heat Kernel Pagerank and Green's Function

The method we present for computing the Green's function of the Laplacian works by computing the sum of random samples of heat kernel pagerank vectors. Heat kernel pagerank was originally introduced as a variant of personalized PageRank of a graph [10,11]. Specifically, it is an exponential sum of random walks generated from a starting vector, $f \in \mathbb{R}^n$. As an added benefit, heat kernel pagerank simultaneously satisfies the heat equation with the rate of diffusion controlled by the parameter t,

$$\frac{\partial u}{\partial t} = -\Delta u$$

where $\Delta = I - P$ and $P = D^{-1}A$ denotes the transition probability matrix for a random walk on the graph G. Namely, if v is a neighbor of u, then $P(u, v) = 1/d_u$ is the probability of moving to v from a vertex u. Let S denote a subset of V. We consider $\Delta_S = I_S - P_S$ where, in general, M_S denotes the submatrix of M restricted to rows and columns indexed by vertices in S. For given $t > 0$ and a preference vector $f \in \mathbb{R}^S$ as a probability distribution on S, the *Dirichlet heat kernel pagerank* is defined to be the exponential sum:

$$\rho_{t,f} = f^T H_{S,t} = f^T e^{-t\Delta_S} = e^{-t} \sum_{k=0}^{\infty} \frac{t^k}{k!} f^T P_S^k \tag{3}$$

where f^T denotes the transpose of f. (Here we follow the notation for random walks and pagerank so that a random walk step is by a right multiplication by P.) We also consider the normalized Laplacian $\mathcal{L} = D^{-1/2}LD^{-1/2}$ which is symmetric version of Δ. The *Dirichlet heat kernel* can be expressed by

$$\check{\mathcal{H}}_t = \mathcal{H}_{S,t} = D_S^{1/2} e^{-t\Delta_S} D_S^{-1/2} = e^{-t\mathcal{L}_S}. \tag{4}$$

We note that as long as the graph G has no isolated vertex and D is invertible, the following two problems of solving linear systems with boundary conditions

$$Lx = b, \quad \text{and} \quad \mathcal{L}x_1 = b_1,$$

are equivalent in the sense that we can set $x_1 = D^{1/2}x$ and $b_1 = D^{-1/2}b$. In addition to L_S, we consider Δ_S since our algorithm and analysis rely heavily on random walks. Furthermore, we need $\mathcal{L}_S$, which is equivalent to Δ_S, since we will discuss the spectral decomposition of $\mathcal{L}_S$ and use the L_2-norm.

Similar to Theorem 1, for solving the linear system $\mathcal{L}x = b$ we have the following.

Theorem 2. *In a graph G, suppose b is a nontrivial vector in $\mathbb{R}^V$ and the induced subgraph on $S = V \setminus supp\ b$ is connected. Suppose x is the solution for a linear system $\mathcal{L}x = b$, satisfying the boundary condition b (i.e. $x_{\delta(S)} = b_{\delta(S)}$). Then the restriction of x to S, denoted by x_S, satisfies*

$$x_S = \mathcal{L}_S^{-1}\big(D_S^{-1/2} A_{\delta S} D_{\delta(S)}^{-1/2} b\big).$$

In the remainder of the paper, we assume that the induced subgraph on S is connected and $\delta(S) \neq \emptyset$. The eigenvalues of $\mathcal{L}_S$ are called Dirichlet eigenvalues, denoted by $\lambda_1 \leq \lambda_2 \leq \ldots \leq \lambda_s$. It is easy to check (see [9]) that $0 < \lambda_i \leq 2$. In fact, $\lambda_1 > 1/|S|^3$. We can write

$$\mathcal{L}_S = \sum_{i=1}^{s} \lambda_i P_i$$

where P_i are the projection to the ith orthonormal eigenvectors.

Let $\mathcal{G}$ denote the inverse of $\mathcal{L}_S$. Namely, $\mathcal{G}\mathcal{L}_S = \mathcal{L}_S\mathcal{G} = I_S$. Then

$$\mathcal{G} = \sum_{i=1}^{s} \frac{1}{\lambda_i} P_i. \tag{5}$$

From (5), we see that

$$\frac{1}{2} \leq ||\mathcal{G}|| \leq \frac{1}{\lambda_1}. \tag{6}$$

Then $\mathcal{G}$ can be related to $\breve{\mathcal{H}} = \mathcal{H}_S$ as follows:

Lemma 1. *Let $\mathcal{G}$ be the Green's function of a connected induced subgraph on $S \subset V$ with $s = |S|$. Let $\breve{\mathcal{H}}$ denote the Dirichlet heat kernel with rows and columns restricted to S. Then*

$$\mathcal{G} = \int_0^\infty \breve{\mathcal{H}}_t \ \mathrm{d}t. \tag{7}$$

Proof. By our definition of the heat kernel,

$$\begin{aligned}\int_0^\infty \check{\mathcal{H}}_t \, \mathrm{d}t &= \int_0^\infty \Big(\sum_{i=1}^{s} e^{-t\lambda_i} P_i \Big) \mathrm{d}t \\ &= \sum_{i=1}^{s} \Big(\int_0^\infty e^{-t\lambda_i} \, \mathrm{d}t \Big) P_i \\ &= \sum_{i=1}^{s} \frac{1}{\lambda_i} P_i \\ &= \mathcal{G}.\end{aligned}$$

We note that $\mathcal{G}$ is related to the Green's function Γ by

$$\mathcal{G} = D_S^{1/2} \Gamma D_S^{1/2}. \tag{8}$$

3 An Efficient Linear Solver Algorithm

Suppose we are given a linear system $\mathcal{L}x = b$, where $\mathcal{L}$ is the normalized Laplacian of a given graph, and b is a specified vector indexed by vertices of the graph. We wish to find the solution x which satisfies the boundary condition b. Let S denote the complement of the support of b. Namely, $S = V \setminus \text{supp } b$. Our approximation algorithm for finding the solution of x, restricted to S, satisfying $x = \mathcal{G}b = \int_0^\infty (\check{\mathcal{H}}_t \; b) \; dt$, is based on a series of approximation and sampling. Here we sketch the ideas. First we bound the tails and approximate $\int_0^\infty (\check{\mathcal{H}}_t \; b) \; dt$ by $\int_0^T (\check{\mathcal{H}}_t \; b) \; dt$ for some appropriate T. Then we approximate this integral by a finite Riemann sum $\sum_{j=1}^{N} \check{\mathcal{H}}_{jT/N} b \frac{T}{N}$ for an appropriate N. Then we approximate by sampling the $\check{\mathcal{H}}_t b$ for sufficiently many values of t. To compute $\check{\mathcal{H}}_t b$, we can use the approximate heat kernel pagerank algorithm as in Section 5 . Here we provide the following definition for an ϵ-approximate heat kernel pagerank vector.

Definition 1. *Let S be a subset of vertices in a graph S. Let $f : V \to \mathbb{R}^S$ be a vector and let $\rho_{t,f} = H_{S,t} f$ be the Dirichlet heat kernel pagerank vector over G according to f. Then we say that $\nu \in \mathbb{R}^S$ is an* ϵ-approximate heat kernel pagerank vector *if*

1. *for every node $v \in V$ in the support of ν, $(1 - \epsilon)(\rho_{t,f}[v] - \epsilon) \le \nu[v] \le (1 + \epsilon)\rho_{t,f}[v]$, and*
2. *for every node with $\nu[v] = 0$, it must be that $\rho_{t,f}[v] \le \epsilon$.*

We note that the error bound for approximated heat kernel pagerank here is much stronger than the approximate pagerank in [3] (by a factor of s). The definition here is similar to that in [6] although the scaling is different there.

To compute the ϵ-approximate heat kernel pagerank vectors requires $O\Big(\frac{\log(1/\epsilon)\log s}{\epsilon^3 \log\log(1/\epsilon)}\Big)$ time provided a unit time allows a random walk step or a sampling from a given distribution (see Section 5). The algorithm for estimating the

Green's function is by taking the sum of r samples of an approximate Dirichlet heat kernel pagerank vector $\hat{\rho}_b(t)$ where t is chosen uniformly over the interval $[0,T]$, $r = O(\epsilon^{-2}\log s)$ and $T = O(s^3(\log(1/\epsilon))$. We use ApproxHK of Section 5 for computing $\hat{\rho}_b(t)$ as a subroutine. Therefore the total complexity for the approximate linear solver algorithm takes time $O\Big(\frac{(\log s)^2(\log(1/\epsilon))^2}{\epsilon^5 \log\log(1/\epsilon)}\Big)$ where the solution is approximated within a multiplicative factor of $(1+\epsilon)$.

Here we give the rather short description for computing an approximate solution x for $\mathcal{L}x = b$ or $x = \mathcal{G}b$, satisfying the boundary condition, for a given graph G, a vector b, a set $S = V \setminus \text{supp } b$ and some $\epsilon > 0$ as the error bound.

Algorithm 1. `GreensSolver`(G, b, ϵ)

---

initialize a 0-vector x of dimension s where $s = |S| = |V \setminus \text{ supp } b|$
$b_1 \leftarrow D_S^{-1/2} A_{\delta(S)} D_{\delta(S)}^{-1/2} b$
$T \leftarrow s^3(\log(1/\epsilon))$
$N \leftarrow T/\epsilon$
$r \leftarrow \epsilon^{-2}(\log s + \log(1/\epsilon))$
for $j = 1$ to r **do**
  $x_j \leftarrow$ `ApproxHK`$(G, b_1, S, kT/N, \epsilon)$ where the integer k is chosen uniformly from $[1, N]$
  $x \leftarrow x + x_j$
end for
return x/r

We will prove the following:

Theorem 3. *Let G be a graph and $\mathcal{L}$ be the Laplacian of G. For the linear system $\mathcal{L}x = b$, the solution x is required to satisfy the boundary condition b, (i.e., $x(v) = b(v)$ for $v \in$ supp b). Let $S = V \setminus$ supp b and $s = |S|$. Suppose the induced subgraph on S is connected and supp b is non-empty. The approximate solution $\hat{x}$ as the output of the algorithm* `GreensSolver`(G, b, ϵ) *satisfies the following:*

1. *For any $a > 0$, the absolute error of $\hat{x}$ is*
$$\|x - \hat{x}\| \le O\big(\epsilon(1 + \|b\|)\big)$$
with probability at least $1 - \epsilon$.
2. *The running time of* `GreensSolver` *is $O\Big(\frac{(\log s)^2(\log(1/\epsilon))^2}{\epsilon^5 \log\log(1/\epsilon)}\Big)$ with additional preprocessing time $O(|\partial(S)|)$.*

The term of $O(|\partial(S)|)$ is from the computation of b_1. We can also compute an approximate solution for $Lx = b$ satisfying the boundary condition. The algorithm is almost identical to Algorithm 1 except for setting $b_1 \leftarrow A_{\delta(S)} b$ and using `ApproxHKPR`$(G, b_1, kT/N, \epsilon)$ instead.

4 Analysis of the Green Linear Solver Algorithm

To prove Theorem 3 we first prove a number of Lemmas. A main tool in our analysis is the following matrix concentration inequality (see [12], also various variations in [1], [15], [30], [21], [36]).

Theorem 4. *Let $X_1, X_2, \ldots, X_m$ be independent random $n \times n$ Hermitian matrices. Moreover, assume that $\|X_i - \mathbb{E}(X_i)\| \leq M$ for all i, and put $v^2 = \|\sum var(X_i)\|$. Let $X = \sum X_i$. Then for any $a > 0$,*

$$\mathbb{P}(\|X - \mathbb{E}(X)\| > a) \leq 2n \exp\left(-\frac{a^2}{2v^2 + 2Ma/3}\right).$$

Theorem 5. *Let G be a graph and $\mathcal{L}$ be the Laplacian of G. Let b denote a vector and for $S = V \setminus supp\ b$, suppose the induced subgraph on S is connected and supp b is non-empty. Then the solution to the linear system $\mathcal{L}x = b$, satisfying the boundary condition, can be computed by sampling $\check{\mathcal{H}}_t b$ for $r = \frac{\log s + \log(1/\epsilon)}{\epsilon^2}$ values and the solution is within a multiplicative factor of $1 + \epsilon$ of the exact solution $x = \mathcal{G}b$ with probability at least $1 - \epsilon$.*

Proof. By Lemma 1, the exact solution is $x = \mathcal{G}b = \int_0^\infty \check{\mathcal{H}}_t b\ \mathrm{d}t$. First, we see that:

$$\begin{aligned} ||\check{\mathcal{H}}_t|| &= ||\sum_i e^{-t\lambda_i} P_i|| \\ &\leq e^{-t\lambda_1} \cdot ||\sum_i P_i|| \\ &= e^{-t\lambda_1} \end{aligned} \tag{9}$$

where λ_i are Dirichlet eigenvalues for the induced subgraph S. So the error incured by taking a definite integral up to $t = T$ is the difference

$$||\int_T^\infty \check{\mathcal{H}}_t\ \mathrm{d}t|| \leq \int_T^\infty e^{-t\lambda_1}\ \mathrm{d}t = \frac{1}{\lambda_1} e^{-T\lambda_1}.$$

It is known that for a general graph on n vertices, $s^{-3} \leq \lambda_1 \leq 1$ (see [9], for example). With this, we restrict the error to ϵ by setting $T = s^3 \log(1/\epsilon)$. Namely,

$$||\mathcal{G} - \int_0^T \check{\mathcal{H}}_t\ \mathrm{d}t|| \leq \epsilon\ ||b||$$

Next, we approximate the definite integral in $[0, T]$ by discretizing it. That is, we divide the interval $[0, T]$ into N intervals of size T/N, and a finite Riemann sum is close to the definite integral:

$$||\int_0^T \check{\mathcal{H}}_t \, \mathrm{d}t - \sum_{j=1}^{N} \check{\mathcal{H}}_{jT/N} \cdot \frac{T}{N}|| \leq \epsilon \, ||\int_0^T \check{\mathcal{H}}_t \, \mathrm{d}t||.$$

for a given ϵ, by choosing choose $N = T/\epsilon = s^3 \log(1/\epsilon)/\epsilon$ so that $T/N \leq \epsilon$. Thus we have

$$||\mathcal{G} - \sum_{j=1}^{N} \check{\mathcal{H}}_{jT/N} \cdot \frac{T}{N}|| \leq 2\epsilon.$$

Let X_i be a random variable which takes on value $T\check{\mathcal{H}}_{jT/N}$ with each $j \in [1, N]$ with probability $1/N$. We consider $X = \sum_{j=1}^{r} X_j$. Then we evaluate the expected value and variance of X as follows:

$$\mathbb{E}(X) = r\mathbb{E}(X_j) = \frac{rT}{N} \sum_{j=1}^{N} \check{\mathcal{H}}_{jT/N}$$

$$\mathrm{Var}(X) = r\mathrm{Var}(X_j) = \frac{rT}{N} \sum_{j=1}^{N} \check{\mathcal{H}}_{2jT/N}$$

Furthermore,

$$||\mathbb{E}(X) - r\mathcal{G}|| \leq 2r\epsilon$$

$$||\mathrm{Var}(X)|| = r||\mathrm{Var}(X_j)|| \leq \frac{r}{2}||\mathcal{G}||$$

We can now apply Theorem 4. By using the above bounds for $||\mathbb{E}(X)||$ and $||\mathrm{Var}(X)||$ as well as the bound for $\mathcal{G}$ in (6), we have

$$\begin{aligned} \mathbb{P}\big(||X - \mathbb{E}(X)|| \geq \epsilon||\mathbb{E}(X)||\big) &\leq 2se^{-\frac{\epsilon^2||\mathbb{E}(X)||^2}{2\mathrm{Var}(X)+\frac{2\epsilon||\mathbb{E}(X)||M}{3}}} \\ &\leq 2se^{-\frac{\epsilon^2 r^2||\mathbb{E}(X_j)||}{r+2\epsilon M/3}} \\ &\leq 2se^{-\frac{\epsilon^2 r^2||\mathcal{G}||}{r+2\epsilon||\mathcal{H}_t||/3}} \\ &\leq 2se^{-\frac{\epsilon^2 r}{2}} \end{aligned} \tag{10}$$

Therefore we have

$$\mathbb{P}\big(||X - \mathbb{E}(X)|| \geq \epsilon||\mathbb{E}(X)||\big) \leq \epsilon$$

if we choose $r \geq \frac{\log s + \log(1/\epsilon)}{\epsilon^2}$. This completes the proof of Theorem 5.

Proof (Proof of Theorem 3). The accuracy of the linear solver algorithm GreensSolver(G, b, ϵ) follows from Theorem 5. For the running time, we rely on Theorem 6. The algorithm makes r calls to ApproxHK, giving a total running time of

$$r \cdot O\left(\frac{\log(1/\epsilon)\log s}{\epsilon^3 \log\log(1/\epsilon)}\right) = O\left(\frac{(\log s)^2(\log(1/\epsilon))^2}{\epsilon^5 \log\log(1/\epsilon)}\right),$$

as claimed.

5 Heat Kernel Pagerank Approximation Algorithm

We first focus on the algorithm ApproxHKPR for approximating a Dirichlet heat kernel pagerank vector for a proper subset S of V in a graph G, using on a preference vector $f \in \mathbb{R}^S$. Here we impose an additional condition that f is a probabilistic function and the induced subgraph on S is connected. The algorithm outputs an ϵ-approximate heat kernel pagerank vector which is denoted by $\hat{\rho}_{t,f}$. The running time of the algorithm is $O\left(\frac{\log(1/\epsilon)\log s}{\epsilon^3 \log\log(1/\epsilon)}\right)$.

The algorithm essentially works by taking a finite sum of truncated random walks. The method and complexity is quite similar to the ApproxRow pagerank algorithm given in [6]. Details of the proof and analysis can be found in [13].

Algorithm 2. ApproxHKPR(G, f, S, t, ϵ)

initialize 0-vector ρ of dimension s, where $s = |S|$.
$r \leftarrow \frac{16}{\epsilon^3}\log s$
$K \leftarrow \frac{\log(1/\epsilon)}{\log\log(1/\epsilon)}$
for r iterations **do**
  choose a starting node v according to the distribution vector f.
  Start
    simulate a $P_S = D_S^{-1}A_S$ random walk where k steps are taken with probability $e^{-t}\frac{t^k}{k!}$ where $k \leq K$, let u be the last node visited in the walk
    $\rho[u] \leftarrow \rho[u] + 1$
  End
end for
return $1/r \cdot \rho$

Theorem 6. *Suppose S is a proper vertex subset in a graph G with $s = |S|$ and the induced subgraph on S is connected. Let f be a preference vector, $f : S \to \mathbb{R}$, $t \in \mathbb{R}$, and $0 < \epsilon < 1$. Then,* ***ApproxHKPR***(G, f, S, t, ϵ) *outputs an ϵ-approximate Dirichlet heat kernel pagerank vector $\hat{\rho}_{f,t}$ with probability at least $1 - o(1)$ and the running time of* ***ApproxHKPR***(G, f, S, t, ϵ) *is* $O\left(\frac{\log(1/\epsilon)\log s}{\epsilon^3 \log\log(1/\epsilon)}\right)$.

Since $\rho_{t,f} = f^T D_S^{-1/2}\mathcal{H}_t D_S^{1/2}$, we modify the above algorithm to approximate $\check{\mathcal{H}}_t g$ for a vector $g \in \mathbb{R}^S$.

Algorithm 3. `ApproxHK`(G, g, S, t, ϵ)

initialize 0-vector y of dimension s, where $s = |S|$.
$g_+ \leftarrow$ the positive portion of g
$g_- \leftarrow$ the negative portion of g so that $g = g_+ - g_-$
$h_1 \leftarrow \|D^{1/2}g_+\|_1$ and $h_2 \leftarrow \|D^{1/2}g_-\|_1$, where $\|\cdot\|$ indicates the L_1-norm.
$r \leftarrow \frac{16}{\epsilon^3}\log s$
$K \leftarrow \frac{\log(1/\epsilon)}{\log\log(1/\epsilon)}$
for r iterations **do**
  choose a starting node v_1 according to the distribution vector $f_+ = D^{1/2}g_+/h_1$.
  Start
    simulate a $P_S = D_S^{-1}A_S$ random walk where k steps are taken with probability $e^{-t}\frac{t^k}{k!}$ where $k \le K$, let u_1 be the last node visited in the walk
    $y[u_1] \leftarrow y[u_1] + h_1/\sqrt{d_{u_1}}$
  End
  choose a starting node v_2 according to the distribution vector $f_- = D^{1/2}g_-/h_2$.
  Start
    simulate a $P_S = D_S^{-1}A_S$ random walk where k steps are taken with probability $e^{-t}\frac{t^k}{k!}$ where $k \le K$, let u_2 be the last node visited in the walk
    $y[u_2] \leftarrow y[u_2] - h_2/\sqrt{d_{u_2}}$
  End
end for
return $1/r \cdot y$

The accuracy of the error estimate for the algorithm `ApproxHK`(G, b, S, t, ϵ) follows immediately from that of algorithm `ApproxHKPR`(G, f, S, t, ϵ) and Theorem 6.

Theorem 7. *Suppose S is a proper vertex subset in a graph G with $s = |S|$, and the induced subgraph on S is connected. Let g be a vector, $g : S \to \mathbb{R}$, $t \in \mathbb{R}$, and $0 < \epsilon < 1$. The algorithm ApproxHK(G, g, S, t, ϵ) outputs an ϵ-approximate vector $\check{\mathcal{H}}_t g$ with probability at least $1-2\epsilon$ and the running time of ApproxHK(G, g, S, t, ϵ) is* $O\Big(\frac{\log(1/\epsilon)\log s}{\epsilon^3 \log\log(1/\epsilon)}\Big)$.

6 Applications

The contributions of this paper have numerous applications. We outline a few areas in which the graph Laplacian arises in linear systems. For scenarios in which the associated networks can be very large, the efficiency of our linear solver is particularly useful.

Effective resistance. Consider a graph as a model of a network of electrical transistors. The vertices correspond to points at which potential is measured, and edges correspond to resistors with weights equal to the inverse of resistance. The *effective resistance* between two vertices v_i, v_j is the difference in potential of v_i, v_j required for one unit of current to flow from v_i to v_j. To measure effective resistance, we require that the in-current of a vertex matches the out-current, which the exception of the injection point and the charge removal point. If $\boldsymbol{p} \in \mathbb{R}^n$ is the vector of potentials and $\boldsymbol{c} \in \mathbb{R}^n$ is the vector of currents, then these values satisfy the equation

$$\boldsymbol{c} = L\boldsymbol{p}.$$

Then effective resistance is the difference of $p(i)$ and $p(j)$ in the solution vector p for the above equation satisfied by the currents

$$c(x) = \begin{cases} 1 & \text{if } x = x_i \\ -1 & \text{if } x = x_j \\ 0 & \text{otherwise} \end{cases}$$

In [8] this problem has been studied in the form of electric networks and flows.

Maximum flow by interior point methods. Maximum flow and minimum cost algorithms for network flows can be solved with an iterative interior point algorithm. Each iteration requires solving a system of linear equations, and this typically dominates the running time. In many applications it is observed that these linear equations can be reduced to restricted Laplacian systems [16]. In one case, Frangioni and Gentile [19] present Prim-based heuristics for generating a support-graph preconditioner for solving linear systems. However, this linear program which iteratively use solutions to linear systems can be improved with our fast direct Laplacian linear solver.

Coupled oscillators. Oscillation is the repetitive variation between states. When the behavior of one oscillator affects that of others, we may reduce the degrees of freedom in variation by coupling these oscillations in the model. These oscillators and the couplings can be described in a network. One question is how the structure or topology of the network affects the synchronization of these oscillators. In [22], it is shown that the Laplacian is closely related to synchronization properties of the oscillator network.

Consider a network of n oscillators symmetrically coupled, modeled as a simple, undirected graph. Let x_i be the oscillator state vector at node i. Say $F(x)$ is a function which determines the uncoupled oscillator dynamics of each node, and $H(x)$ specifies the coupling of the vector fields. Then the equations of motion for the oscillator state vector x_i at each node i are linearly described by:

$$\frac{\mathrm{d}x_i}{\mathrm{d}t} = F(x_i) - \sigma \sum_{j=1}^{n} L_{ij} H(x_j),$$

where σ is the coupling strength. Solutions to the above equation are said to be synchronized if $x_i(t) = x_j(t)$ for all nodes i and j in the network.

Consensus of mult-agent networks. In a network of dynamic agents which have a communication topology represented by a weighted graph, a group consensus value can be modeled by a differential equation in the graph Laplacian (see [29], [28]). If each node in a graph represents an agent and x is a vector of decision values over the agents, then the value $x(i) - x(j)$ measures the level of disagreement among agents i and j. The goal of consensus is to minimize disagreement among agents, which is achieved by minimizing the quadratic form. If the network of integrator agents applies the distributed linear protocol given by

$$u_i = \sum_{j \in N_i} (x_j - x_i)$$

with each agent governed by dynamics $\dot{x}_i = u_i$, then the evolution of the dynamic system is the solution to the differential equation

$$\dot{x} = -Lx.$$

Further, the rate of convergence is given in terms of the second eigenvalue of L.

References

1. Ahlswede, R., Winter, A.: Strong converse for identification via quantum channels. IEEE Trans. Inform. Theory 48(3), 569–579 (2002)
2. Andersen, R., Chung, F.: Detecting sharp drops in pageRank and a simplified local partitioning algorithm. In: Cai, J.-Y., Cooper, S.B., Zhu, H. (eds.) TAMC 2007. LNCS, vol. 4484, pp. 1–12. Springer, Heidelberg (2007)
3. Andersen, R., Chung, F., Lang, K.: Local graph partitioning using pagerank vectors. In: FOCS, pp. 475–486 (2006)
4. Boman, E.G., Hendrickson, B.: On spanning tree preconditioners, Sandia National Laboratories (2001) (manuscript)
5. Boman, E.G., Hendrickson, B.: Support theory for preconditioning. SIAM Journal on Matrix Analysis and Applications 25(3), 694–717 (2003)
6. Borgs, C., Brautbar, M., Chayes, J., Teng, S.-H.: A sublinear time algorithm for pagerank computations. In: Bonato, A., Janssen, J. (eds.) WAW 2012. LNCS, vol. 7323, pp. 41–53. Springer, Heidelberg (2012)
7. Brin, S., Page, L.: The anatomy of a large-scale hypertextual web search engine. Computer Networks and ISDN Systems, 107–117 (1998)
8. Christiano, P., Kelner, J.A., Madry, A., Spielman, D.A., Teng, S.-H.: Electrical flows, laplacian systems, and faster approximation of maximum flow in undirected graphs. In: STOC, pp. 273–282 (2011)
9. Chung, F.: Spectral graph theory. American Mathematical Society (1997)
10. Chung, F.: The heat kernel as the pagerank of a graph. Proceedings of the National Academy of Sciences 104(50), 19735–19740 (2007)
11. Chung, F.: A local graph partitioning algorithm using heat kernel pagerank. Internet Mathematics 6(3), 315–330 (2009)
12. Chung, F., Radcliffe, M.: On the spectra of general random graphs. The Electronic Journal of Combinatorics 18, P215 (2011)
13. Chung, F., Simpson, O.: Local graph partitioning using heat kernel pagerank, `http://cseweb.ucsd.edu/~osimpson/research.html`
14. Chung, F., Zhao, W.: A sharp pagerank algorithm with applications to edge ranking and graph sparsification. In: Kumar, R., Sivakumar, D. (eds.) WAW 2010. LNCS, vol. 6516, pp. 2–14. Springer, Heidelberg (2010)

15. Cristofides, D., Markström, K.: Expansion properties of random cayley graphs and vertex transitive graphs via matrix martingales. Random Structures Algs. 32(8), 88–100 (2008)
16. Daitch, S.I., Spielman, D.A.: Faster approximate lossy generalized flow via interior point algorithms. In: STOC, pp. 451–460 (2008)
17. Doyle, P.G., Snell, J.L.: Random walks and electric networks, vol. 22. Math. Ass. of America (1984)
18. Forsythe, G.E., Leibler, R.A.: Matrix inversion by a monte carlo method. Mathematical Tables and Other Aids to Computation 4(31), 127–129 (1950)
19. Frangioni, A., Gentile, C.: Prim-based support-graph preconditioners for min-cost flow problems. Computational Optimization and Applications 36(2-3), 271–287 (2007)
20. Golub, G.H., Overton, M.L.: The convergence of inexact chebyshev and richardson iterative methods for solving linear systems. Numerische Mathematik 53(5), 571–593 (1988)
21. Gross, D.: Recovering low-rank matrices from few coefficients in any basis. IEEE Trans. Inform. Theory 57, 1548–1566 (2011)
22. Hagberg, A., Schult, D.A.: Rewiring networks for synchronization. Chaos: An Interdisciplinary Journal of Nonlinear Science 18(3), 037105 (2008)
23. Joshi, A.: Topics in optimization and sparse linear systems. PhD Thesis (1996)
24. Kelner, J.A., Orecchia, L., Sidford, A., Zhu, Z.A.: A simple, combinatorial algorithm for solving sdd systems in nearly-linear time. In: STOC, pp. 911–920 (2013)
25. Koutis, I., Miller, G.L., Peng, R.: Approaching optimality for solving sdd linear systems. In: FOCS, pp. 235–244 (2010)
26. Koutis, I., Miller, G.L., Peng, R.: A nearly-m log n time solver for sdd linear systems. In: FOCS, pp. 590–598 (2011)
27. Lee, Y.T., Sidford, A.: Efficient accelerated coordinate descent methods and faster algorithms for solving linear systems. In: FOCS (2013)
28. Olfati-Saber, R., Fax, J.A., Murray, R.M.: Consensus and cooperation in networked multi-agent systems. Proceedings of the IEEE 95(1), 215–233 (2007)
29. Olfati-Saber, R., Murray, R.M.: Consensus protocols for networks of dynamic agents. In: Proceedings of the American Control Conference 2003, vol. 2, pp. 951–956 (2003)
30. Oliveira, R.I.: Concentration of the adjacency matrix and of the laplacian in random graphs with independent edges, arXiv preprint arXiv:0911.0600 (2009)
31. Reif, J.H.: Efficient approximate solution of sparse linear systems. Computers & Mathematics with Applications 36(9), 37–58 (1998)
32. Sachdeva, S., Vishnoi, N.K.: Matrix inversion is as easy as exponentiation, arXiv preprint arXiv:1305.0526 (2013)
33. Spielman, D.A.: Algorithms, graph theory, and linear equations in laplacian matrices. In: Proceedings of the International Congress of Mathematicians, vol. 4, pp. 2698–2722 (2010)
34. Spielman, D.A., Teng, S.-H.: Solving sparse, symmetric, diagonally-dominant linear systems in time $o(m^{1.31})$. In: FOCS, pp. 416–427 (2003)
35. Spielman, D.A., Teng, S.-H.: Nearly-linear time algorithms for graph partitioning, graph sparsification, and solving linear systems. In: STOC, pp. 81–90 (2004)
36. Tropp, J.A.: User-friendly tail bounds for sums of random matrices. Foundations of Computational Mathematics 12(4), 389–434 (2012)
37. Vaidya, P.M.: Solving linear equations with symmetric diagonally dominant matrices by constructing good preconditioners. A talk based on this manuscript 2(3.4), 2–4 (1991)

Anarchy Is Free in Network Creation*

Ronald Graham[1], Linus Hamilton[2], Ariel Levavi[1], and Po-Shen Loh[2]

[1] Department of Computer Science and Engineering, University of California, San Diego, La Jolla, CA 92093
[2] Department of Mathematical Sciences, Carnegie Mellon University, Pittsburgh, PA 15213

Abstract. The Internet has emerged as perhaps the most important network in modern computing, but rather miraculously, it was created through the individual actions of a multitude of agents rather than by a central planning authority. This motivates the game theoretic study of network formation, and our paper considers one of the most-well studied models, originally proposed by Fabrikant et al. In it, each of n agents corresponds to a vertex, which can create edges to other vertices at a cost of α each, for some parameter α. Every edge can be freely used by every vertex, regardless of who paid the creation cost. To reflect the desire to be close to other vertices, each agent's cost function is further augmented by the sum total of all (graph theoretic) distances to all other vertices.

Previous research proved that for many regimes of the (α, n) parameter space, the total social cost (sum of all agents' costs) of every Nash equilibrium is bounded by at most a constant multiple of the optimal social cost. In algorithmic game theoretic nomenclature, this approximation ratio is called the price of anarchy. In our paper, we significantly sharpen some of those results, proving that for all constant non-integral $\alpha > 2$, the price of anarchy is in fact $1 + o(1)$, i.e., not only is it bounded by a constant, but it tends to 1 as $n \to \infty$. For constant integral $\alpha \geq 2$, we show that the price of anarchy is bounded away from 1. We provide quantitative estimates on the rates of convergence for both results.

Keywords: Network creation, price of anarchy, algorithmic game theory.

1 Introduction

Networks are of fundamental importance in modern computing, and substantial research has been invested in network design and optimization. However, one of the most significant networks, the Internet, was not created "top-down" by a central planning authority. Instead, it was constructed through the cumulative actions of countless agents, many of whom built connections to optimize their individual objectives. To understand the dynamics of the resulting system, and to answer the important question of how much inefficiency is introduced through

* Research supported by NSF grants DMS-1201380 and DMS-1041500, an NSA Young Investigators Grant and a USA-Israel BSF Grant.

A. Bonato, M. Mitzenmacher, and P. Prałat (Eds.): WAW 2013, LNCS 8305, pp. 220–231, 2013.
© Springer International Publishing Switzerland 2013

the selfish actions of the agents, it is therefore natural to study it through the lens of game theory.

In this paper, we focus on a well-studied game-theoretic model of network creation, which was formulated by Fabrikant et al. in [1]. There are n agents, each corresponding to a vertex. They form a network (graph) by laying down connections (edges) between pairs of vertices. For this, each agent v has an individual strategy, which consists of a subset S_v of the rest of the vertices that it will connect to. The resulting network is the disjoint union of all (undirected) edges between vertices v and vertices in their S_v. Note that in this formulation, an edge may appear twice, if v lays a connection to w and w lays a connection to v. Let α be an arbitrary real parameter, which represents the cost of making a connection. In order to incorporate each agent's desire to be near other vertices, the total cost to each agent is defined to be:

$$\text{cost}(v) = \alpha|S_v| + \sum_w \text{dist}(v, w),$$

where the sum is over all vertices in the graph, and $\text{dist}(v, w)$ is the number of edges in the shortest path between v and w in the graph, or infinity if v and w are disconnected. The *social cost* is defined as the total of the individual costs incurred by each agent. This cost function summarizes the fact that v must pay the construction cost for the connections that it initiates, but v also prefers to be graph-theoretically close to the other nodes in the network. This model also encapsulates the fact that, just as in the Internet, once a connection is made, it can be shared by all agents regardless of who paid the construction cost.

The application of approaches from algorithmic game theory to the study of networks is not new. The works [2–7] all consider network design issues such as load balancing, routing, etc. Numerous papers, including [8–14] and the surveys [15, 16], have considered network formation itself, by formulating and studying network creation games. From a game-theoretic perspective, a (pure) *Nash equilibrium* is a tuple of deterministic strategies S_v (one per agent) under which no individual agent can strictly reduce its cost by unilaterally changing its strategy assuming all other agents maintain their strategies. If every unilateral deviation strictly increases the deviating agent's cost, then the Nash equilibrium is *strict*.

To quantify the cumulative losses incurred by the lack of coordination, the key ratio is called the *price of anarchy*, a term coined by Koutsoupias and Papadimitriou [17]. It is defined as the maximum social cost incurred by any Nash equilibrium, divided by the minimum possible social cost incurred by any tuple of strategies. Note that the minimizer, also known as the *social optimum*, is not necessarily a Nash equilibrium itself. The central questions in this area are thus to understand the price of anarchy, and to characterize the Nash equilibria.

1.1 Previous Work

To streamline our discussion, we will represent a tuple of strategies with a directed graph, whose underlying undirected graph is the resulting network, and where each edge vw is oriented from v to w if it was constructed by v's strategy ($w \in S_v$). This is well-defined because it is clear that the social optimum and all Nash equilibria will avoid multiple edges, and so each edge is either not present at all, or present with a single orientation.

The problem is trivial for $\alpha < 1$, because all Nash equilibria produce complete graphs, as does the social optimum, and therefore the price of anarchy is 1 in this range. For $\alpha \geq 1$, a new Nash equilibrium arises: the star with all edges oriented away from the central vertex. Indeed, the central vertex has no incentive to disconnect any of the edges which it constructed, as its individual cost function would rise to infinity, and no other vertex has incentive to add more connections, because a new connection would cost an additional $\alpha \geq 1$, and reduce at most one of the pairwise distances by 1. Yet, as observed in the original paper of Fabrikant et al. [1], when $\alpha < 2$, the social optimum is a clique, and they calculate the price of anarchy to be $\frac{4}{2+\alpha} + o(1)$, where the error term tends to 0 as $n \to \infty$. This ranges from $\frac{4}{3}$ to 1 as α varies in that interval.

For $\alpha \geq 2$, the social optimum is the star. Various bounds on the price of anarchy were achieved, with particular interest in constant bounds, which were derived in many ranges of the parameter space. From the point of view of approximation algorithms, these show that in those ranges of α, the Nash equilibria that arise from the framework of selfish agents still are able to approximate the optimal social cost to within a constant factor. The current best bounds are summarized in Table 1.

Table 1. Previous upper bounds on the price of anarchy. The last bound above is due to Mihalák et al. [14], and the other bounds are due to Demaine et al. [13].

Regime	Upper bound on price of anarchy
General α	$2^{O(\sqrt{\log n})}$
$2 \leq \alpha < \sqrt[3]{n/2}$	4
$\sqrt[3]{n/2} \leq \alpha < \sqrt{n/2}$	6
$\alpha = O(n^{1-\epsilon})$	$O(1)$
$\alpha > 273n$	$O(1)$

1.2 Our Contribution

Much work had been done to achieve constant upper bounds on the price of anarchy in various regimes of α, because those imply the satisfying conclusion that selfish agents fare at most a constant factor worse than optimally coordinated agents. Perhaps surprisingly (or perhaps reassuringly), it turns out that the price of anarchy is actually $1 + o(1)$ for most constant values of α. In other words, the lack of coordination has negligible effect on the social cost as n grows.

Theorem 1. *For non-integral $\alpha > 2$, and $n > \alpha^3$, the price of anarchy is at most*

$$1 + \frac{150\alpha^6}{(\alpha - \lfloor \alpha \rfloor)^2} \sqrt{\frac{\log n}{n}} = 1 + o(1) .$$

On the other hand, for each integer $\alpha \geq 2$, the price of anarchy is at least

$$\frac{3}{2} - \frac{3}{4\alpha} + o(1),$$

and it is achieved by the following construction. Start with an arbitrary orientation of the complete graph on k vertices. For each vertex v of the complete graph, add $\alpha - 1$ new vertices, each with a single edge oriented from v.

2 Proof for Non-integral α

Suppose that $\alpha > 2$ is fixed, and is not an integer. Assume that we are given a Nash equilibrium. In this section, we prove that its total social cost is bounded by $1 + o(1)$ times the social optimum, as stated in Theorem 1. Throughout this proof, we impose a structure on the graph as follows: select a vertex v, and partition the remainder of the graph into sets based on their distance from v. Let N_1 denote the set of vertices at distance 1 from v, let N_2 denote the set of vertices at distance 2 from v, etc., as diagrammed in Figure 1. Since the graph in every Nash equilibrium is obviously connected, every vertex falls into one of these sets.

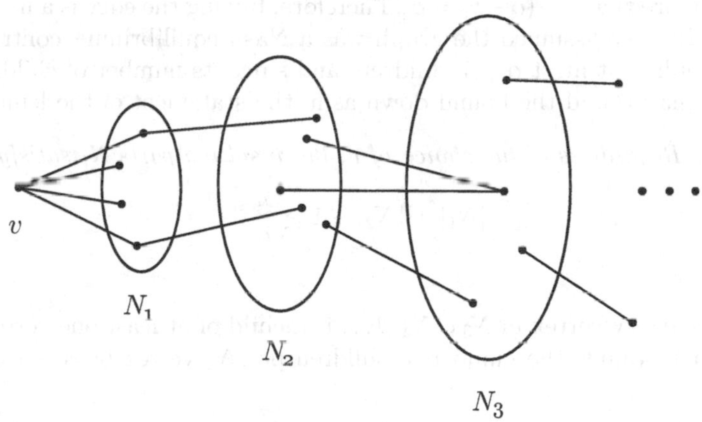

Fig. 1. Partitioning the graph into sets

Consider any vertex $v_i \in N_i$ where $i \geq 3$. Since the graph is connected, we can always find a path $v_i v_{i-1} v_{i-2} \ldots v_2 v_1 v$, where $v_j \in N_j$ for all $1 \leq j \leq i$. In this case, we will call v_i a *child* of v_2. (Note that v_i may be a child of more than one vertex, but is always a child of at least one vertex.) This is diagrammed in Figure 2.

Lemma 1. *No matter which vertex is used as v to construct the vertex partition, every vertex in N_2 has at most $\lfloor \alpha - 1 \rfloor$ children.*

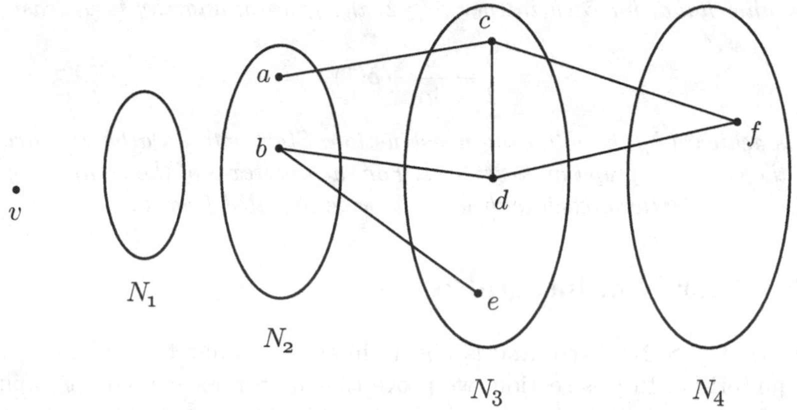

Fig. 2. Here, d and e are children of b; c is a child of a, but not a child of b; and f is a child of both a and b

Proof. Suppose $w \in N_2$ has more than $\alpha - 1$ children. Consider what happens if v buys an edge to w. Although v pays α for the edge, it gets one step closer to w and all of its children, and so the distance component of v's cost function reduces by more than $1+(\alpha-1) = \alpha$. Therefore, buying the edge is a net positive gain for v. But we assumed the graph was a Nash equilibrium—contradiction. Therefore, w has at most $\alpha - 1$ children, and since its number of children is an integer, we may round the bound down as in the statement of the lemma.

Lemma 2. *Regardless of the choice of v, the resulting parts N_i satisfy:*

$$|N_1| + |N_2| + 1 \geq \frac{n}{\alpha}.$$

Proof. Since every vertex in $N_3 \cup N_4 \cup \ldots$ is a child of at least one vertex of N_2, but Lemma 1 bounds the number of children per N_2-vertex by $\alpha - 1$, we must have

$$(\alpha - 1)|N_2| \geq |N_3 \cup N_4 \cup \ldots| = (n - 1 - |N_1| - |N_2|)$$
$$\alpha|N_2| + |N_1| + 1 \geq n,$$

which implies the desired result.

Lemma 3. *If x has degree at least α, then every vertex is at most distance 3 from it.*

Proof. If some vertex w is distance at least 4 from x, then w can buy an edge to x. Vertex w will pay α for the edge, and get 3 steps closer to x, as well as at least 1 step closer to all of x's immediate neighbors, for a net gain. Hence this cannot appear in a Nash equilibrium.

Corollary 1. *If n is sufficiently large ($n > \alpha^3$), then the graph has diameter at most 4.*

Proof. Consider an arbitrary pair of vertices v, w. Lemma 2 implies that for n sufficiently large ($n > \alpha^3$ suffices), either v has degree at least α, or one of v's neighbors has degree at least α. In either case, we can travel from v to a vertex with degree at least α in at most one step, and then by Lemma 3, travel to w in at most 3 more steps. Therefore, v and w are at distance at most 4.

Remark. From now on, we will assume $n > \alpha^3$, and so for any initial choice of v, the resulting partition will only have N_1, N_2, N_3, and N_4.

Lemma 4. *Consider the partition constructed from an arbitrary initial vertex v. Select any $w \in N_2$, and let d be the number of edges w pays for which connect to other vertices in N_2. Then $d \leq |N_1| \cdot \frac{\alpha}{\alpha - \lfloor \alpha \rfloor}$.*

Proof. Consider the following strategy for w: disconnect those d edges, and instead connect to every vertex in N_1. We will carefully tally up the potential gain for this amendment.

- *Paying for edges:* w **saves at least $(d - |N_1|)\alpha$** in terms of paying for edges. (The "at least" is because w might already be connected to some vertices in N_1.)
- *Connectedness to v and N_1:* w obviously can't get farther away from v or any vertices in N_1.
- *Connectedness within N_2:* w gets farther away from all d vertices it disconnected from, but remains at distance 2 from all of N_2, since every vertex in N_2 is connected to some vertex in N_1. This results in a maximum **increased cost of d** in terms of distances to other vertices within N_2.
- *Connectedness to N_3 and N_4:* When disconnecting from a vertex $x \in N_2$, w might get farther away from all of x's children in N_3 and N_4. However, remember that w is still distance 2 from all of N_2. Hence, w is still distance 3 from all of N_3 and distance 4 from all of N_4. Therefore, w can only get 1 step farther from x's children, and doesn't get any farther from vertices in N_3 and N_4 that aren't x's children. By Lemma 1, every N_2-vertex has at most $\lfloor \alpha - 1 \rfloor$ children. Therefore, in disconnecting from d vertices, w gets 1 step farther from at most $d\lfloor \alpha - 1 \rfloor$ vertices in N_3 and N_4, for a **cost increase of at most $d\lfloor \alpha - 1 \rfloor$**.

Adding, w's net cost savings total to at least $(d - |N_1|)\alpha - d - d\lfloor \alpha - 1 \rfloor$, which must be ≤ 0 since we are at a Nash equilibrium. Rearranging, $d \leq |N_1| \cdot \frac{\alpha}{\alpha - \lfloor \alpha \rfloor}$, as desired.

Lemma 5. *If $|N_1|$ is $o(n)$, then so is $|N_3 \cup N_4|$. Quantitatively, $|N_3 \cup N_4| < |N_1| \cdot \frac{5\alpha^3}{\alpha - \lfloor \alpha \rfloor}$.*

Proof. Let P be the number of pairs of vertices (x, y), such that $x \in N_3 \cup N_4$ and y is at most distance 2 from x. We will bound this number in two ways.

First, Lemma 2 tells us that for any vertex in the graph, the number of vertices at most distance 2 from it is at least $\frac{n}{\alpha}$. Therefore, $P \geq |N_3 \cup N_4| \cdot \frac{n}{\alpha}$.

For the second way, we will find an upper bound for the number of ways to start at a vertex $x \in N_3 \cup N_4$, and then travel along at most two edges in some way. This is an overcount for P, so it will give an upper bound. To count the number of these paths, we do casework on the various ways to start at a vertex in $N_3 \cup N_4$ and then travel along at most two edges.

Case 1: The path stays inside $N_3 \cup N_4$. Any vertex in $N_3 \cup N_4$ can be connected to at most $\alpha - 1$ other vertices in $N_3 \cup N_4$ (otherwise v would gain from connecting to it directly), so the number of paths for us to count for each starting vertex is at most $1 + (\alpha - 1) + (\alpha - 1)^2 \leq \alpha^2$. Therefore, the total number of paths of this type is at most $\boldsymbol{|N_3 \cup N_4|\alpha^2}$.

Case 2: The path travels from $N_3 \cup N_4$ *to* N_3 *to* N_2*, or is a length 1 path traveling from* N_3 *to* N_2. We count these backwards, starting from N_2. The number of edges from N_2 to N_3 is at most $\alpha|N_2|$ by Lemma 1, and again, every vertex in N_3 is connected to at most α vertices in $|N_3 \cup N_4|$, if including itself. Therefore, the number of paths here is at most $\boldsymbol{\alpha^2|N_2|}$.

Case 3: The path travels from N_3 *to* N_2 *to* N_3. We can count these by looking at the vertex in N_2 first, and then picking 2 of its children in N_3. Thus, the number of such paths is at most $\boldsymbol{|N_2|\alpha^2}$.

Case 4: The path travels from N_3 *to* N_2 *to* N_1. Similarly to Case 2, the number of such paths is at most $\boldsymbol{\alpha|N_2||N_1|}$.

Case 5: The path travels from N_3 *to* N_2 *to* N_2. By Lemma 4, the number of edges inside N_2 is at most $|N_2||N_1|\frac{\alpha}{\alpha - \lfloor\alpha\rfloor}$. Each such path consists of one of these edges, together with an edge to N_3 from one of its two endpoints. Therefore, the number of paths for us to count is at most $\boldsymbol{|N_2||N_1|\frac{2\alpha^2}{\alpha - \lfloor\alpha\rfloor}}$.

Total: summing over all cases, we have:

$$\begin{aligned}
P &\leq |N_3 \cup N_4|\alpha^2 + 2\alpha^2|N_2| + |N_1||N_2|\alpha + |N_1||N_2|\frac{2\alpha^2}{\alpha - \lfloor\alpha\rfloor} \\
&< 2\alpha^2 n + |N_1|n\left(\alpha + \frac{2\alpha^2}{\alpha - \lfloor\alpha\rfloor}\right) \\
&< |N_1|n\left(\frac{5\alpha^2}{\alpha - \lfloor\alpha\rfloor}\right).
\end{aligned}$$

But $P \geq |N_3 \cup N_4|\frac{n}{\alpha}$ from above, so:

$$\begin{aligned}
|N_3 \cup N_4|\frac{n}{\alpha} &< |N_1|n\left(\frac{5\alpha^2}{\alpha - \lfloor\alpha\rfloor}\right) \\
|N_3 \cup N_4| &< |N_1| \cdot \frac{5\alpha^3}{\alpha - \lfloor\alpha\rfloor}.
\end{aligned}$$

Lemma 6. *If every vertex has degree more than $\sqrt{n \log n}$, then the graph is asymptotically socially optimal: the total social cost is at most $2n^2 + \alpha n^{3/2}\sqrt{\log n}$.*

Proof. Suppose we have a Nash equilibrium where all vertices have degree greater than $\sqrt{n \log n}$. We give a strategy for an arbitrary vertex to achieve an individual cost of at most $\alpha\sqrt{n \log n} + 2n$, by changing only its own behavior. Since this is a Nash equilibrium, we will then be able to conclude that every vertex must have had individual cost at most $\alpha\sqrt{n \log n} + 2n$, proving this claim.

Specifically, we show that for any vertex w, the strategy "undo all edges you are currently paying for, and connect to $\sqrt{n \log n}$ vertices at random" has a positive probability of bringing it within distance ≤ 2 from every other vertex in the graph. Indeed, if w does this, then for any other vertex x,

$$\begin{aligned}
&\mathbb{P}\left[x \text{ is now distance} > 2 \text{ from } w\right] \\
&\leq \mathbb{P}\left[w \text{ didn't choose any of } x\text{'s neighbors}\right] \\
&\leq \left(1 - \frac{\sqrt{n \log n}}{n}\right)^{\sqrt{n \log n}} \leq e^{-\log n} = \frac{1}{n}.
\end{aligned}$$

Since there are only $n-1$ other vertices $x \neq w$ to consider, a union bound shows that the probability of failure is at most $(n-1)\frac{1}{n} < 1$, and therefore there is a way for w to attain an individual cost of at most $\alpha\sqrt{n \log n} + 2n$, as desired.

Lemma 7. *Even if there is a vertex of degree at most $\sqrt{n \log n}$, the graph is still asymptotically socially optimal: the total social cost is at most $2n^2 + n^{3/2}\sqrt{\log n} \cdot \frac{290\alpha^6}{(\alpha - \lfloor\alpha\rfloor)^2}$.*

Proof. Let v be a vertex of degree at most $\sqrt{n \log n}$, and construct the vertex partition N_1, N_2, N_3, N_4. We already know $|N_1|$ is at most $\sqrt{n \log n} = o(n)$, so by Lemma 5, $|N_3 \cup N_4|$ is at most $\sqrt{n \log n} \cdot \frac{5\alpha^3}{\alpha - \lfloor\alpha\rfloor} = o(n)$. By Lemma 4, the total number of edges inside N_2 is at most $|N_2||N_1|\frac{\alpha}{\alpha - \lfloor\alpha\rfloor} \leq n^{3/2}\sqrt{\log n} \cdot \frac{\alpha}{\alpha - \lfloor\alpha\rfloor} = o(n^2)$. Also, the total number of edges not completely inside N_2 is at most $n \cdot (1 + |N_1| + |N_3 \cup N_4|) \leq n^{3/2}\sqrt{\log n} \cdot \frac{6\alpha^3}{\alpha - \lfloor\alpha\rfloor} = o(n^2)$. Therefore, the total number of edges is the whole graph is at most $n^{3/2}\sqrt{\log n} \cdot \frac{7\alpha^3}{\alpha - \lfloor\alpha\rfloor} = o(n^2)$.

Next, we calculate a bound on the total sum of distances in the graph. Using Lemma 5 on every vertex in the graph, and the fact that all distances are at most 4 (Corollary 1), we get:

$$\begin{aligned}
&[\text{total sum of distances in the graph}] \\
&\leq 2n^2 + 4[\#\text{ of distances in the graph that are 3 or 4}] \\
&= 2n^2 + 4\sum_w [\#\text{ of vertices at distance 3 or 4 from } w] \\
&< 2n^2 + 4\sum_w \deg(w) \cdot \frac{5\alpha^3}{\alpha - \lfloor\alpha\rfloor}.
\end{aligned}$$

The degree sum is precisely twice the total number of edges in the graph, a quantity which we just bounded above. Putting everything together, the total sum of distances is at most:

$$2n^2 + 8n^{3/2}\sqrt{\log n} \cdot \frac{7\alpha^3}{\alpha - \lfloor\alpha\rfloor} \cdot \frac{5\alpha^3}{\alpha - \lfloor\alpha\rfloor} = 2n^2 + n^{3/2}\sqrt{\log n} \cdot \frac{280\alpha^6}{(\alpha - \lfloor\alpha\rfloor)^2} \, .$$

Adding α times the number of edges to compute the total social cost, we obtain the desired bound.

Lemmas 6 and 7 cover complementary cases, so we now conclude that the total social cost of every Nash equilibrium is at most the bound obtained in Lemma 7. As was observed by previous authors [1], the social optimum for $\alpha \geq 2$ is the star, achieving a social cost of at least $2n(n-1)$. Dividing, we find that the price of anarchy is at most

$$1 + \frac{150\alpha^6}{(\alpha - \lfloor\alpha\rfloor)^2}\sqrt{\frac{\log n}{n}} = 1 + o(1) \, ,$$

proving the first part of Theorem 1.

3 Integral α

There is one catch in our bound above. Namely, when α is only slightly greater than an integer (e.g. 4.0001), the terms of the form $\frac{\star}{\alpha - \lfloor\alpha\rfloor}$ all blow up, giving the final $o(n^2)$ terms for our bound a large constant factor. Even worse, when α is an exact integer, the proof fails completely. Perhaps surprisingly, this is not an artifact of the proof. In this section, we construct a counterexample when α is an integer. Let $v_1, v_2, \ldots, v_k$ be a large clique with edges oriented arbitrarily. In addition, each vertex v_i in the clique also pays for edges to $\alpha - 1$ separate leaves $l_{i:1}, l_{i:2}, \ldots, l_{i:\alpha-1}$. This graph also appears in [9], as an example of a Nash equilibrium which does not correspond to a tree, but its social cost is not calculated there.

Lemma 8. *In this graph, no single vertex has a better strategy than the one it is currently using.*

Proof. First, consider any leaf, say $l_{1:1}$. This leaf is not currently paying for any edges, so its only option is to pay for some set of edges. Notice that purely choosing some set of edges to pay for, without being able to delete any edges, is an instance of convex optimization. Therefore, by convexity, if any vertex in any graph can improve its station purely by adding some set of edges S, then it can also do this by adding some single edge $s \in S$. By observation, the leaf can only break even by adding one edge, so it can only break even overall.

Next, consider a members of the clique, say v_1. This vertex cannot delete its connections to its leaves, because that would disconnect the graph, making the distance component of its cost infinite. If v_1 remains neighbors with v_i and also buys an edge to some leaf $l_{i:j}$, then this is suboptimal: the edge to $l_{i:j}$ costs α but only gets v closer to one vertex. If v_1 deletes its edge to v_i but buys an edge to some leaf $l_{i:j}$, this is unnecessary: v_1 can move the edge from $l_{i:j}$ to v_i, switching its distances to those two vertices and not increasing the distance to any other vertex. Therefore, it is unnecessary for v_1 to consider strategies involving connecting to other vertices' leaves.

Thus, similarly to the previous case, v_1 only needs to consider strategies involving purely deleting edges. Again, by convexity, this reduces to considering strategies involving deleting a single edge. But again, v_1 can only break even by deleting an edge, so it can only break even overall.

Therefore, the graph is indeed a weak Nash equilibrium. Let n be the number of vertices in the graph. The size of the clique is $k = \frac{n}{\alpha}$, and so the cost of all of the edges is

$$\alpha\left[\binom{k}{2} + (\alpha - 1)k\right] = \alpha\left[(1+o(1))\frac{n^2}{2\alpha^2} + \frac{\alpha - 1}{\alpha}n\right] = (1+o(1))\frac{n^2}{2\alpha}.$$

Every clique vertex is distance 1 from the rest of the clique, as well as its leaves, and distance 2 from every other vertex; therefore, each clique vertex sees a distance sum of

$$(1+o(1))n\left(2 - \frac{1}{\alpha}\right).$$

Since there are $\frac{n}{\alpha}$ clique vertices, these contribute a total of

$$(1+o(1))n^2\left(\frac{2}{\alpha} - \frac{1}{\alpha^2}\right).$$

Every leaf vertex is distance 2 from almost all of the clique, and distance 3 from almost all of the leaves, and so it sees a distance sum of

$$(1+o(1))n\left(3 - \frac{1}{\alpha}\right).$$

Since there are $n(1 - \frac{1}{\alpha})$ leaves, these contribute a total distance sum of

$$(1+o(1))n^2\left(3 - \frac{4}{\alpha} + \frac{1}{\alpha^2}\right).$$

Putting everything together, we find that this graph has a total social cost of

$$(1+o(1))n^2\left(3 - \frac{3}{2\alpha}\right),$$

giving a price of anarchy at least $\frac{3}{2} - \frac{3}{4\alpha} + o(1)$, as claimed in the second part of Theorem 1.

4 Concluding Remarks

It is interesting that the price of anarchy converges to 1 for non-integral $\alpha > 2$, but is bounded away from 1 for integer $\alpha \geq 2$. Our convergence rate is non-uniform in the sense that it slows down substantially when α is slightly more than an integer. On the other hand, when α is slightly less than an integer, the convergence rate is still relatively rapid. It would be nice to prove a uniform convergence rate for all non-integral α.

References

1. Fabrikant, A., Luthra, A., Maneva, E., Papadimitriou, C.H., Shenker, S.: On a network creation game. In: Proc. 22nd Symposium on Principles of Distributed Computing, PODC 2003, pp. 347–351 (2003)
2. Anshelevich, E., Dasgupta, A., Kleinberg, J., Tardos, E., Wexler, T., Roughgarden, T.: The price of stability for network design with fair cost allocation. In: Proc. 45th IEEE Symposium on Foundations of Computer Science, FOCS 2004, pp. 295–304. IEEE Computer Society, Washington, DC (2004)
3. Anshelevich, E., Dasgupta, A., Tardos, E., Wexler, T.: Near-optimal network design with selfish agents. In: Proc. 35th ACM Symposium on Theory of Computing, STOC 2003, pp. 511–520. ACM, New York (2003)
4. Czumaj, A., Vöcking, B.: Tight bounds for worst-case equilibria. In: Proc. 13th ACM-SIAM Symposium on Discrete Algorithms, SODA 2002 (2002)
5. Papadimitriou, C.: Algorithms, games, and the internet. In: Proc. 33rd ACM Symposium on Theory of Computing, STOC 2001, pp. 749–753. ACM, New York (2001)
6. Roughgarden, T.: The price of anarchy is independent of the network topology. In: Proc. 34th ACM Symposium on Theory of Computing, STOC 2002, pp. 428–437 (2002)
7. Roughgarden, T.: Selfish Routing and the Price of Anarchy. MIT Press (2005)
8. Albers, S.: On the value of coordination in network design. In: Proc. 19th ACM-SIAM Symposium on Discrete Algorithms, SODA 2008, pp. 294–303. Society for Industrial and Applied Mathematics, Philadelphia (2008)
9. Albers, S., Eilts, S., Even-dar, E., Mansour, Y., Roditty, L.: On Nash equilibria for a network creation game. In: Proc. 16th ACM-SIAM Symposium on Discrete Algorithms, SODA 2006, pp. 89–98 (2006)
10. Alon, N., Demaine, E.D., Hajiaghayi, M., Leighton, T.: Basic network creation games. In: Proc. 22nd ACM Symposium on Parallelism in Algorithms and Architectures, SPAA 2010, pp. 106–113. ACM, New York (2010)
11. Andelman, N., Feldman, M., Mansour, Y.: Strong price of anarchy. In: Proc. 18th ACM-SIAM Symposium on Discrete Algorithms, SODA 2007, pp. 189–198. Society for Industrial and Applied Mathematics, Philadelphia (2007)
12. Corbo, J., Parkes, D.: The price of selfish behavior in bilateral network formation. In: Proc. 24th ACM Symposium on Principles of Distributed Computing, PODC 2005, pp. 99–107. ACM, New York (2005)
13. Demaine, E.D., Hajiaghayi, M., Mahini, H., Zadimoghaddam, M.: The price of anarchy in network creation games. ACM Transactions on Algorithms 8(2), Paper 13 (April 2012)
14. Mihalák, M., Schlegel, J.C.: The price of anarchy in network creation games is (mostly) constant. Theory Comput. Syst. 53(1), 53–72 (2013)

15. Jackson, M.O.: A survey of network formation models: stability and efficiency. In: Demange, G., Wooders, M. (eds.) Group Formation in Economics; Networks, Clubs, and Coalitions. Cambridge University Press (2005)
16. Tardos, E., Wexler, T.: Network formation games and the potential function method. In: Nisan, N., Roughgarden, T., Tardos, E., Vazirani, V.V. (eds.) Algorithmic Game Theory, pp. 487–516. Cambridge University Press (2007)
17. Koutsoupias, E., Papadimitriou, C.: Worst-case equilibria. In: Meinel, C., Tison, S. (eds.) STACS 1999. LNCS, vol. 1563, pp. 404–413. Springer, Heidelberg (1999)

Author Index